T0335593

Graduate Texts in Mathematics 128

Graduate Texts in Mathematics

continued after index

Jeffrey Rauch

Partial Differential Equations

With 42 Illustrations

Springer

Jeffrey Rauch
Department of Mathematics
University of Michigan
Ann Arbor, MI 48109
USA

Mathematics Subject Classifications (1991): 35-01, 35AXXX

Library of Congress Cataloging-in-Publication Data
Rauch, Jeffrey.
Partial differential equations / Jeffrey Rauch.
p. cm. — (Graduate texts in mathematics ; 128)
Includes bibliographical references and index.
ISBN 0-387-97472-5 (alk. paper)
1. Differential equations, Partial. I. Title. II. Series.
QA374.R38 1991
515'.353—dc20 90-19680
CIP

Printed on acid-free paper.

Typeset by Asco Trade Typesetting Ltd., Hong Kong.
Printed and bound by R.R. Donnelley & Sons, Harrisonburg, Virginia.
Printed in the United States of America.

9 8 7 6 5 4 3 2 (Corrected second printing, 1997)

ISBN 0-387-97472-5 Springer-Verlag New York Berlin Heidelberg
ISBN 3-540-97472-5 Springer-Verlag Berlin Heidelberg New York SPIN 10558487

Preface

This book is based on a course I have given five times at the University of Michigan, beginning in 1973. The aim is to present an introduction to a sampling of ideas, phenomena, and methods from the subject of partial differential equations that can be presented in one semester and requires no previous knowledge of differential equations. The problems, with hints and discussion, form an important and integral part of the course.

In our department, students with a variety of specialties—notably differential geometry, numerical analysis, mathematical physics, complex analysis, physics, and partial differential equations—have a need for such a course.

The goal of a one-term course forces the omission of many topics. Everyone, including me, can find fault with the selections that I have made.

One of the things that makes partial differential equations difficult to learn is that it uses a wide variety of tools. In a short course, there is no time for the leisurely development of background material..Consequently, I suppose that the reader is trained in advanced calculus, real analysis, the rudiments of complex analysis, and the language of functional analysis. Such a background is not unusual for the students mentioned above. Students missing one of the "essentials" can usually catch up simultaneously.

A more difficult problem is what to do about the Theory of Distributions. The compromise which I have found workable is the following. The first chapter of the book, which takes about nine fifty-minute hours, does not use distributions. The second chapter is devoted to a study of the Fourier transform of tempered distributions. Knowledge of the basics about $\mathscr{D}(\Omega)$, $\mathscr{E}(\Omega)$, $\mathscr{D}'(\Omega)$, and $\mathscr{E}'(\Omega)$ is assumed at that time. My experience teaching the course indicates that students can pick up the required facility. I have provided, in an appendix, a short crash course on Distribution Theory. From Chapter 2 on, Distribution Theory is the basic language of the text, providing a good setting for reinforcing the fundamentals. My experience in teaching this course is that students have less difficulty with the distribution theory than with geometric ideas from advanced calculus (e.g. $d\varphi$ is a one-form which annihilates the tangent space to $\{\varphi = 0\}$).

There is a good deal more material here than can be taught in one semester. This provides material for a more leisurely two-semester course and allows the reader to browse in directions which interest him/her. The essential core is the following:

Chapter 1. Almost all. A selection of examples must be made.

Chapter 2. All but the L^p theory for $p \neq 2$. Some can be left for students to read.

Chapter 3. The first seven sections. One of the ill-posed problems should be presented.

Chapter 4. Sections 1, 2, 5, 6, and 7 plus a representative sampling from Sections 3 and 4.

Chapter 5. Sections 1, 2, 3, 10, and 11 plus at least the statements of the standard Elliptic Regularity Theorems.

These topics take less than one semester.

An introductory course should touch on equations of the classical types, elliptic, hyperbolic, parabolic, and also present some other equations. The energy method, maximum principle, and Fourier transform should be used. The classical fundamental solutions should appear. These conditions are met by the choices above.

I think that one learns more from pursuing examples to a certain depth, rather than giving a quick gloss over an enormous range of topics. For this reason, many of the equations discussed in the book are treated several times. At each encounter, new methods or points of view deepen the appreciation of these fundamental examples.

I have made a conscious effort to emphasize qualitative information about solutions, so that students can learn the features that distinguish various differential equations. Also the origins in applications are discussed in conjunction with these properties. The interpretation of the properties of solutions in physical and geometric terms generates many interesting ideas and questions.

It is my impression that one learns more from trying the problems than from any other part of the course. Thus I plead with readers to attempt the problems.

Let me point out some omissions. In Chapter 1, the Cauchy–Kowaleskaya Theorem is discussed, stated, and much applied, but the proof is only indicated. Complete proofs can be found in many places, and it is my opinion that the techniques of proof are not as central as other things which can be presented in the time gained. The classical integration methods of Hamilton and Jacobi for nonlinear real scalar first-order equations are omitted entirely. My opinion is that when needed these should be presented along with symplectic geometry. There is a preponderance of linear equations, at the expense of nonlinear equations. One of the main points for nonlinear equations is their differences with the linear. Clearly there is an order in which these things should be learned. If one includes the problems, a reasonable dose of nonlinear examples and phenomena are presented. With the exception of the elliptic theory, there is a strong preponderance of equations with constant coefficients, and especially Fourier transform techniques. The reason for this choice is that one can find detailed and interesting information without technical complexity. In this way one learns the ideas of the theory of partial differential equations at minimal cost. In the process, many methods are introduced which work for variable coefficients and this is pointed out at the appropriate places.

Compared to other texts with similar level and scope (those of Folland, Garabedian, John, and Treves are my favorites), the reader will find that the present treatment is more heavily weighted toward initial value problems. This, I confess, corresponds to my own preference. Many time-independent problems have their origin as steady states of such time-dependent problems and it is as such that they are presented here.

A word about the references. Most are to textbooks, and I have systematically referred to the most recent editions and to English translations. As a result the dates do not give a good idea of the original publication dates. For results proved in the last 40 years, I have leaned toward citing the original papers to give the correct chronology. Classical results are usually credited without reference.

I welcome comments, critiques, suggestions, corrections, etc. from users of this book, so that later editions may benefit from experience with the first.

So many people have contributed in so many different way to my appreciation of partial differential equations that it is impossible to list and thank them all individually. However, specific influences on the structure of this book have been P.D. Lax and P. Garabedian from whom I took courses at the level of this book; Joel Smoller who teaches the same course in a different but related way; and Howard Shaw whose class notes saved me when my own lecture notes disappeared inside a moving van. The integration of problems into the flow of the text was much influenced by the *Differential Topology* text of Guillemin and Pollack. I have also benefited from having had exceptional students take this course and offer their criticism. In particular, I would like to thank Z. Xin whose solutions, corrections, and suggestions have greatly improved the problems. Chapters of a preliminary version of this text were read and criticized by M. Beals, J.L. Joly, M. Reed, J. Smoller, M. Taylor, and M. Weinstein. Their advice has been very helpful. My colleagues and coworkers in partial differential equations have taught me much and in many ways. I offer a hearty thank you to them all.

The love, support, and tolerance of my family were essential for the writing of this book. The importance of these things to me extends far beyond professional productivity, and I offer my profound appreciation.

[illegible] geometry, there is a great number [illegible] of the subject [illegible] equations. One of the main points for [illegible] is the difference with [illegible] in order to [illegible] should [illegible] the [illegible] examples and [illegible] are presented. With the [illegible] theory [illegible] correspondence [illegible] techniques. The [illegible] of this choice [illegible] one can find [illegible] and interest [illegible] without [illegible]

[illegible] of the theory of [illegible] equations [illegible] [illegible] [illegible] which [illegible] the [illegible] in the [illegible]

Compared to other [illegible] level and [illegible] the [illegible] and Treves [illegible], the reader [illegible] the present treatment [illegible] [illegible] devices, corresponding [illegible] problems [illegible] and [illegible] that they are presented here.

[illegible] about the references. Most are [illegible] referred to the [illegible] and to [illegible] expositions. As a result [illegible] of the original publication [illegible] the results [illegible] I have [illegible] papers [illegible] the [illegible] credit [illegible]

I welcome comments, notes, suggestions, corrections [illegible] from users of this book, so that later editions [illegible] from experience with the first.

So many people have [illegible] many different ways to my approach and [illegible] that it is impossible to [illegible] and thank them all individually. However, specific influences on the structure of the book have been [illegible] from whom I [illegible] of this book; Joel Smoller who [illegible] the same course [illegible] and Howard [illegible] [illegible] when [illegible] [illegible] [illegible] of the [illegible] and [illegible] I have [illegible] benefited [illegible] the course and offer [illegible] to thank [illegible] whose [illegible] comments and suggestions have greatly improved the [illegible] chapters [illegible] and [illegible] [illegible] [illegible] helped. My [illegible] in partial differential equations [illegible] [illegible] thank you [illegible] all.

The [illegible] support [illegible] family were essential to the writing of this book. The importance of these [illegible] to [illegible] and [illegible] my [illegible] approach [illegible]

Contents

CHAPTER 1

Power Series Methods

§1.1. The Simplest Partial Differential Equation

It takes a little time and a few basic examples to develop intuition. This is particularly true of the subject of partial differential equations which has an enormous variety of technique and phenomena within its confines. This section describes the simplest nontrivial partial differential equation

$$u_t(t, x) + cu_x(t, x) = 0, \qquad t, x \in \mathbb{R}, \quad c \in \mathbb{C}. \tag{1}$$

The equation is of first order, is linear with constant coefficients, and involves derivatives with respect to both variables. The unknown is a possibly complex valued function u of two real variables. This example reveals one of the fundamental dichotomies of the subject, the equation is hyperbolic if $c \in \mathbb{R}$ and elliptic otherwise. The equation is radically different in these two cases in spite of the similar appearance.

The use of "t" is meant to suggest time. One can use the equation to march forward in time as follows. Given u at time t, $u(t, \cdot)$, one can compute the value of

$$u_t(t, \cdot) = -c\partial_x u(t, \cdot),$$

and then advance the time using

$$u(t + \Delta t, \cdot) \approx u(t, \cdot) + u_t(t, \cdot)\Delta t = (1 - c\Delta t \partial_x)u(t, \cdot). \tag{2}$$

This marching algorithm suggests that the *initial value problem* or *Cauchy problem* is appropriate. Thus, given $g(x)$ we seek u satisfying (1) and the initial condition

$$u(0, \cdot) = g(\cdot). \tag{3}$$

For $g \in C^\infty(\mathbb{R})$ and $n \in \mathbb{N}$ we may choose a time step $\Delta t = 1/n$ and find approximate values

$$u\left(\frac{k}{n}, \cdot\right) \approx \left(1 - \frac{c}{n}\partial_x\right)^k g(\cdot). \tag{4}$$

Since the approximation (2) improves as Δt decreases to zero it is not unreasonable to think that as $n, k \to \infty$ with $k/n = t$ fixed, the approximations on the right approach the values $u(t, \cdot)$ of a solution.

With $t = k/n$, (4) reads

$$u(t, \cdot) \approx \left(1 - \frac{tc\partial_x}{k}\right)^k g(\cdot). \tag{5}$$

Letting k tend to infinity suggests the formal identity

$$u(t, \cdot) = \exp(-ct\partial_x)g. \tag{6}$$

For polynomial g, formally expanding the exponential and using Taylor's Theorem yields

$$u(t, \cdot) = \sum \frac{(-ct)^n g^n(\cdot)}{n!} = g(\cdot - ct). \tag{7}$$

It is easy to verify that for polynomial g, $g(x - ct)$ is indeed a solution of the initial value problem and is also the limit of the approximations (4). In fact, if g is the restriction to $\mathbb{R}$ of an analytic function on $|\text{Im } x| < R$, then one has convergence for $|t| < R/|c|$ to the solution $g(x - ct)$.

If c is real, then the formula $g(x - ct)$ still provides a solution even when g does not have an analytic continuation to a neighborhood of the real axis. However, the approximations (5) will not converge if the derivatives of g grow faster than those of an analytic function.

Finally, if c is complex then the formula suggests that g must have a natural extension from real to complex values of x in order for there to be a solution.

The ideas suggested by the formal computations are next verified by examining the initial value problem (1), (3) following a different and easier route.

For real c, the differential equation (1) asserts that the directional derivative of u in the direction $(1, c)$ vanishes (Figure 1.1.1). Thus $u \in C^1(\mathbb{R}^2)$ is a solution if and only if u is constant on each of the lines $x - ct =$ constant. These lines, integral curves of the vector field $\partial/\partial t + c\partial/\partial x$, are called *characteristic lines* or *rays*. This observation yields the following result.

Theorem 1. *If c is real and $g \in C^1(\mathbb{R})$, there is a unique solution $u \in C^1(\mathbb{R}^2)$ to the initial value problem* (1), (3). *The solution is given by the formula $u(t, x) = g(x - ct)$. If $g \in C^k(\mathbb{R})$ with $k > 1$, then $u \in C^k(\mathbb{R})$.*

The solution u represents undistorted wave propagation with speed c. The characteristic lines have slope dt/dx equal to $1/c$ and speed dx/dt equal to c. The value of u at $\bar{t}, \bar{x}$ is determined by g at $\bar{x} - c\bar{t}$. This illustrates the ideas of *domain of determinacy* and *domain of influence*. The domain of determinacy of $\bar{t}, \bar{x}$ is the point $(0, \bar{x} - c\bar{t})$ on the line $t = 0$. The domain of influence of the point $(0, \underline{x})$ on the initial line is the characteristic $x - ct = \underline{x}$ (Figure 1.1.2).

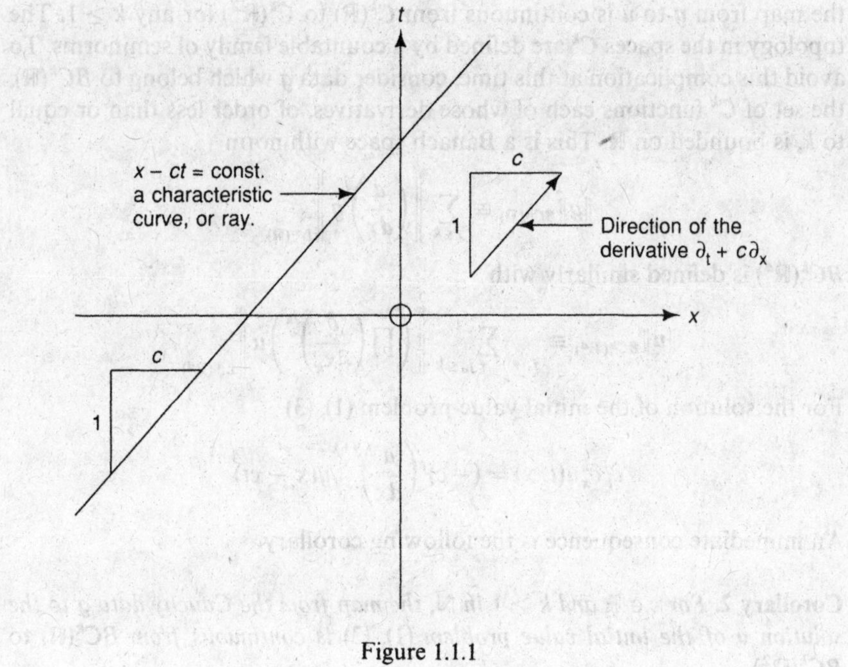

Figure 1.1.1

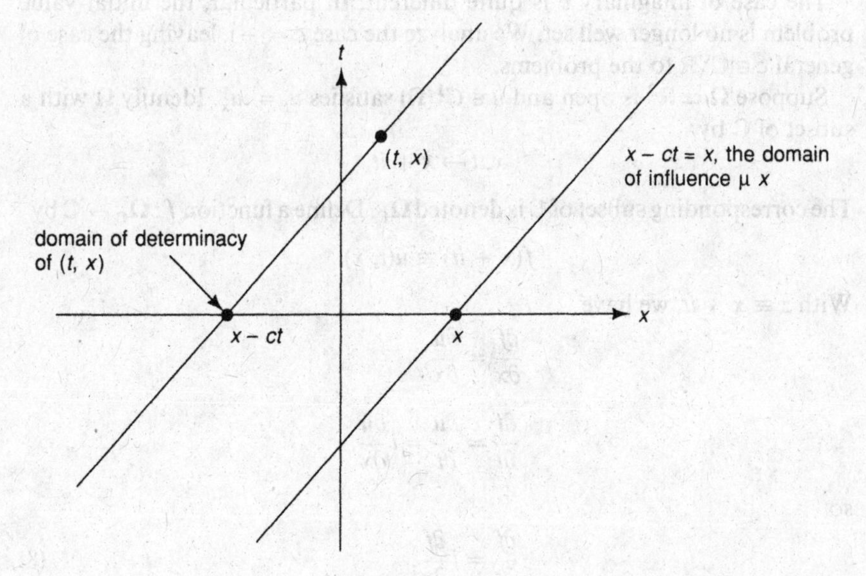

Figure 1.1.2

Nearby initial data g yield nearby solutions u. A precise statement is that the map from g to u is continuous from $C^k(\mathbb{R})$ to $C^k(\mathbb{R}^2)$ for any $k \geq 1$. The topology in the spaces C^k are defined by a countable family of seminorms. To avoid this complication at this time, consider data g which belong to $BC^k(\mathbb{R})$, the set of C^k functions each of whose derivatives, of order less than or equal to k, is bounded on $\mathbb{R}$. This is a Banach space with norm

$$\|g\|_{BC^k(\mathbb{R})} \equiv \sum_{j \leq k} \left\| \left(\frac{d}{dx}\right)^j g \right\|_{L^\infty(\mathbb{R})}.$$

$BC^k(\mathbb{R}^d)$ is defined similarly with

$$\|u\|_{BC^k(\mathbb{R}^d)} \equiv \sum_{j_1 + \cdots + j_d \leq k} \left\| \left(\prod \left(\frac{\partial}{\partial x_k}\right)^{j_k}\right) u \right\|_{L^\infty(\mathbb{R}^d)}.$$

For the solution of the initial value problem (1), (3)

$$\partial_t^j \partial_x^k u(t, x) = (-c)^j \left(\frac{d}{dx}\right)^{j+k} g(x - ct).$$

An immediate consequence is the following corollary.

Corollary 2. *For $c \in \mathbb{R}$ and $k \geq 1$ in $\mathbb{N}$, the map from the Cauchy data g to the solution u of the initial value problem* (1), (3) *is continuous from $BC^k(\mathbb{R})$ to $BC^k(\mathbb{R}^2)$.*

The case of imaginary c is quite different. In particular, the initial value problem is no longer well set. We analyze the case $c = -i$, leaving the case of general $c \in \mathbb{C}\backslash\mathbb{R}$ to the problems.

Suppose $\Omega \subset \mathbb{R}^2$ is open and $u \in C^1(\Omega)$ satisfies $u_t = iu_x$. Identify Ω with a subset of $\mathbb{C}$ by

$$x, t \mapsto x + it.$$

The corresponding subset of $\mathbb{C}$ is denoted $\Omega_{\mathbb{C}}$. Define a function $f: \Omega_{\mathbb{C}} \to \mathbb{C}$ by

$$f(x + it) \equiv u(t, x).$$

With $z \equiv x + it$, we have

$$\frac{\partial f}{\partial x} = \frac{\partial u}{\partial x},$$

$$\frac{\partial f}{\partial t} = \frac{\partial u}{\partial t} = i\frac{\partial u}{\partial x},$$

so

$$\frac{\partial f}{\partial t} = i\frac{\partial f}{\partial x}. \tag{8}$$

Equation (8) is called the *Cauchy–Riemann equation*. In elementary function theory one shows that the solutions, called *holomorphic* or *analytic functions* of $z = x + it$, are infinitely differentiable. Moreover, if $p \in \Omega_{\mathbb{C}}$, then f is equal to the sum of a convergent power series in $z - p$,

$$f = \sum_{n=0}^{\infty} a_n(z - p)^n, \qquad |z - p| < \operatorname{dist}(p, \partial\Omega_{\mathbb{C}}).$$

Differentiating term by term shows that the converse is also true, that is, convergent power series in $z - p$ are solutions.

Theorem 3. *$u \in C^1(\Omega)$ satisfies $u_t - iu_x = 0$ if and only if it defines a holomorphic function on $\Omega_{\mathbb{C}}$.*

Next consider the initial value problem $u_t - iu_x = 0$, $u(0, \cdot) = g$. If there is a solution on a neighborhood of $(0, \underline{x})$, then u is holomorphic. Thus

$$u = \sum a_n(z - \underline{x})^n,$$

so

$$g = \sum a_n(x - \underline{x})^n,$$

is given by a convergent power series. Such a function is called *real analytic*. Conversely, if g is real analytic at $\underline{x}$ then the above formula defines u holomorphic near $\underline{x} + i0$.

Warning. If the a_n are complex, such real analytic functions need not be real valued. They are defined on a real domain, hence the name.

Theorem 4. *The initial value problem*

$$u_t - iu_x = 0, \qquad u(0, \cdot) = g(\cdot),$$

has a C^1 solution on a neighborhood of $(0, \underline{x})$ if and only if g is the restriction to $\mathbb{R}$ of a holomorphic function defined on a neighborhood of $\underline{x}$, that is, if and only if g is real analytic at $\underline{x}$.

As a consequence, we see that if g is C^∞ but not real analytic at $\underline{x}$, then the approximation scheme 2, 4 cannot converge to a solution of the initial value problem on a neighborhood of $(0, \underline{x})$. It is not difficult to show that one does have convergence for real analytic g.

Summary

(i) For $c \in \mathbb{R}$, the initial value problem is nicely solvable.

(ii) For $c \in \mathbb{C}\backslash\mathbb{R}$, $u_t + cu_x = 0$ has only real analytic solutions. The initial value problem is not solvable unless the data are real analytic.

(iii) For $c \in \mathbb{R}$ the equation is *hyperbolic*. For $c \in \mathbb{C}\backslash\mathbb{R}$, it is *elliptic*. These terms will be defined later and describe two of the most important classes of partial differential equations.

Problems

1. If $c \in \mathbb{C}\backslash\mathbb{R}$, $g \in C^1(\mathbb{R})$, then the initial value problem
$$u_t + cu_x = 0, \qquad u(0, x) = g(x),$$
has a C^1 solution on a neighborhood of the origin if and only if g is real analytic on a neighborhood of the origin.

2. Prove that if $c \in \mathbb{R}$ and u is a $C^1(\mathbb{R}^2)$ solution of the equation $\partial_t u + c\partial_x u = 0$, then
$$\{(t, x) \in \mathbb{R}^2 \colon u \in C^k \text{ on a neighborhood of } t, x\}$$
is a union of rays.
Discussion. This elementary result is typical. Solutions of partial differential equations inherit a great deal of structure from the equation they satisfy. This result asserts *propagation of singularities* and *propagation of regularity* along rays.

3. Prove that if $u \in C^\infty(\mathbb{R}^2)$ satisfies $\partial_t u + c\partial_x u = 0$ with c real, and k is a nonnegative integer, then
$$\{(t, x) \colon u \text{ vanishes to order } k \text{ at } (t, x)\}$$
is a union of rays. For any closed set $\Gamma \subset \mathbb{R}^2$ which is a union of rays, prove that there is a u as above such that Γ is exactly the set where u vanishes.
Discussion. Contrast this to the case where c is not real. Then, if a solution vanishes on any open set it must vanish identically.

4. Show that for $c \in \mathbb{R}$ and $f \in C^1(\mathbb{R}^2)$ there is one and only one solution to the initial value problem
$$u_t + cu_x = f, \qquad u(0, x) = 0.$$
Find a formula for the solution. Find an $f \in C^1(\mathbb{R}^2)$ such that the solution is not in $C^2(\mathbb{R}^2)$.
Discussion. This may be surprising since "first derivatives in C^1 indicate $u \in C^2$". However, the partial differential equation contains only a linear combination of first derivatives. Nevertheless, when $c \in \mathbb{C}\backslash\mathbb{R}$ the equation is elliptic and, in a sense, controls all derivatives. In that case, $u_t + cu_x \in C^\infty$ implies that $u \in C^\infty$. Also u has one more derivative than $u_t + cu_x$, but not in the sense of the classical spaces C^k (see Propositions 2.4.5 and 5.9.1, and Problem 5.9.3).

5. For the nonlinear initial value problem,
$$u_t + cu_x + u^2 = 0, \qquad u(0, x) = g(x), \qquad c \in \mathbb{R},$$
show that if $g \in C_0^\infty(\mathbb{R})$, g not identically zero, there is a local solution $u \in C^\infty(\{-\delta < t < \delta\} \times \mathbb{R})$ but that the solution does not extend to a C^∞ solution on all of $\mathbb{R}^2$.
Discussion. This blow-up of solutions is just like that for the nonlinear ordinary differential equation $dy/dt = y^2$. Nonlinear partial differential equations have more subtle blow-up mechanisms too. See the formation of shocks discussed in §1.9.

§1.2. The Initial Value Problem for Ordinary Differential Equations

Many of the first steps in studying the initial value problem have direct ancestors in the theory of ordinary differential equations. For that reason, we begin with a quick review. Consider an ordinary differential equation of order m solved for the derivative of highest order

$$\frac{d^m}{dt^m}u(t) = G\left(t, u(t), \frac{du(t)}{dt}, \ldots, \frac{d^{m-1}u(t)}{dt^{m-1}}\right). \tag{1}$$

Simple examples from applications are the equation $du/dt = au$ modeling radioactive decay if $a < 0$ and the Malthusian population explosion if $a > 0$. Equally elementary is the equation of the damped spring

$$mu'' + au' + k^2u = 0, \qquad m, k > 0 \quad \text{and} \quad a \geq 0.$$

More generally, Newton's second law of motion reads

$$mu'' = G(t, u(t), u'(t)),$$

where we have supposed that the force on the particle at time t is determined by t and the position and velocity of the particle. A complicated example is

$$u'' = ((1 + t^2)(u')^2)^{1/2}.$$

The equation for population growth or radioactive decay has solution $u = u(0)\exp(at)$ which is uniquely determined once the initial state is known. For Newton's law initial position and velocity are required. More generally, the correct initial value problem is the following.

Cauchy Problem. Given $u_0, u_1, \ldots, u_{m-1} \in \mathbb{R}$ find a solution u to the ordinary differential equation (1) which satisfies the initial conditions

$$\frac{d^ju}{dt^j}(0) = u_j, \qquad j = 0, 1, \ldots, m - 1. \tag{2}$$

That it is reasonable to expect to determine u is indicated by the following calculation of Cauchy. Given the initial conditions one computes

$$\frac{d^mu}{dt^m}(0) = G(0, u_0, \ldots, u_{m-1}),$$

thus, $d^\nu u/dt^\nu(0)$ is determined for $\nu \leq m$. Inductively, we determine all derivatives at $t = 0$ as follows.

Differentiate (1) $k - m + 1$ times to find

$$\frac{d^{k+1}u}{dt^{k+1}} = \left(\frac{d}{dt}\right)^{k-m+1} G(t, u, \ldots, u^{(m-1)}).$$

Using Leibniz' rule shows that the right-hand side is a function $G_k(t, u, \ldots, u^{(k)})$.

Suppose $u^{(\nu)}(0)$ is determined for $\nu \le k$, $k \ge m-1$. Setting $t = 0$ determines $u^{(k+1)}(0) = G_k(0, u(0), \ldots, u^{(k)}(0))$ completing the induction.

Once $u^{(\nu)}(0)$ is determined for all ν, then

$$\sum_0^\infty \left(\frac{u^{(\nu)}(0)}{\nu!}\right) t^\nu$$

is a good candidate for a solution if the series converges. At any rate, it is the Taylor series of any infinitely differentiable solution.

Example. Consider the initial value problem for a hard spring

$$u'' - u^3 = 0, \qquad u(0) = 1, \qquad u'(0) = 0.$$

To find the Taylor series at $t = 0$, compute

$$u'' = u^3, \qquad @t = 0, \quad u''(0) = 1.$$

$$u''' = (u^3)' = 3u^2u', \qquad @t = 0, \quad u'''(0) = 0.$$

$$u^{(4)} = (3u^2u')' = 6u(u')^2 3u^2u'', \qquad @t = 0, \quad u^4(0) = 3.$$

$$u \approx 1 + \frac{t^2}{2!} + \frac{3t^4}{4!} + \cdots.$$

Recall that a C^∞ function defined on an open set in $\mathbb{R}^d$ is called *real analytic*, if on a neighborhood of every point it is equal to the sum of its Taylor series. We denote by $C^\omega(\Omega)$ the class of real analytic functions on Ω. Since the Taylor expansion of a C^∞ solution is uniquely determined, we have the following uniqueness result in the real analytic category.

Theorem 1. *If G is infinitely differentiable, then the initial value problem* (1), (2) *can have at most one real analytic solution.*

Example. If $m = 1$, $G = G(t) \in C^\infty$, but is not real analytic at $t = 0$, then

$$\frac{du}{dt} = G(t), \qquad u(0) = u_0,$$

does not have a solution given by a convergent power series, since if it did then $G(t) = du/dt$ would be given by a convergent power series. An example is given by

$$G(t) = \begin{cases} e^{-1/t}, & t > 0, \\ 0, & t \le 0. \end{cases}$$

Cauchy's algorithm yields $u^{(\nu)}(0) = 0$ for all $\nu \ge 0$, so the Taylor series converges but not to a solution. If one chooses $G \in C^\infty$ with divergent Taylor series, then Cauchy's recipe will construct a divergent power series for u. For real analytic data, we state Cauchy's positive result.

Theorem 2 (Cauchy). *If G is real analytic on a neighborhood of $(0, u_0, \ldots, u_{m-1})$, then the Taylor series computed above converges on a neighborhood of $t = 0$ to a real analytic solution to the initial value problem* (1), (2).

A second approach to constructing a solution is to march forward in time in steps $\Delta t \equiv h$, and then take the limit $h \to 0$. More precisely, given $h > 0$, let

$$t_n \equiv nh, \qquad n = 0, 1, \ldots.$$

With h fixed we construct an approximation to $u(nh)$. At the same time, we construct approximations to the derivatives $u^{(\nu)}(nh)$ for $\nu = 1, 2, \ldots, m - 1$. The notation U_n^ν is used for the approximation to $u^\nu(nh)$. The values of U_{n+1}^ν are computed from the values of U_n^ν according to Euler's scheme:

$$U_{n+1}^\nu = U_n^\nu + hU_n^{\nu+1}, \qquad \nu \le m - 2,$$
$$U_{n+1}^{m-1} = U_n^{m-1} + hG(t_n, U_n^0, \ldots, U_n^{m-1}).$$

The last expression comes from the approximation

$$u^{(m)}(nh) = G(t_n, u^0(nh), \ldots, u^{m-1}(nh)) \cong G(t_n, U_n^0, \ldots, U_n^{m-1}).$$

Note that to continue this process one needs to know that $t_n, U_n^0, \ldots, U_n^{m-1}$ remains in the domain of definition of G. Thus, the U may only be defined for a finite set of n. For $h = \Delta t$ fixed, let $g_h(t)$ be the piecewise linear function which is linear on each interval $[t_n, t_{n+1}]$, and is equal to $G(t_n, U_n^0, \ldots, U_n^{m-1})$ at time t_n. Then g_h is an approximation to $u^m(t)$. Let $I\colon C([0, \infty[: \mathbb{R}) \to C^1([0, \infty[: \mathbb{R})$ be the integration operator

$$(If)(t) \equiv \int_0^t f(s)\, ds.$$

A reasonable approximation for u is then

$$u_h(t) \equiv \sum_0^{m-1} \frac{u^{(\nu)}(0)t^\nu}{\nu!} + I^m(g_h).$$

One hopes or expects that, as h tends to zero, $\lim u_h(t)$ exists and gives a solution. Note that u_h is defined on an h-dependent interval, so part of this expressed optimism is that the interval does not shrink to $\{0\}$ as h decreases. In fact, all goes well.

Existence Theorem 3 (Peano). *If $G \in C(\Omega)$, then there exists $T > 0$ and a sequence $h_n \to 0$ such that, as $n \to \infty$, $u_{h_n}(t)$ converges in $C^m([0, T] : \mathbb{R})$ to a solution of the initial value problem* (1), (2).

To guarantee uniqueness of solutions, G must be more regular. Lipshitz continuity as a function of its arguments is sufficient.

Uniqueness Theorem 4 (Picard). *Suppose Ω is an open neighborhood of $(0, u_0, \ldots, u_{m-1})$ in $\mathbb{R}^{m+1}$ and $G \in C^1(\Omega)$. If u and $v \in C^m([0, T] : \mathbb{R})$ are solutions* *of the initial value problem* (1), (2), *then* $u = v$ *provided that* $(t, u, u^{(1)}, \ldots, u^{(m-1)})$ *and* $(t, v, v^{(1)}, \ldots, v^{(m-1)})$ *lie in* Ω *for* $0 \le t \le T$.

Picard proved both existence and uniqueness in this setting by recasting the initial value problem in the form of a fixed point equation, $Mu = u$, where M is the operator

$$Mv(t) \equiv \sum_{0}^{m-1} \frac{u^{\nu}(0)t^{\nu}}{\nu!} + I^{m}(G(t, v(t), \ldots, v^{m-1}(t))).$$

Picard's proof marked a watershed in the theory of differential equations as it established existence and uniqueness in cases where no reasonable formula for a solution exists. His argument is a model for all later results. Existence is proved by demonstrating the convergence of a sequence of approximate solutions, called "Picard iterates". These are defined by $u_{n+1} \equiv Mu_n$. This idea, called fixed point iteration, is an effective numerical method, though for this initial value problem there are much better techniques. Picard's proof is now the industry standard and can be found in many texts on ordinary differential equations as well as in Picard's elegant *Traité d'Analyse* [P].

Euler's method relies on a finite difference replacement of the differential equation based on

$$\frac{u^{m-1}(t_{n+1}) - u^{m-1}(t_n)}{h} \cong u^{m}(t_n) \cong G(t_n, u(t_n), \ldots, u^{m-1}(t_n)),$$

$$\frac{u^{j}(t_{n+1}) - u^{j}(t_n)}{h} \cong u^{j+1}(t_n), \qquad j = 0, 1, \ldots, m-2.$$

Experience from §1.1 should have left you wary of such algorithms, but in this circumstance, it converges to a solution (see Problem 2).

When $G \in C^1$, the error in Euler's method is $O(h)$ in C^m norm. Proofs can be found in texts on numerical analysis which address the approximate solution of ordinary differential equations. The best approximate methods are refinements of Euler's method. One can also find in such texts a discussion of fixed point iteration, as a method for solving linear and nonlinear equations.

PROBLEMS

1. Show that the initial value problem $(u')^2 - u^2 = 0$, $u(0) = 1$, has exactly two real analytic solutions on a neighborhood of $t = 0$.
   DISCUSSION. This sort of problem, in the partial differential equations category, is the subject of §1.4.

2. For the two simple initial value problems
   (i) $u' = g(t)$, $u(0) = 0$, $g \in C(\mathbb{R})$,
   (ii) $u' = u$, $u(0) = 1$,
   verify that the approximations defined by Euler's scheme converge uniformly on $[0, 1]$ to a solution.

§1.3. Power Series and the Initial Value Problem for Partial Differential Equations

Our goal is to investigate through two examples the partial differential equation analogue of Cauchy's Theorem. The upshot is the theorem of Cauchy–Kowaleskaya.

EXAMPLE. Consider the initial value problem

$$u_t - iu_x = 0, \qquad u(0, \cdot) = g(\cdot), \tag{1}$$

which we know from §1.1 cannot be solved unless g is real analytic. Nevertheless, for any solution, the differential equation and initial condition determine $\partial_t^j \partial_x^k u(0, 0)$ and therefore the Taylor series

$$\sum \frac{\partial_t^j \partial_x^k u(0, 0)}{j!\, k!} t^j x^k.$$

To see this, observe that

$$u_t = iu_x,$$

$$u_{tt} = i\partial_x \partial_t u = i\partial_x i\partial_x u,$$

$$\partial_t^j \partial_x^k u = (i\partial_x)^j \partial_x^k u,$$

$$\partial_t^j \partial_x^k u(0, 0) = i^j \left(\frac{d}{dx}\right)^{j+k} g(0),$$

$$u \sim \sum \frac{i^j g^{(j+k)}(0)}{j!\, k!} t^j x^k.$$

The Taylor series for g is $\sum g^j(0)x^j/j!$. If it converges for $|x| \leq R$, then $|g^j(0)|R^j/j! \leq C$. Thus, the series for u is dominated by

$$C \sum \frac{(j + k)!}{j!\, k!} \left|\frac{t}{R}\right|^j \left|\frac{x}{R}\right|^k.$$

In Problem 2, you are asked to prove that this series converges on a neighborhood of (0, 0), reproving the existence part of Theorem 1.1.4.

Proposition 1. *If g is real analytic at $\bar{x}$, then the initial value problem* (1) *has a unique real analytic solution on a neighborhood of* $(0, \bar{x})$.

Virtually the same argument works for $u_t + cu_x = 0$ for any $c \in \mathbb{C}$.

For partial differential equations there is a wide variety of "mixtures" of orders corresponding to the large number of distinct partial derivatives. Here are some examples:

$$u_{tt} + u_{xx} = 0, \quad \text{Laplace's equation,}$$

$$u_{tx} = 0, \quad 45^\circ \text{ wave equation,}$$

$$u_{tt} + u_{xxxx} = 0, \quad \text{linearized beam equation,}$$

$$u_t - u_{xx} = 0, \quad \text{heat equation,}$$

$$u_{tt} - u_x = 0, \quad \text{sideways heat equation.}$$

To make it easier to manage the bookkeeping of the possible partial derivatives we use the multi-index notation of L. Schwartz. For $\alpha \in \mathbb{N}^d$,

$$\partial^\alpha \equiv \left(\frac{\partial}{\partial x_1}, \frac{\partial}{\partial x_2}, \ldots, \frac{\partial}{\partial x_d}\right)^\alpha \equiv \left(\frac{\partial}{\partial x_1}\right)^{\alpha_1}\left(\frac{\partial}{\partial x_2}\right)^{\alpha_2}\cdots\left(\frac{\partial}{\partial x_d}\right)^{\alpha_d},$$

$$x^\alpha = (x_1, x_2, \ldots, x_d)^\alpha \equiv x_1^{\alpha_1} x_2^{\alpha_2} \ldots x_d^{\alpha_d},$$

$$\alpha! \equiv \alpha_1!\,\alpha_2! \ldots \alpha_d! \quad \text{where} \quad 0! \equiv 1,$$

$$|\alpha| \equiv |\alpha_1| + |\alpha_2| + \cdots + |\alpha_d|.$$

EXAMPLES. 1. The most general partial derivative of order m is ∂^α, $|\alpha| = m$. Equality of mixed partials is assumed here.

2. The most general linear partial differential operator of order m with constant coefficients is

$$\sum_{|\alpha| \le m} a_\alpha \partial^\alpha, \qquad a_\alpha \in \mathbb{C}.$$

The principal part, consisting of terms of order exactly m, is the sum over terms with $|\alpha| = m$.

3. Taylor's series in several variables takes the elegant form

$$f(x) \sim \sum \frac{\partial^\alpha f(\underline{x})(x - \underline{x})^\alpha}{\alpha!}.$$

4. The most general partial derivative of order m in t and x is $\partial_t^j \partial_x^\alpha$ with $j + |\alpha| = m$. Equivalently, it is $(\partial_t, \partial_x)^\beta$ with β an $\mathbb{N}^{1+d}$ multi-index with $|\beta| = m$.

For a partial differential operator of order m in t, the derivatives which occur are $\partial_t^j \partial_x^\alpha$, $j \le m$. The highest time derivative possible is ∂_t^m. In analogy with equation (1.2.1), we begin by considering a partial differential equation which is solved for this highest derivative. The equation then takes the form

$$\partial_t^m u = G(t, x, \partial_t^j \partial_x^\alpha u; j \le m - 1). \tag{2}$$

The notation means that G is a function of the variables t, x and the partial derivatives of order $\le m - 1$ in t.

EXAMPLE. The operator $u_{tx} = 0$ is *not* of the form (2), but $u_t = u_{xx}$ is.

Proposition 2. *If u is a smooth solution of a partial differential equation (2), then knowing*

$$\partial_t^\nu u(0, \cdot) = g_\nu(\cdot), \qquad \nu = 0, 1, \ldots, m-1,$$

on a neighborhood of $0 \in \mathbb{R}_x^d$ determines all the derivatives of u at $(0, 0)$.

PROOF. From the initial data compute $\partial_t^\nu \partial_x^\alpha u(0, \cdot) = \partial_x^\alpha g(\cdot)$ for $0 \le \nu \le m-1$. If $k \ge m$ and $\partial_t^\nu \partial_x^\alpha u(0, \cdot)$ is known for $\nu \le k-1$ and all α, then

$$\begin{aligned}\partial_t^k \partial_x^\alpha u &= \partial_t^{k-m} \partial_x^\alpha (G(t, x, \partial_t^j \partial_x^\alpha u; j \le m-1)) \\ &\equiv G_{k\alpha}(t, x, \partial_t^j \partial_x^\alpha u; j \le k-1).\end{aligned}$$

When $t = 0$, the arguments of $G_{k\alpha}$ are known on an $\mathbb{R}_x^d$ neighborhood of 0 by the inductive hypothesis. □

We have seen by example that:

(1) For real analytic ordinary differential equations with real analytic data the Taylor series converges (Cauchy's Theorem).
(2) For the same class of equations, the series need not converge if the data are not real analytic.
(3) For $(\partial_t - i\partial_x)u = 0$, real analytic data yields a series which converges.

This leads naturally to the question: Does the Taylor series of u always converge if G and g_ν are real analytic?

EXAMPLE (The Heat Equation). This is one of the fundamental partial differential equations of mathematical physics. In addition, it is the equation which guides our intuition about the class of *parabolic* equations.

We begin by presenting a derivation based on physical arguments. Suppose $\Omega \subset \mathbb{R}^3$ is occupied by homogeneous ($\equiv$ local physical properties translation invariant), isotropic ($\equiv$ local physical properties invariant under rotations), materials like air, water, jello, steel, etc. Let $u(t, x)$ denote the temperature at time t and place $x \in \Omega$. The second important physical quantity is the *heat current*, $\underline{J}(t, x)$, which gives the direction and speed at which heat is flowing at the point (t, x). The interpretation of $\underline{J}$ is that the flux per unit time through a piece of surface Σ is

$$\int_\Sigma \underline{J} \cdot n \, dA.$$

Thus, the rate, per unit time, at which heat leaves a volume V is $\int_{\partial V} \underline{J} \cdot n \, dA$. Using the Divergence Theorem yields

$$\text{Flux out of } V = \int_{\partial V} \underline{J} \cdot n \, dA = \int_V \operatorname{div} \underline{J} \, dx.$$

Two simple physical laws lead to an equation of motion for u.

The first fundamental law asserts that heat flows from hot to cold at a rate proportional to the temperature gradient. Thus the vector heat current is given by

$$\underline{J} \equiv \text{heat current} = -k \operatorname{grad}_x u.$$

The proportionality constant is called the *heat conductivity*.

The second law expresses the idea that a small volume, δV, of material heats up by an amount proportional to the quantity of heat which flows into it

$$c\frac{\partial u}{\partial t}|\delta V| \sim \text{rate at which heat flows into } \delta V,$$

where c is called the *heat capacity per unit volume*. The error in this approximation is no larger than

$$c\left(\operatorname*{osc}_V \frac{\partial u}{\partial t}\right)|\delta V| = o(1)|\delta V|$$

as the size of the δV tends to zero. Summing over small volumes comprising V yields

$$-\int_V c\frac{\partial u}{\partial t}\,dx = \text{rate at which heat flows out of } V.$$

Using our expression for the flux out of V yields

$$\int_V \left(\operatorname{div} \underline{J} + c\frac{\partial u}{\partial t}\right) dx = 0$$

for all nice subsets $V \subset \Omega$. It follows that we must have $c\partial_t u = -\operatorname{div} \underline{J}$ throughout Ω. Using the formula for $\underline{J}$ yields

$$c\frac{\partial u}{\partial t} = \operatorname{div}(k \operatorname{grad} u).$$

If k is constant this simplifies to

$$cu_t = k \sum \frac{\partial^2 u}{\partial x_i^2} \equiv k\Delta u.$$

Thus with $v \equiv k/c$, we have $u_t = v\Delta u$.

In many problems the hypothesis that c and k are constant is quite good. In others they may depend on t, x, u or even the derivatives of u or the values of u in the past (materials with memory). In any event, the case $v = \text{constant}$ is the starting point for any analysis.

The heat equation is not only of intrinsic interest but it serves as a test case for the question raised above. Consider the one-dimensional heat equation which is the equation for solutions u which do not depend on y, z, namely, $u_t = vu_{xx}$, $x \in \mathbb{R}$. The initial condition is $u(0, \cdot) = g$. The derivatives of u at $t = 0$ are computed as follows:

$$u_t = vu_{xx}, \qquad @t = 0, \qquad = vg_{xx},$$

$$u_{tt} = vu_{xxt} = v^2 u_{xxxx}, \qquad @t = 0, \qquad = (v\partial_x^2)^2 g,$$

$$\partial_t^{j_1}\partial_x^{j_2} u(0, \bar{x}) = (v\partial_x^2)^{j_1}(\partial_x)^{j_2} g(\bar{x}).$$

The Taylor expansion is

$$u \approx \sum \frac{(\nu\partial_x^2)^{j_1}(\partial_x)^{j_2} g(\underline{x})}{j_1!\,j_2!} t^{j_1}(x - \underline{x})^{j_2}.$$

This will usually *not* converge even if g has convergent Taylor series. The cause of the problem is that the coefficient of the jth power of t contains a derivative of order $2j$ of g. Problem 4 gives an example.

We have seen two "obstructions" to convergence:

(1) If either G or g_j is not real analytic, then the series need not converge.
(2) If the partial differential equation is not of highest order in ∂_t, that is, ∂_t^m is not the highest-order derivative that occurs, then the series may not converge.

If neither obstruction is present, the series does converge. This is the celebrated theorem of Cauchy–Kowaleskaya. The theorem concerns the initial value problem

$$\begin{cases} \partial_t^m u = G(t, x, \partial_t^j \partial_x^\alpha u;\ 0 \le j \le m-1, j + |\alpha| \le m), \\ \partial_t^j u(0, \cdot) = g_j(\cdot), \qquad 0 \le j \le m-1. \end{cases} \tag{3}$$

Theorem 3 (Cauchy–Kowaleskaya). *Suppose that g_j is real analytic on a neighborhood of $\underline{x} \in \mathbb{R}_x^d$ and that G is real analytic on a neighborhood of*

$$(0, \underline{x}, \partial_t^j \partial_x^\alpha g_j(\underline{x});\ j \le m-1, j + |\alpha| \le m).$$

Then there is a real analytic solution to (3) *defined on an $\mathbb{R}_t \times \mathbb{R}_x^d$ neighborhood of* $(0, \underline{x})$. *The solution is unique in the sense that if u and v are real analytic solutions of* (3) *defined on a connected neighborhood of* $(0, \underline{x})$, *then $u = v$.*

Proof. We have seen that (3) determines all the derivatives at $(0, \underline{x})$ of any solution. Thus, two real analytic solutions must agree to infinite order at $(0, \underline{x})$, and therefore must agree on any connected open set containing $(0, \underline{x})$ on which they are real analytic.

For the existence proof, one shows that the Taylor series computed above converges. Cauchy's method of majorants yields an elegant though lengthy proof. See the texts of Folland [Fo], Garabedian [Gara], or John [J] for details. The method of proof is, in my opinion, atypical within partial differential equations and if one is forced to omit things from a short introduction here is one place to start. □

Examples. The theorem applies to the first four equations but not to the last two:

$$u_t + iu_x = 0, \qquad \text{Cauchy–Riemann equation},$$

$$u_{tt} + \Delta u = 0, \qquad \text{Laplace equation},$$

$$u_{tt} - \Delta u = 0, \qquad \text{wave equation},$$

$$u_{tt} + u_x = 0 \qquad \text{sideways heat equation,}$$

$$u_t - \nu u_{xx} = 0, \qquad \text{heat equation,}$$

$$u_{tt} + u_{xxxx} = 0, \qquad \text{linearized beam equation.}$$

PROBLEMS

1. Prove the following elegant identities involving multi-index notation:

$$(x_1 + x_2 + \cdots + x_d)^m = \sum_{|\alpha|=m} \frac{m!\, x^\alpha}{\alpha!},$$

$$(1 - (x_1 + x_2 + \cdots + x_d))^{-1} = \sum \frac{|\alpha|!\, x^\alpha}{\alpha!}.$$

2. For $u_t - iu_x = 0$, $u(0, \cdot) = g(\cdot)$, g real analytic at zero, the Taylor series for u was dominated by

$$c \sum \frac{(j + k)!\, (Bt)^j (Bx)^k}{j!\, k!}.$$

Show that this power series converges on a neighborhood of $(0, 0)$. Prove that u, given by its convergent Taylor series, solves the initial value problem.

3. Consider the heat equation, $u_t = \nu u_{xx}$, $\nu > 0$, with initial value $g(x)$, a polynomial in x. Show that the Taylor series solution u has radius of convergence $R = \infty$. Show that for each t, u is a polynomial in x. Is u polynomial in t?

4. For the heat equation, $u_t = \nu u_{xx}$, $\nu > 0$, with real analytic initial data $g(x) = 1/(1 - ix)$, show that the Taylor series

$$\sum \frac{(\partial_{t,x})^\alpha u(0, 0)(t, x)^\alpha}{\alpha!}, \qquad \alpha \in \mathbb{N} \times \mathbb{N},$$

converges for no t, x with $t \neq 0$.

5. Suppose that $P(\partial_t, \partial_x) = \partial_t^m + \sum A_j(\partial_x)\partial_t^{m-j}$ where the A's are constant coefficient differential operators *of any order*. Generalizing Problem 3, show that if $g_j(x)$ is a polynomial in x for $0 \leq j \leq m - 1$, then the initial value problem $Pu = 0$, $\partial_t^j u(0, \cdot) = g_j(\cdot)$, $j \leq m - 1$, has a unique real analytic solution u defined on all of $\mathbb{R}_t \times \mathbb{R}_x^d$. Is u polynomial in t?

   DISCUSSION. If P is of order higher than m, then this solution will not be unique in the C^∞ category (see Problems 1.7.1 and 1.7.2, and §3.9). This is in contrast to Holmgren's Theorem to be studied shortly.

6. Use the Cauchy–Kowaleskaya Theorem to show that the initial value problem

$$u_t u_x = f(t, x, u), \qquad u(0, x) = g(x),$$

has a real analytic solution on a neighborhood of $(0, 0)$, provided that f is real analytic on a neighborhood of $(0, 0, g(0))$ and g is real analytic on a neighborhood of 0, and $g'(0) \neq 0$.

   Construct an example with $g'(0) = 0$, $g''(0) \neq 0$, g and f real analytic, and such that the initial value problem does not have even a C^1 solution on a neighborhood of $(0, 0)$.

7. Show that if the initial value problem $u_{tt} + u_{xx} = 0$, $u(0, \cdot) = 0$, $u_t(0, \cdot) = f(\cdot)$, has a C^2 solution on a neighborhood of (0, 0), then f and u must be real analytic on a neighborhood of (0, 0). *Hint*. Use the Schwarz reflection principle and the fact that harmonic functions are real analytic. For harmonic functions on $\mathbb{R}^2$, this can be proved by constructing a *harmonic conjugate* v satisfying $dv = u_t\,dx - u_x\,dt$. Then $u + iv$ is a holomorphic function of $x + it$, so its real part is C^∞.

§1.4. The Fully Nonlinear Cauchy–Kowaleskaya Theorem

The previous section was devoted to the Cauchy problem for nonlinear equations of order m which are solved for ∂_t^m in the sense of (1.3.2). The general case presents some additional phenomena.

EXAMPLE. For $t, x \in \mathbb{R} \times \mathbb{R}$, consider the initial value problem

$$u_t^2 + u_x^2 = 1, \qquad u(0, \cdot) = g(\cdot) \text{ real valued.}$$

First, observe that at $t = 0$

$$u_t(0, \cdot)^2 = 1 - g_x^2(\cdot). \tag{1}$$

If one seeks a real valued solution one must have $|g'| \le 1$. For complex solutions this constraint is not needed.

Second, note that (1) does *not* determine u_t. For $u_t(0, 0)$ there are two possibilities

$$u_t(0, 0) = \pm(1 - g_x(0)^2)^{1/2}.$$

Once the value of $u_t(0, 0)$ is chosen the rest follows, since near (0, 0) one solves

$$u_t^2 + u_x^2 = 1, \qquad u_t \approx u_t(0, 0),$$

uniquely by

$$u_t = \pm(1 - u_x^2)^{1/2}, \qquad \pm \text{ following the choice at } (0, 0).$$

Then the Cauchy–Kowaleskaya Theorem of the last section applies provided $(g_x)^2 \ne 1$. One finds two solutions. They are real if $|g'(0)| < 1$.

Consider next the general nonlinear equation of order m, where the derivative ∂_t^m plays a distinguished role

$$F(t, x, \partial_t^m u, \partial_t^j \partial_x^\alpha u; j \le m - 1, j + |\alpha| \le m) = 0. \tag{2}$$

Once $\partial_t^j u(0, \underline{x}) = g_j(\underline{x})$ for $j \le m - 1$ are known, $\partial_t^m u(\underline{x})$ must be determined by solving the nonlinear equation

$$F(t, x, \partial_t^m u, \partial_x^\alpha g_j) = 0.$$

As in the above example, there may be several solutions. Suppose that γ is a solution at $(0, \underline{x})$

$$F(0, \underline{x}, \gamma, \partial_x^\alpha g_j(\underline{x})) = 0. \tag{3}$$

To solve (2) for $\partial_t^m u$ with $\partial_t^m u \sim \gamma$ near $(0, \underline{x})$, the Implicit Function Theorem shows that if

$$\frac{\partial}{\partial s} F(0, \underline{x}, s, \partial_x^\alpha g_\gamma(\underline{x}))|_{s=\gamma} \neq 0, \tag{4}$$

then for $t, x \sim 0, \underline{x}$ and $\partial_t^m u \sim \gamma$, (2) is equivalent to

$$\partial_t^m u = G(t, x, \partial_t^j \partial_x^\alpha u; j \leq m - 1, j + |\alpha| \leq m)$$

with G real analytic if F is. The result of the last section immediately gives the fully nonlinear version of the Cauchy–Kowaleskaya Theorem.

Theorem 1 (Cauchy–Kowaleskaya). *Suppose that F and g_j are real analytic near $(0, \underline{x}, \gamma, \partial_x^\alpha g_j(\underline{x}))$ and $\underline{x}$, respectively, and that γ is a solution of* (3). *If in addition* (4) *is satisfied, then, on a neighborhood of* $(0, \underline{x})$, *there is a real analytic solution u to* (1) *with*

$$\partial_t^j u(0, \cdot) = g(\cdot), \qquad 0 \leq j \leq m - 1,$$

and

$$\partial_t^m u(0, 0) = \gamma. \tag{5}$$

Two such solutions defined on a connected neighborhood of $(0, \underline{x})$ *must be equal.*

The condition $\partial F/\partial(\partial_t^m u)\ (0, x, \partial_t^m u(0, x), \partial_t^j \partial_{t,x}^\alpha u(0, x)) \neq 0$ is very important. When it holds we say that the surface $t = 0$ is *noncharacteristic at* $(0, x)$ *on the solution u of the partial differential equation $F = 0$.*

The rest of this section is devoted to discussing several interpretations of this condition.

EXAMPLES. 1. For the equation $xu_t = u_x^2$, the surface $t = 0$ is noncharacteristic at all points $(0, x)$, $x \neq 0$.

2. For the equation $u_t^2 = u_x^2$, the surface $t = 0$ is noncharacteristic on the solution u at all x such that $u_t(0, x) \neq 0$.

3. For the equation $uu_t = u_x^2$, the surface $t = 0$ is noncharacteristic at $(0, x)$ on the solution u if and only if $u(0, x) \neq 0$.

4. If F is a linear partial differential operator

$$F = \sum_{|\alpha|+j \leq m} a_{j,\alpha}(t, x) \partial_x^\alpha \partial_t^j u - f(t, x).$$

Then $\partial F/\partial(\partial_t^m u) = a_{m,0}(t, x)$ the coefficient of ∂_t^m, and

$$F = a_{m,0}(t, x)\partial_t^m + \text{terms lower order in } \partial_t.$$

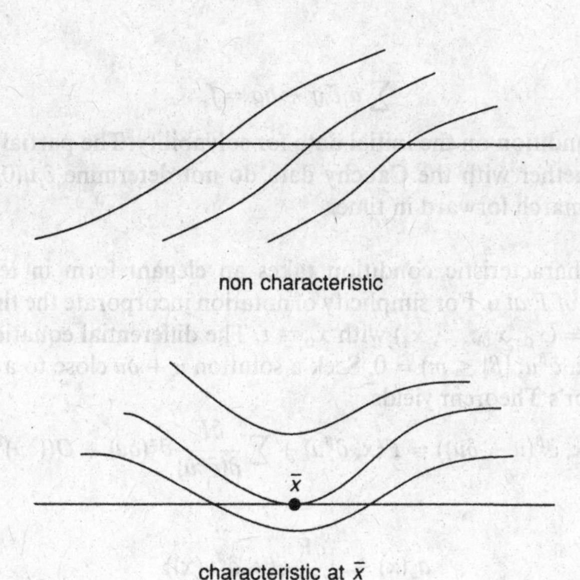

Figure 1.4.1

The noncharacteristic condition is then that $a_{m,0}(0, x) \neq 0$, in which case it is obvious that one can solve for ∂_t^m. In the linear case, the condition depends only on the equation and not on the solution. No Implicit Function Theorem is needed.

5. Even more special is when F is linear and of order $m = 1$,

$$F = a_0 \partial_t u + \sum a_i \partial_i u + bu - f(t, x).$$

The noncharacteristic condition is $a_0 \neq 0$. If a_0, a_i are real, this is equivalent to the condition that the vector field $a_0 \partial_t + \sum a_i \partial_i$ is transverse to $\{t = 0\}$ (Figure 1.4.1).

In the real noncharacteristic case, we can find C^k solutions of the initial value problem

$$F = a_0 \partial_t u + \sum a_i \partial_i u + bu - f(t, x) = 0,$$

$$u(0, \cdot) = g(\cdot),$$

for arbitrary $f, g \in C^k$, $k \geq 1$. The proof is by integrating along the integral curves of $a_0 \partial_t + \sum a_i \partial_i$. These are the characteristic curves and this is a simple case of the *method of characteristics*, generalizing the analysis of §1.1.

In the complex case, for example, $\partial_t - i\partial_x$, we have seen in §1.1 that real analyticity is indispensable for the solution of the initial value problem.

The surface $t = 0$ is characteristic at $(0, x)$ if and only if $a_0 = 0$. In that case, the partial differential operator involves only differentiations parallel to the initial surface. The differential equation, restricted to $t = 0$, yields

$$\sum a_i \partial_i g + bg = f,$$

which is a condition on the initial data for solvability. The partial differential equation together with the Cauchy data do not determine $\partial_t u(0, \cdot)$, so that you cannot march forward in time.

The noncharacteristic condition takes an elegant form in terms of the *linearization of F at u*. For simplicity of notation incorporate the time variable in x, thus $x = (x_0, x_1, \ldots, x_d)$ with $x_0 \equiv t$. The differential equation (2) takes the form $F(x, \partial^\beta u; |\beta| \le m) = 0$. Seek a solution $u + \delta u$ close to a given solution u. Taylor's Theorem yields

$$F(x, \partial^\beta(u + \delta u)) = F(x, \partial^\beta u) + \sum \frac{\partial F}{\partial(\partial^\alpha u)} \partial^\alpha(\delta u) + O((\delta u)^2).$$

Let

$$a_\alpha(x) \equiv \frac{\partial F}{\partial(\partial^\alpha u)}(x, \partial^\beta u(x)). \tag{6}$$

We are led to the equation $P(\delta u) = 0$, where $P = \sum a_\alpha(x)\partial^\alpha$ is the linear operator with coefficients a_α.

Definition. If $F(x, \partial^\beta u; |\beta| \le m) = 0$, the *linearization of F at u* is the linear partial differential operator $P(x, \partial) \equiv \sum a_\alpha(x)\partial^\alpha$.

If $P(x, \partial)v = 0$, then

$$F(x, \partial^\beta(u + v)) = O(v^2),$$

and

$$F(x, \partial^\beta(u + \varepsilon v)) = O(\varepsilon^2).$$

Thus $u + \varepsilon v$ satisfies the equation $F = 0$ to first order. Equivalently, $u + \varepsilon v$ is a solution in *first-order perturbation theory* (see also Problem 1). The equation $Pv = 0$ is sometimes called the *equation of variation* or *perturbation equation.*

These ideas are now illustrated with the *inviscid Burgers equation*

$$u_t + uu_x = 0.$$

This equation, for real valued u, arises in the study of the motion of fluids of very small viscosity. Air and water have small viscosity compared to honey and molasses. The linearization is the partial differential operator P defined by $Pv = (\partial_t + u\partial_x + u_x)v$ (Problem 2).

Suppose that $\underline{u}$ is a solution of the inviscid Burgers equation, and consider the perturbed initial value problem

$$u_t + uu_x = 0, \qquad u(0, x) = \underline{u}(0, x) + \varepsilon\varphi(x),$$

then a first approximation is given by $\underline{u} + \varepsilon v$ where v satisfies the pertubation

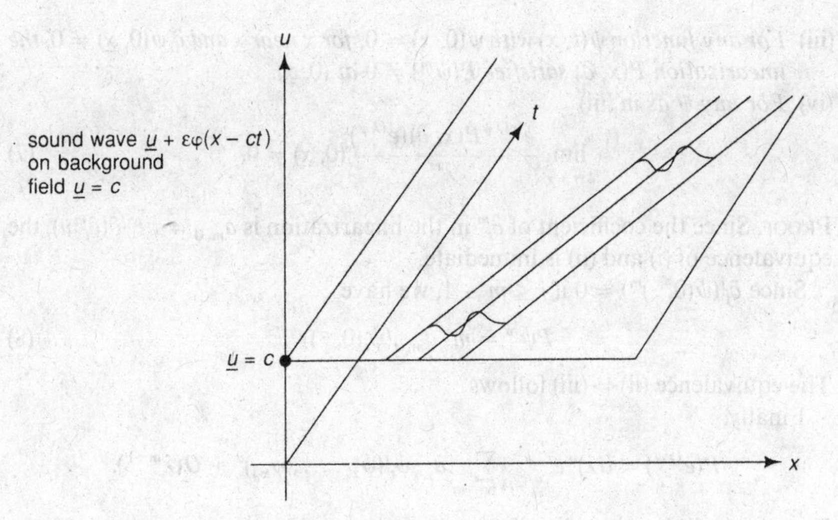

Figure 1.4.2

equation

$$v_t + \underline{u}v_x + \underline{u}_x v = 0, \qquad v(0, x) = \varphi(x).$$

This linear equation is much simpler than the nonlinear Burgers equation. In fact, along each integral curve of the vector field, $\partial_t + \underline{u}\partial_x$, it is a linear ordinary differential equation for v.

For example, if $\underline{u} = c \in \mathbb{R}$, then the perturbation equation is exactly the simple equation $v_t + cv_x = 0$ from §1.1 and $v = \varphi(x - ct)$. To first order in ε, small perturbations of real constant solutions are rigidly propagated at speed c. These small linearly propagating disturbances are called *sound waves* (Figure 1.4.2). Eventually, nonlinear effects predominate and this lineared approximation is inappropriate.

If u is a solution of the inviscid Burgers equation, then a C^1 curve Γ is characteristic at $p \in \Gamma$ if and only if the vector field $\partial_t + u\partial_x$ is tangent to Γ at p. Γ is called a *characteristic curve* if it is characteristic at all points. These are the same curves along which v, from the previous paragraph, satisfied an ordinary differential equation. They will reappear in §1.6.

For such a curve, the differential equation $u_t + uu_x = 0$ implies that u is constant on all components of Γ. Suppose that Γ is connected. The vector field $\partial_t + u\partial_x$ is then constant along Γ and also tangent to Γ which implies that Γ is a straight-line segment. These remarks will permit us to describe nonlinear effects alluded to two paragraphs back (see §1.9).

The last two conditions of the next theorem give coordinate invariant versions of the noncharacteristic condition.

Theorem 2. *The following are equivalent:*

(i) $\{t = 0\}$ *is noncharacteristic at* $(0, \underline{x})$ *for the solution* u *to* $F(x, \partial^\beta u; |\beta| \le m) = 0$.

(ii) $\{t = 0\}$ *is noncharacteristic at* $(0, \underline{x})$ *for the linearization of* F *at* u.

(iii) *For any function* $\psi(t, x)$ *with* $\psi(0, x) = 0$, *for* x *near* $\underline{x}$ *and* $\partial_t \psi(0, \underline{x}) \neq 0$, *the linearization* $P(x, \partial)$ *satisfies* $P(\psi^m) \neq 0$ *at* $(0, \underline{x})$.

(iv) *For any* ψ *as in* (iii)

$$\lim_{\lambda \to +\infty} \frac{e^{-i\lambda\psi} P(x, \partial)(e^{i\lambda\psi})}{\lambda^m}(0, \underline{x}) \neq 0. \tag{7}$$

PROOF. Since the coefficient of ∂_t^m in the linearization is $a_{m,0} = \partial F/\partial(\partial_t^m u)$, the equivalence of (i) and (ii) is immediate.

Since $\partial_t^j(\psi(0, \cdot)^m) = 0$ if $j \leq m - 1$, we have

$$P\psi^m = m!\, a_{m,0}(\psi_t(0, \cdot))^m. \tag{8}$$

The equivalence (ii) $\Leftrightarrow$ (iii) follows.

Finally,

$$P(e^{i\lambda\psi}) = (i\lambda)^m e^{i\lambda\psi} \sum_{j+|\alpha|=m} a_{j,\alpha} \psi_t^j (\psi_{x_1}, \ldots, \psi_{x_d})^\alpha + O(\lambda^{m-1}).$$

For our ψ all the $\psi_{x_i} = 0$ at $(0, \underline{x})$, whence

$$P(e^{i\lambda\psi})(0, x) = (i\lambda)^m e^{i\lambda\psi} a_{m,0}(0, x) \psi_t(0, x)^m + O(\lambda^{m-1}). \tag{9}$$

The equivalence of the previous conditions with (iv) follows. □

The formulas (8), (9) show that if (iii) or (iv) holds for one such ψ, then it holds for all of them.

PROBLEMS

1. Suppose that I is an interval in $\mathbb{R}$, $0 \in I$, and that $u(\sigma, x)$ is a smooth one-parameter family of solutions of $F(x, \partial_x^\beta u; |\beta| \leq m) = 0$, that is, $F(x, \partial_x^\beta u(\sigma, x)) = 0$ for all $\sigma \in I$. Let P be the linearization of F at $u(0, \cdot)$ and let $v = \partial u/\partial\sigma(0, \cdot)$. Prove that $Pv = 0$.
DISCUSSION. Since $u(\sigma, x) = u(0, x) + \sigma v + O(\sigma^2)$, we see for the second time that the linearization describes first-order changes in solutions of $F(x, \partial_x^\beta u) = 0$.

2. Show that the linearizations of $u_t + uu_x = 0$ and $u_t + (u_x)^2 = 0$ are $Pv = \partial_t v + u\partial_x v + u_x v$ and $Pv = \partial_t v + 2u_x \partial_x v$, respectively.

3. Consider again the initial value problem $u_t + uu_x = 0$, $u(0, x) = c + \varepsilon\varphi(x)$. We found $\underline{u} + \varepsilon v$ which satisfied the initial condition and $u_t + uu_x = o(\varepsilon)$. Find a corrected expansion $\underline{u} + \varepsilon v + \varepsilon^2 w$ which improves the error to $o(\varepsilon^2)$. *Hint.* Plug $\underline{u} + \varepsilon v + \varepsilon^2 w$ into the equation and set leading terms in ε equal to zero. This gives an independent derivation of the perturbation equation at the same time.
DISCUSSION. This is an example of higher order perturbation theory.

§1.5. Cauchy–Kowaleskaya with General Initial Surfaces

In many situations, initial value problems are natural but a distinguished time variable t is not available. For example, the wave operator $\partial_0^2 - \partial_1^2 - \partial_2^2 - \partial_3^2$ is Lorentz invariant (§4.6 begins with a discussion of invariant operators).

Here, planes $\sum a_i x_i = 0$ are candidates for initial hypersurfaces with corresponding time variable $t = \sum a_i x_i$. All planes with $a_0^2 - a_1^2 - a_2^2 - a_3^2 > 0$ are equivalent by Lorentz transformations. The principle of special relativity implies that all such time functions should be treated on an equal footing. For nonlinear $t(x_0, \ldots, x_3)$, the condition becomes $(\partial_0 t)^2 - (\partial_1 t)^2 - (\partial_2 t)^2 - (\partial_3 t)^2 > 0$, and the equivalence of all such is in the spirit of general relativity.

Another example arises in searching for isometric embeddings $M_2 \to \mathbb{R}^3$ of a Riemanian two-manifold M of negative scalar curvature (Spivak, [Sp]). One solves an "initial" value problem on the manifold but there is no natural time variable or initial curve. These examples suggest the importance of the following.

Problem. For a partial differential equation of order m

$$F(x, \partial^\alpha u; |\alpha| \le m) = 0, \tag{1}$$

and a smooth hypersurface, Σ in $\mathbb{R}^d_x$ study the Cauchy problem with initial data given on Σ.

As one no longer has a time, t, it is no longer reasonable to prescribe $\partial_t^j u$, $0 \le j \le m - 1$. If there were a distinguished variable t, the data $\partial_t^j u(0, \cdot) = g_j(\cdot)$, $0 \le j \le m - 1$, would determine all derivatives of u of order $\le m - 1$. Thus one could hope to give as data all derivatives of u up to order $m - 1$ along the surface Σ. However, the functions $\partial^\alpha u|_\Sigma = g_\alpha$ are not independent. There are compatibility conditions. For example, if $\Sigma = \{x_1 = 0\}$ and $\alpha = (0, \alpha_2, \ldots, \alpha_n)$ and $|\alpha + \beta| \le m - 1$, then $\partial^\alpha(\partial^\beta u|_\Sigma) = \partial^{\alpha+\beta} u|_\Sigma$.

A common formulation of the Cauchy problem involving the "normal derivatives" $(\partial/\partial n)^j u|_\Sigma$ is not correct (see Problem 1).

A good way to account automatically for the compatibility relations among the derivatives is to ask that the derivatives of u be equal to the derivatives of a given function.

> Given an $m - 1$ times differentiable function v defined on a neighborhood of Σ, find a solution u to (1) such that
>
> $$\partial^\alpha u = \partial^\alpha v \quad \text{on } \Sigma \quad \text{for all } |\alpha| \le m - 1. \tag{2}$$

Knowing all the derivatives of order $\le m - 1$ determines all but one of the derivatives of order $\le m$. If $\Sigma = \{x_1 = 0\}$ the missing derivative is $\partial_1^m u$. For nonlinear problems, one must supply that additional derivative at one point of Σ as in the Cauchy–Kowaleskaya Theorem (1.4.1). The general case follows that pattern once the notion of noncharacteristic is defined.

Definition. The *linearization* of F at a solution u to $F(x, \partial^\beta u; |\beta| \le m) = 0$ is the linear operator

$$P = \sum a_\alpha(x) \partial^\alpha, \qquad a_\alpha(x) \equiv \frac{\partial F}{\partial(\partial^\alpha u)}(x, \partial^\beta u(x)).$$

Definition. If $F(\bar{x}, \partial^\beta u(\bar{x}); |\beta| \le m) = 0$, then the hypersurface Σ is *noncharacteristic* for F on u at $\bar{x}$, if and only if the following equivalent conditions hold:

(A) For any real valued C^∞ function ψ defined on a neighborhood of $\bar{x}$ with $\psi|_\Sigma = 0$, $d\psi|_\Sigma \neq 0$, we have $P\psi^m(\bar{x}) \neq 0$.
(B) For any ψ as in (A)

$$\lim_{\lambda\to\infty} \frac{e^{-i\lambda\psi}P(x,\partial)e^{i\lambda\psi}}{\lambda^m} \neq 0 \quad \text{at } \bar{x}.$$

Remarks. 1. $d\psi = \sum(\partial\psi/\partial x_j)\,dx_j$ is the differential of ψ.

2. If (A) or (B) holds for one ψ it hold for any such (exercise).

3. To check if Σ is noncharacteristic at $\bar{x}$ it is sufficient to know $\partial^\alpha u(\bar{x})$, for all $|\alpha| \leq m$. One does *not* need to know u on a neighborhood of $\bar{x}$.

4. In the special case $\Sigma = \{t = 0\}$, (A) and (B) become conditions (iii) and (iv) of Theorem 1.4.2.

Theorem 1. *Suppose that:*

(1) $\bar{x} \in \Sigma \subset \mathbb{R}^d_x$ *and* Σ *is a real analytic hypersurface;*
(2) v *is real analytic on a neighborhood of* $\bar{x} \in \mathbb{R}^d$, *and that* $F(x, \partial^\beta v(x)) = 0$ *for* x *in* Σ;
(3) Σ *is noncharacteristic for* F *on* v *at* $\bar{x}$;
(4) F *is real analytic on a neighborhood of* $(\bar{x}, \partial^\beta v(\bar{x}))$.

Then there is a u, *real analytic on a neighborhood* Ω *of* $\bar{x}$, *such that*

$$F(x, \partial^\alpha u(x)) = 0 \quad \text{in } \Omega,$$

$$\partial^\alpha u|_{\Sigma\cap\Omega} = \partial^\alpha v|_{\Sigma\cap\Omega} \qquad \text{for all} \quad |\alpha| \leq m-1,$$

$$\partial^\alpha u(\bar{x}) = \partial^\alpha v(\bar{x}) \qquad \text{for all} \quad |\alpha| = m,$$

Two such solutions defined on a connected neighborhood of $\bar{x}$ *must be equal.*

PROOF. Introduce new real analytic variables so that $\Sigma = \{t = 0\}$. Theorem 1.4.1 immediately implies the above result, once one notes that the hypotheses of Theorem 1 are expressed in a coordinate independent way, and that they reduce to the hypotheses of Theorem 1.4.1 in coordinates so that $\Sigma = \{t = 0\}$. □

PROBLEMS

It is common to pose the Cauchy problem as follows:

Find a function u such that

$$F(x, \partial^\beta u) = 0, \qquad \left(\frac{\partial}{\partial n}\right)^j u = g_j \quad \text{on } \Sigma, \qquad 0 \leq j \leq m-1,$$

where g_j are given functions on Σ, and $\partial/\partial n = \sum n_i(x)\,\partial/\partial x_i$ is the derivative in the direction of the unit normal to Σ.

There are two serious problems with this formulation. First, in order to choose a normal along Σ, one needs a Riemannian metric. More telling is that, even in the Riemannian case, $(\partial/\partial n)^2$ is not meaningful. To see this, note that

$$\left(\sum n_i(x)\partial/\partial x_i\right)\left(\sum n_i(x)\partial/\partial x_i\right)u$$

involves all of the partials of $n(x)$. As n is defined only on Σ, only tangential derivatives exist. Thus for some j, $\partial n/\partial x_j$ is not meaningful. One solutions is to extend the $n_j(x)$ so as to define a vector field on a neighborhood of Σ. The results depend on the extension.

1. Construct an example showing that the value of $(\partial/\partial n)^2 u$ may be different for different extensions.

A way to avoid both difficulties is to drop the idea of normal derivatives and settle for differentiations transverse to Σ (that is, nowhere tangent to Σ). This leads to the following formulation of the Cauchy problem.

> Given V, a smooth vector field defined on a neighborhood of (and transverse to) Σ, find a function u such that
>
> $$F(x, \partial^\beta u) = 0, \qquad (V)^j u = g_j \quad \text{on } \Sigma, \qquad 0 \le j \le m - 1.$$

The next problem shows that this formulation is equivalent to prescribing consistently all derivatives of order $\le m - 1$. The present formulation is more appealing geometrically, but requires a choice of V which is not canonical.

2. Given $x \in \Sigma$ and $g_j \in C^\infty(\omega)$, $0 \le j \le m - 1$, $\omega \subset \Sigma$ a neighborhood of x, prove that there is a smooth v defined on Ω an $\mathbb{R}^d$ neighborhood of x so that $(V)^j v = g_j$ on $\Sigma \cap \Omega$, $0 < j < m - 1$. Show that the g_j determine all the derivatives of v up to order $m - 1$ by proving that if w is a second such function, then $\partial^\alpha(v - w) = 0$ on $\Sigma \cap \Omega$ whenever $|\alpha| \le m - 1$.
3. Suppose that $P(x, \partial)$ is a linear partial differential operator with coefficients $a_\alpha(x)$ real analytic on a neighborhood of $\underline{x}$. Suppose, in addition, that the principal part at $\underline{x}$, $\sum_{|\alpha|=m} a_\alpha(\underline{x})\partial^\alpha$, is nonzero. Prove that for any $f(x)$ real analytic on a neighborhood of $\underline{x}$, there is a possibly smaller neighborhood Ω of $\underline{x}$ and $u \in C^\omega(\Omega)$ such that $Pu = f$ in Ω. *Hint*. Show that there is a hyperplane which is noncharacteristic at $\underline{x}$, then solve an initial value problem.
   DISCUSSION. This result shows that linear P are locally solvable in the real analytic category. In particular, this shows that there is no obstruction to the solvability of $Pu = f$ comparable, for example, to the condition $d\gamma = 0$ as the solvability condition of $d\omega = \gamma$.

It came as a surprise to the mathematical community when H. Lewy found, in 1956, a P as above, such that $Pu = f$ is not locally solvable at $\underline{x}$ for most $f \in C^\infty$ (see Garabedian [Gara] or Folland [Fo]).

§1.6. The Symbol of a Differential Operator

Given a solution of a nonlinear partial differential equation $F(x, \partial^\beta u; |\beta| \le m) = 0$, we have seen how to associate an mth order linear operator, the *linearization of F at u.* The recipe is

$$P(x, \partial) = \sum a_\alpha(x)\partial^\alpha,$$

$$a_\alpha(x) = \frac{\partial F}{\partial(\partial^\alpha u)}(x, \partial^\beta u(x)).$$

To $P(x, \partial)$ we associate a function $P(x, i\xi)$ of x and $\xi \in \mathbb{C}^d$, by replacing $(\partial_1, \ldots, \partial_d)$ by $(i\xi_1, \ldots, i\xi_d)$,

$$P(x, i\xi) = \sum a_\alpha(x)(i\xi)^\alpha.$$

The function $P(x, i\xi)$ is called the *complete symbol* of the differential operator $P(x, \partial)$. It is a polynomial in ξ of degree m whose coefficients depend on x. The regularity of the coefficients depends on the regularity of F and u. The reason for the i will be clear later. Let

$$D_j \equiv \frac{1}{i}\partial_j, \qquad D \equiv \frac{1}{i}(\partial_1, \ldots, \partial_d).$$

Then $c_\alpha(x)D^\alpha$ has symbol $c_\alpha(x)\xi^\alpha$.

Definition. The *principal symbol* of $P = \sum a_\alpha(x)D^\alpha$ is the function

$$P_m(x, \xi) \equiv \sum_{|\alpha|=m} a_\alpha(x)\xi^\alpha,$$

P_m is a homogeneous polynomial of degree m in ξ. One of the basic themes in partial differential equations is to associate properties of the operator $P(x, D)$ with algebraic/geometric properties of the symbols of P which can in principle be verified.

Examples of Symbols

Operator	Principal symbol
$\partial_t + c\partial_x$	$i\tau + ic\xi$
Δ_x	$-\xi_1^2 - \xi_2^2 - \cdots - \xi_d^2$
$\partial_{tt} - \Delta_x$	$-\tau^2 + \xi_1^2 + \cdots + \xi_d^2$
$\partial_t - \Delta_x$	$\xi_1^2 + \cdots + \xi_d^2$

In proving Theorem 1.4.2(iv), we evaluated $D^\alpha e^{i\lambda\varphi}$, noting that to get highest order in λ the derivatives must all fall on the exponent. Thus, for smooth real φ,

$$D^\alpha e^{i\lambda\varphi} = \lambda^{|\alpha|}e^{i\lambda\varphi}\left(\frac{\partial\varphi}{\partial x_1}, \ldots, \frac{\partial\varphi}{\partial x_d}\right)^\alpha + O(\lambda^{|\alpha|-1}).$$

This yields the fundamental asymptotic expansion.

Theorem 1 (Fundamental Asymptotic Expansion). *If φ is a smooth real valued function, then as $|\lambda| \to \infty$*

$$\begin{aligned} e^{-i\lambda\varphi}P(x, D)e^{i\lambda\varphi} &= \lambda^m P_m\left(x, \frac{\partial\varphi}{\partial x_1}, \ldots, \frac{\partial\varphi}{\partial x_d}\right) + O(\lambda^{m-1}) \\ &= \lambda^m P_m(x, d\varphi) + O(\lambda^{m-1}). \end{aligned} \tag{1}$$

This result shows that the principal symbol is particularly important when considering highly oscillatory functions.

In new coordinates, $y = Y(x)$ with inverse $x = X(y)$, a linear partial differential operator in ∂_x is transformed to a linear partial differential operator in ∂_y. Precisely, if u is a function of x, then $\tilde{u}(y) \equiv u(X(y))$ is the corresponding function of y. Similarly, $(Pu) \circ X$ is the function of y corresponding to Pu. Thus the operator P viewed in the y coordinates is the map sending $u \circ X$ to $(Pu) \circ X$. For example, the operator ∂_j viewed in the y variables is given by the familiar law

$$\frac{\partial}{\partial x_j} = \sum \frac{\partial y_k}{\partial x_j} \frac{\partial}{\partial y_k}.$$

It follows that the map P viewed in the y variables is a differential operator which we denote by $\widetilde{P}(y, D_y)$.

EXAMPLE. $\Delta \equiv \partial_1^2 + \partial_2^2$ in polar coordinates r, ϑ is equal to

$$\frac{1}{r}\partial_r r \partial_r + \frac{1}{r^2}\partial_{\vartheta\vartheta}.$$

The relation between P and $\widetilde{P}$ is

$$(P(x, D_x)u) \circ X = \widetilde{P}(y, D_y)(u \circ X). \tag{2}$$

Many interesting analytic properties of P have expressions which are independent of coordinates. For example,

"The Cauchy problem with data on Σ is solvable."
"All solutions of $Pu = 0$ are C^∞."

If we expect that these correspond to properties of the symbol, then the symbol itself should have reasonable transformation properties under change of coordinates. A natural question is What is the relation between the symbol of the transformed operator and that of the original? Using formula (1) for $P_m(x, d\varphi)$ yields

$$P_m(x, d_x\varphi) = \lim_{\lambda\to\infty} \frac{e^{-i\lambda\varphi(x)}P(x, D)e^{i\lambda\varphi(x)}}{\lambda^m},$$

$$\widetilde{P}_m(y, d_y\tilde{\varphi}) = \lim_{\lambda\to\infty} \frac{e^{-i\lambda\tilde{\varphi}(y)}\widetilde{P}(y, D_y)e^{i\lambda\tilde{\varphi}(y)}}{\lambda^m}.$$

Equation (2), applied with $u = e^{i\lambda\varphi}$, shows that the right-hand sides are equal at corresponding points x and $Y(x)$. We next interpret this important conclusion.

The differential $d\varphi = \sum (\partial\varphi/\partial x_j)\, dx_j$ is a one-form. Equivalently, $(\partial\varphi/\partial x_1(x), \ldots, \partial\varphi/\partial x_d(x))$ transforms as a covector, that is, an element of the dual, $T_x^*(\mathbb{R}^d)$, of the tangent space $T_x(\mathbb{R}^d)$. This part of advanced calculus is sometimes unfamiliar. Here is a brief exposition (see Spivak [Sp] or Loomis and Sternberg [LS] for detailed treatment). The goal is a geometric foundation for differential calculus so that invariants under nonlinear coordinate changes are easily recognized.

A *tangent vector* v to $\mathbb{R}^d$ at x (i.e. $v \in T_x(\mathbb{R}^d)$) is visualized as a vector with tail at x and/or as the tangent vector to a curve passing through x. The set of all tangent vectors at x is called the *tangent space* at x and is denoted $T_x(\mathbb{R}^d)$. The set of all pairs x, v with $v \in T_x(\mathbb{R}^d)$ is the *tangent bundle* $T(\mathbb{R}^d)$. Under a change of coordinates, $y = Y(x)$, v is transformed to the vector $Y_* v = Y'(x)v$ with tail at $Y(x)$, that is, $Y_* v \in T_{y(x)}(\mathbb{R}^d)$. Here Y' is the Jacobian matrix $\partial y_i/\partial x_j$ of Y. The map $x, v \mapsto y, Y_* v$ is the transformation law for tangent vectors. If $\gamma(t)$ is a curve with $\gamma(0) = x$, $\gamma'(0) = v$, then $Y(\gamma(t))'|_{t=0} = Y_* v$ is tangent to the curve $Y \circ \gamma$ corresponding to γ (Figure 1.6.1).

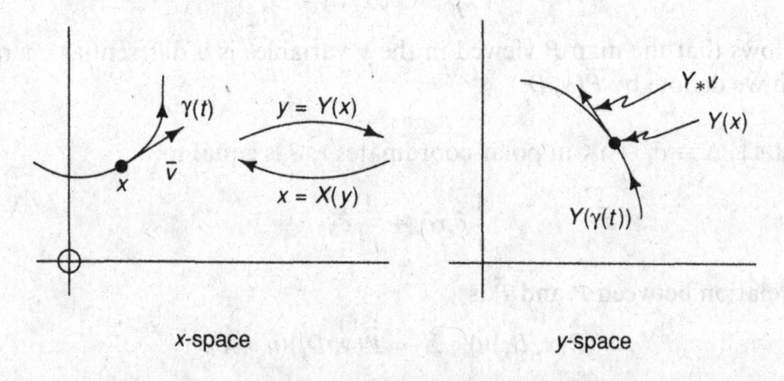

Figure 1.6.1

The differential of φ acts on tangent vectors by $d\varphi(x)(v) = \sum(\partial\varphi(x)/dx_j)v_j = d\varphi(\gamma(t))/dt|_{t=0}$. One can think of this as measuring the rate at which $\gamma(t)$ or the vector v at x cuts the level surfaces of φ (Figure 1.6.2). The fact that $d\varphi$ transforms as a one-form means that computing $d\varphi$ in x coordinates on v gives the same answer as computing $d\varphi$ in y coordinates on $Y_* v$. This is clear from the level surface interpretation. More formally, one has $d\varphi(x)(v) = d\tilde{\varphi}(y(x))(Y_* v)$ where $\tilde{\varphi} \equiv \varphi \circ X$ denotes the function corresponding to φ in the y coordinates. Written out, the identity is

$$\sum \frac{\partial\varphi(x)}{dx_j} v_j = \sum \frac{\partial\tilde{\varphi}(y)}{dy_j} (Y_* v)_j.$$

This can be verified by brute calculation using the chain rule. Alternatively, note that $\varphi(\gamma(t)) = \tilde{\varphi}(Y(\gamma(t)))$. Differentiating with respect to t at $t = 0$ proves it.

The set of all points ξ in the dual of the vector space $T_x(\mathbb{R}^d)$ is called the

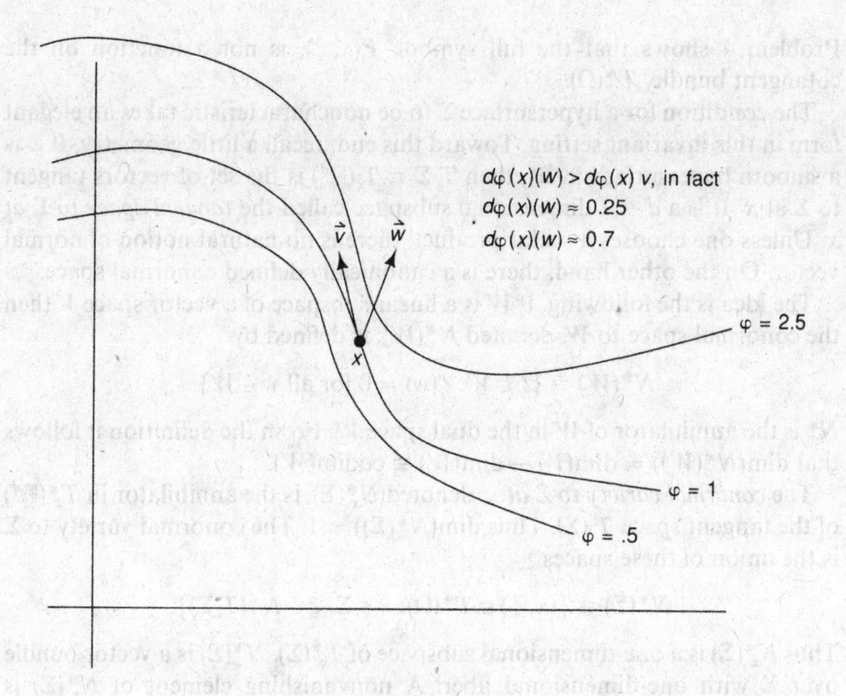

Figure 1.6.2

cotangent space at x and is denoted $T_x^*(\mathbb{R}^d)$. An element $\xi \in T_x^*(\mathbb{R}^d)$ can be visualized by imagining the level sets of ξ which are a family of parallel hypersurfaces in $T_x(\mathbb{R}^d)$ given by the equations $\xi(v) = \text{const}$. Then $\xi(v)$ "counts the number of level surfaces cut by v." The set of all pairs x, ξ with $\xi \in T_x^*(\mathbb{R}^d)$ is called the *cotangent bundle* $T^*(\mathbb{R}^d)$. The computations of the previous paragraph show that the pair x, $d\varphi(x)$ transforms as an element of the cotangent bundle.

Just as the functions $u, \tilde{u}$ take the same value at corresponding points x and y, we have shown that the principal symbols of P and $\tilde{P}$ take on the same values at x, $d_x\varphi(x)$ and y, $d_y\tilde{\varphi}(y)$ which are corresponding points of the cotangent bundle $T^*(\mathbb{R}^d)$. This proves the following theorem.

Theorem 2. *If $P = P(x, D)$ is a linear partial differential operator defined in Ω, then its principal symbol P_m is a well-defined function on the cotangent bundle, $T^*(\Omega)$.*

To find $P_m(\bar{x}, \bar{\xi})$ for $\bar{x}, \bar{\xi} \in T^*(\Omega)$, one need only choose a real valued φ with $d\varphi(\bar{x}) = \bar{\xi}$. Then

$$P_m(\bar{x}, \bar{\xi}) = \lim_{\lambda \to \infty} \frac{e^{-i\lambda\varphi} P e^{i\lambda\varphi}}{\lambda^m}(\bar{x}).$$

Note that this recipe does not depend on the particular coordinate system.

Problem 4 shows that the full symbol, $P(x, \xi)$, is not a function on the cotangent bundle, $T^*(\Omega)$.

The condition for a hypersurface Σ to be noncharacteristic takes an elegant form in this invariant setting. Toward this end, recall a little geometry. If Σ is a smooth hypersurface in $\mathbb{R}^d$, then $T_x\Sigma \subset T_x(\mathbb{R}^d)$ is the set of vectors tangent to Σ at x. It is a $d - 1$ dimensional subspace called the *tangent space to Σ at x*. Unless one chooses a scalar product, there is no natural notion of normal vector. On the other hand, there is a canonically defined conormal space.

The idea is the following. If W is a linear subspace of a vector space V then the conormal space to W, denoted $N^*(W)$, is defined by

$$N^*(W) \equiv \{\ell \in V' : \ell(w) = 0 \text{ for all } w \in W\}.$$

N^* is the annihilator of W in the dual space V'. From the definition it follows that $\dim(N^*(W)) = \dim(V) - \dim(W) \equiv \operatorname{codim}(W)$.

The *conormal variety to Σ at x*, denoted $N_x^*(\Sigma)$, is the annihilator in $T_x^*(\mathbb{R}^d)$ of the tangent space $T_x(\Sigma)$. Thus $\dim(N^*(\Sigma)) = 1$. The conormal variety to Σ is the union of these spaces

$$N^*(\Sigma) \equiv \{(x, \xi) \in T^*(\Omega) : x \in \Sigma, \xi \in N^*(T_x\Sigma)\}.$$

Thus $N_x^*(\Sigma)$ is a one-dimensional subspace of $T_x^*(\Sigma)$. $N^*(\Sigma)$ is a vector bundle over Σ with one-dimensional fiber. A nonvanishing element of $N_x^*(\Sigma)$ is called a *conormal to Σ at x*. If $\varphi \in C^\infty(\Omega : \mathbb{R})$ and $\varphi|_\Sigma = 0$, then for $x \in \Sigma$, $d\varphi(x) \in N_x^*(\Sigma)$.

Recall that Σ is noncharacteristic at x if for such a φ with $d\varphi(x) \neq 0$, $\lim e^{-i\lambda\varphi} P e^{i\lambda\varphi}/\lambda^m \neq 0$ at x. That is, Σ is noncharacteristic if and only if $P_m(x, d\varphi(x)) \neq 0$. As $(x, d\varphi(x))$ generates $N_x^*(\Sigma)$, this proves the following result.

Proposition 3. *The hypersurface Σ is noncharacteristic at $x \in \Sigma$ for $P(x, D)$ if and only if $P_m \neq 0$ on $N_x^*(\Sigma)\backslash 0$.*

Definition. A smooth hypersurface is *characteristic at* x if and only if $P_m(x, \xi) = 0$ for all $\xi \in N_x^*(\Sigma)$. A surface which is characteristic (resp. noncharacteristic) at all points is called a *characteristic surface* (resp. *noncharacteristic surface*).

In case $d = 2$, hypersurfaces have dimension 1 and the name characteristic curves is natural. We met such a situation in §1.1 and §1.4.

If a surface is given by $\varphi(x) = 0$ with φ real valued and satisfying $d\varphi(x) \neq 0$, then Σ is characteristic at x if and only if $P_m(x, d\varphi(x)) = 0$. This is called the *eikonal equation* for φ.

For solutions of nonlinear equations, whether or not a surface is characteristic, depends not only on the surface but also on the solution. One applies the above criteria with P equal to the linearization of F at u.

Examples. 1. Find all the characteristic lines for $\partial_t + c\partial_x$.

Solution. For $\tau, \xi \in \mathbb{R}^2\backslash 0$ the line $L = \{(t, x)\colon \tau t + \xi x = \text{constant}\}$ has conormal variety

$$N^*(L) = \{(t, x, \lambda\tau, \lambda\xi)\colon t, x \in L \text{ and } \lambda \in \mathbb{R}\}.$$

In order to be characteristic the principal symbol of P must vanish at these points, thus

$$P_1(\tau, \xi) = i\tau + ic\xi = 0, \qquad \text{that is,} \quad \tau = -c\xi.$$

If $c \in \mathbb{C}\backslash\mathbb{R}$, no real solution exists and therefore all lines are noncharacteristic. Otherwise, $(\tau, \xi) = \text{constant}(-c, 1)$ is the general solution, so the characteristic lines have equation, $x - ct = \text{constant}$. Note that this recovers the same lines which played such an important role in the analysis of §1.1. □

2. Find all characteristic hyperplanes for the wave operator

$$u_{tt} - c^2\Delta u \equiv Pu, \qquad c \in \mathbb{R}_+.$$

Solution. The hyperplane with equation $\tau t + \langle \xi, x\rangle = \text{constant}$ has conormal τ, ξ. The equation $P(\tau, \xi) = 0$ reads, $-\tau^2 + c^2|\xi|^2 = 0$, which has general solution, $\tau = \pm c|\xi|$. Multiplying ξ and τ by the same nonzero constant leaves the hyperplane unchanged. Since there are no nontrivial solutions with $\xi = 0$, it suffices to consider those ξ with $|\xi| = 1$. The most general characteristic hyperplane has equation

$$\text{constant} = \langle x, \xi\rangle \pm ct, \qquad \xi \in \mathbb{R}^d, \quad |\xi| = 1.$$

Planes are noncharacteristic if and only if $\tau^2 \neq c^2|\xi|^2$. □

3. Find all the characteristic curves at a real solution u of the inviscid Burgers equation.

Solution. The linearized operator is $\partial_t + u\partial_x + u_x$ which has principal symbol equal to $i\tau + u(t, x)i\xi$. If Γ is a curve which is characteristic at t, x, then its conormal τ, ξ at t, x must satisfy $\tau + u(t, x)\xi = 0$. It follows that the tangent to Γ at t, x is parallel to the vector with components $(1, u(t, x))$. Thus $\partial_t + u\partial_x$ must be tangent to Γ, so the characteristic curves are exactly the integral curves of Γ which played an important role in the perturbation theory in §1.4. □

Definition. A linear partial differential operator $P(x, D)$ is called *elliptic* at x if and only if $P_m(x, \xi) \neq 0$ for all $\xi \in \mathbb{R}^d\backslash 0$. It is elliptic on $\Omega \subset \mathbb{R}^d$ if it is elliptic at each point of Ω.

EXAMPLES. 1. If $c(x) \notin \mathbb{R}$, then $\partial_t + c\partial_x$ is elliptic at x.

2. Δ is elliptic.

3. $\partial_t + c(t, x)\partial_x$ is *not* elliptic at t, x if $c(t, x) \in \mathbb{R}$.

Definition. For $\underline{x}$ fixed the set of real $\xi \neq 0$ with $P_m(\underline{x}, \xi) = 0$ is called the *characteristic variety of P at $\underline{x}$*. If P is defined on a open set Ω, then the characteristic variety of P is the set

$$\operatorname{char}(P) \equiv \{(x, \xi) \in T^*(\Omega)\backslash 0: P_m(x, \xi) = 0\}.$$

Here $T^*(\Omega)\backslash 0$ denotes the set of x, ξ with $\xi \neq 0$.

The characteristic variety at x is invariant under multiplication by nonzero constants and is closed in $T^*_{\underline{x}}\backslash 0$. If the coefficients of P are continuous, $\operatorname{char}(P)$ is a closed subset of $T^*(\Omega)\backslash 0$ invariant under multiplication by nonzero constants in the second, or fiber, variable. An operator is elliptic in Ω if and only if its characteristic variety is the empty set.

A point $\underline{x}$, ξ is in the characteristic variety if and only if $P_m(\underline{x}, D)e^{ix\xi} = 0$. Here the "$i$" convention in D is convenient and by $P(\underline{x}, D)$ we mean the constant coefficient operator whose coefficients are $a_\alpha(\underline{x})$.

Problems

1. For each of the following partial differential operators on $\mathbb{R}^2$, find all characteristic lines:
(a) ∂_t; (b) ∂_x; (c) $\partial_t\partial_x$; (d) $\partial_t^2 + \partial_t\partial_x$; (e) $\partial_t^2 + \partial_t\partial_x + \partial_x^2$;
(f) $i\partial_t - \partial_x^2$; (g) $\partial_{tt} + \partial_{xx}$.

2. If $P(D)$ is a nonzero partial differential operator of degree m, show that the set of $\xi \in \mathbb{R}^d$ such that $P(\xi) = 0$ is a closed set of measure zero. Applied to P_m, this shows that most planes are noncharacteristic. *Hint*. Choose $\eta \in \mathbb{R}^d$ with $P_m(\eta) \neq 0$. Choose V a linear subspace of $\mathbb{R}^d$ complimentary to $\mathbb{R}\eta$. Show that for each $v \in V$, $\{s \in \mathbb{R}: P(s\eta + v) = 0\}$ is a set of measure zero. Then apply Fubini's Theorem.
Discussion. The set $\{P(\xi) = 0\}$ is a real algebraic variety, the intersection of an algebraic variety with the real space $\mathbb{R}^d$. As such, one can say a good deal more than that it is of measure zero. However, the intersection with $\mathbb{R}^d$ renders the description far less detailed than the known properties of complex algebraic varieties.
The variety, $\{P_m(\xi) = 0\}$, is conic in the sense that it is invariant under $\xi \to a\xi$ for all $a \in \mathbb{R}\backslash 0$.

3. If P is a homogeneous partial differential operator, prove that the intersection of the characteristic variety with any sphere, $\{|\xi| = R\}$, is a subset of the sphere with $d - 1$ dimensional measure equal to zero.

4. The principal symbol $P_m(x, \xi)$ is a well-defined function on the cotangent bundle. The same is not true of the complete symbol $P(x, \xi)$. Construct an example of a linear partial differential operator $P(x, D)$, a change of variable $y = y(x)$, and corresponding points x, ξ and y, η in T^*, such that $P(x, \xi) \neq \tilde{P}(y, \eta)$ where $\tilde{P}(y, D_y)$ is the expression for P in the new coordinates. *Hint*. Almost any P and any nonlinear change of coordinates works.

The next sequence of problems concerns the solvability of the Cauchy problem up to errors which vanish to infinite order at $\Sigma = \{x_1 = 0\}$. This is the *Infinitesimal Cauchy Problem*. The problem is equivalent, as we will see, to solving the Cauchy problem on the level of formal power series. The latter question can even be raised for operators whose coefficients are formal power series. The resulting circle of ideas yield another perspective on the noncharacteristic condition in the Cauchy–Kowaleskaya Theorem.

Let $x = (x_1, x')$, $x' \equiv (x_2, \ldots, x_d)$. For ω open in $\mathbb{R}^{d-1}_{x'}$, let $\mathscr{G}(x_1 : \omega)$ be the ring of formal power series in x_1 with coefficients in $C^\infty(\omega)$. Equivalently, $\mathscr{G}$ is the ring of germs of smooth functions on a neighborhood of $\{0\} \times \omega \subset \mathbb{R}^d$. If $P(x, D)$ is a linear differential operator with coefficients smooth on a neighborhood of ω, construct a formal operator $\mathscr{P}(x, D)$ with coefficients in $\mathscr{G}(x_1 : \omega)$ by replacing the coefficients of P by their Taylor expansions in x_1. By definition, a formal operator with coefficients in $\mathscr{G}(x_1 : \omega)$ maps $\mathscr{G}(x_1 : \omega)$ to itself. The plane $\{x_1 = 0\}$ is noncharacteristic at (0, 0) for P if and only if the coefficient of $(\partial/\partial x_1)^m$ in $\mathscr{P}$ does not vanish at (0, 0).

5. Prove

Theorem. *The following are equivalent:*
(i) $\{x_1 = 0\}$ *is noncharacteristic at* (0, 0) *for* P.
(ii) *There is an open neighborhood* ω *of* 0 *in* $\mathbb{R}^d_x$, *such that for any* $f \in \mathscr{G}(x_1 : \omega)$ *and* $g_j \in C^\infty(\omega)$, $0 \le j \le m - 1$, *there is a unique* $u \in \mathscr{G}(x_1 : \omega)$ *solution of* $\mathscr{P}(x, D)u = f$, $\partial_1^j u(0, \cdot) = g_j(\cdot)$, $j \le m - 1$.
(iii) *For any* f *and* g_j, $j \le m - 1$, *smooth on a neighborhood of* 0 *in* $\mathbb{R}^d$ *and* $\mathbb{R}^{d-1}_{x'}$, *respectively, there is a neighborhood* Ω *of* $0 \in \mathbb{R}^{d-1}_{x'}$ *and a function* u *smooth on a neighborhood of* $\{0\} \times \Omega \subset \mathbb{R}^d$, *such that for* $j \le m - 1$, $\partial_t^j u(0, \cdot) = g_j(\cdot)$ *on* Ω *and* $Pu - f$ *vanishes to infinite order on* Ω.

Hints. Prove (i) ⇒ (ii) ⇒ (iii) ⇒ (i). To prove (ii), use Borel's Theorem which asserts that for any element γ of $\mathscr{G}(x_1 : \omega)$ there is a smooth function on a neighborhood of $\{0\} \times \omega$ which has γ as Taylor series in x_1.

6. Find necessary and sufficient conditions on the real constants a, b, c so that $a\partial_1^2 + b\partial_1\partial_2 + c\partial_2^2$ is elliptic on $\mathbb{R}^2$.

7. (a) Prove that if the coefficients of P are continuous then the set of points x, such that P is elliptic at x, is open.
(b) Prove that $P(x, D)$ is elliptic at $\underline{x}$ if and only if there are constants $c_1 > 0$ and c_2 such that

$$|P(\underline{x}, \xi)| \ge c_1|\xi|^m - c_2|\xi|^{m-1}.$$

§1.7. Holmgren's Uniqueness Theorem

If u_1 and u_2 are two local solutions of the mth order *linear* Cauchy problem

$$Pu = f,$$

$$\partial^\alpha u = \partial^\alpha v \quad \text{on } \Sigma \qquad \text{for all} \quad |\alpha| \le m - 1, \quad v \text{ given},$$

then the difference is a solution of

$$\begin{cases} Pu = 0, \\ \partial^\alpha u = 0 \quad \text{on } \Sigma \qquad \text{for all} \quad |\alpha| \le m - 1. \end{cases} \tag{1}$$

Thus, to prove local uniqueness one must show that solutions to (1) on a neighborhood of $\bar{x} \in \Sigma$ must vanish on a neighborhood of $\bar{x}$. When Σ is noncharacteristic, solutions of (1) must vanish to infinite order at Σ. Thus, if u is real analytic it must vanish on a neighborhood of Σ. Holmgren's Theorem asserts that the same conclusion is valid if the coefficients of P are real analytic, Σ is noncharacteristic, and u is only supposed to be C^m.

The strategy of the proof is motivated by an abstract result. Suppose that X and Y are normed linear spaces and that $T: X \to Y$ is a continuous linear map. The transpose $T^t: Y' \to X'$ of T is a linear map, defined on the dual spaces by

$$\langle T^t y', x \rangle \equiv \langle y', Tx \rangle \quad \text{for all} \quad y' \in Y' \quad \text{and} \quad x \in X.$$

Proposition 1. *If range*(T^t) *is dense in* X', *then* $\ker(T) = \{0\}$, *that is,* $u = 0$ *is the only* $u \in X$ *satisfying* $Tu = 0$.

Proof. For all $y' \in Y'$, $\langle y', Tu \rangle = 0$. The definition of transpose gives $\langle T^t y', u \rangle = 0$ for all y'.

Since $\{T^t y' : y' \in Y'\}$ is dense in X', we conclude that $\langle x', u \rangle = 0$ for all $x' \in X'$. The Hahn–Banach Theorem implies that $u = 0$. □

The point is that existence of solutions of the transposed equation $T^t v = g$, for a dense set of right-hand sides g, proves uniqueness for solutions of $Tu = f$. The idea of Holmgren is to use the Cauchy–Kowaleskaya Theorem for the existence part, the real analytic functions being dense.

Motivated by this idea, we begin by defining the transpose of a linear partial differential operator on Ω

$$P(x, \partial) = \sum a_\alpha(x) \partial^\alpha, \qquad a_\alpha \in C^\infty(\Omega).$$

If $u \in C^m(\Omega)$ satisfies $Pu = 0$ in Ω, and $v \in C_0^m(\Omega)$, then

$$\int_\Omega (Pu)v = 0.$$

One then integrates by parts passing the operator P to the function v. The resulting operator is called the *transpose of* P and is denoted P^t. To see the form of P^t consider the individual terms

$$\int (a_\alpha \partial^\alpha u) v \, dx = (-1)^{|\alpha|} \int u \partial^\alpha (a_\alpha v) \, dx.$$

Thus if

$$P^t(x, \partial) v \equiv \sum (-\partial)^\alpha (a_\alpha v),$$

then

$$\int_\Omega (Pu) v \, dx = \int_\Omega u P^t v \, dx, \qquad \text{for all} \quad u \in C^m(\Omega), \quad v \in C_0^m(\Omega).$$

Note that the principal symbol of P^t satisfies

$$P_m^t(x, \xi) = (-1)^m P_m(x, \xi), \tag{2}$$

so Σ is noncharacteristic at x for P if and only if it is noncharacteristic for P^t.

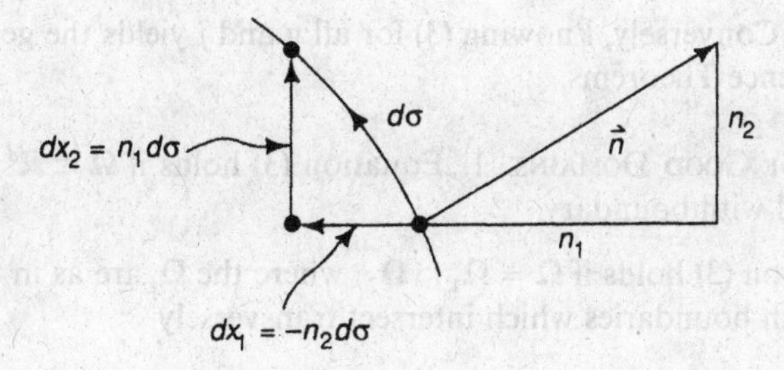

Figure 1.7.1

Next consider the boundary terms which appear when the supports of u and v reach $\partial\Omega$. We suppose $u, v, a_\alpha \in C^m(\bar{\Omega})$ and $\bar{\Omega}$ is sufficiently regular so that the Fundamental Theorem of Calculus is valid. That is,

$$\int_\Omega \partial_j u \, dx = \int_{\partial\Omega} n_j u \, d\sigma, \qquad \text{for all} \quad u \in C_0^1(\mathbb{R}^d), \tag{3}$$

where $n = (n_1, n_2, \ldots, n_d)$ is the unit outward normal to Σ and $d\sigma$ is the element of $d-1$ dimensional surface area on Σ. In the language of differential forms, the right-hand side is the integral of the $d-1$ form

$$\omega \equiv (-1)^{j-1} u \, dx_1 \wedge \cdots \wedge \widehat{dx_j} \wedge \cdots \wedge dx_d \qquad (\text{factor } dx_j \text{ omitted}), \tag{4}$$

over Σ with Σ oriented as the boundary of Ω. This means that $v_1, \ldots, v_{d-1}$ is an oriented basis for the tangent space to Σ if and only if $n, v_1, \ldots, v_{d-1}$ is an oriented basis for the tangent space to $\mathbb{R}^d$. These two observations for $d = 2$ are illustrated in Figure 1.7.1 and Figure 1.7.2. Note that

$$d\omega = \frac{\partial u}{\partial x_j} dx_1 \wedge \cdots \wedge dx_d$$

so (3) follows from Stokes' Theorem. Similarly, (3) is a consequence of the Divergence Theorem applied to the vector field $(0, 0, \ldots, u, \ldots, 0)$ with u in

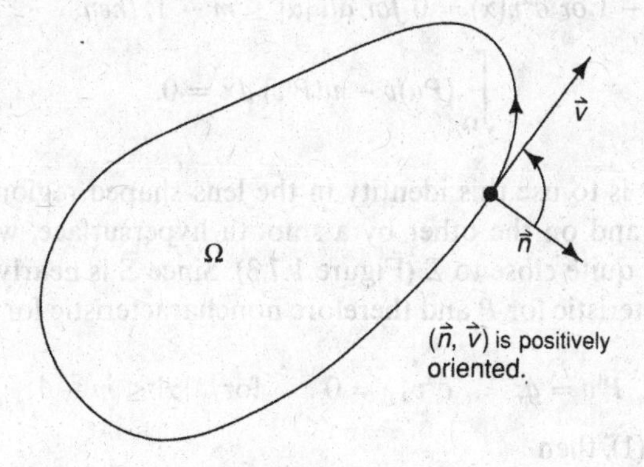

Figure 1.7.2

the jth slot. Conversely, knowing (3) for all u and j yields the general Stokes and Divergence Theorems.

Examples of Good Domains. 1. Equation (3) holds if $\Omega \subset \mathbb{R}^d$ is a smooth submanifold with boundary.

2. Equation (3) holds if $\Omega = \Omega_1 \cap \Omega_2$, where the Ω_j are as in the previous example with boundaries which intersect transversely.

To investigate the boundary terms from our integrations by parts, write ∂^α as a product of partial derivatives of first order

$$\partial^\alpha = \prod_{j=1}^{|\alpha|} \partial_{k_j}.$$

Then moving ∂_{k_1} from u to av by an integration by parts yields

$$\int_\Omega \partial^\alpha u\, av\, dx = -\int_\Omega \left(\prod_{j=2}^{|\alpha|} \partial_{k_j}\right) u\, (\partial_{k_1} av)\, dx + \int_{\partial\Omega} \left(\prod_{j=2}^{|\alpha|} \partial_{k_j}\right) u\, n_{k_1} av\, d\sigma.$$

Next move the second derivate, ∂_{k_2}, to find

$$-\int_\Omega \left(\prod_{j=2}^{|\alpha|} \partial_{k_j}\right) u\, (\partial_{k_1} av)\, dx = \int_\Omega \left(\prod_{j=3}^{|\alpha|} \partial_{k_j}\right) u\, (\partial_{k_2}\partial_{k_1} av)\, dx$$
$$- \int_{\partial\Omega} \left(\prod_{j=3}^{|\alpha|} \partial_{k_j}\right) u\, n_{k_2}\partial_{k_1} av\, d\sigma.$$

Continuing in this fashion yields an identity

$$\int_\Omega (Pu)v - u(P^t v)\, dx = \sum_{|\beta|+|\gamma|\le m-1} \int_{\partial\Omega} a_{\beta\gamma}(x)\partial^\beta u \partial^\gamma v\, d\sigma. \tag{5}$$

The following lemma is a consequence.

Lemma 2. *If u and v belong to $C^m(\bar{\Omega})$, and for each $x \in \partial\Omega$ either $\partial^\gamma u(x) = 0$ for all $|\gamma| \le m - 1$ or $\partial^\alpha v(x) = 0$ for all $|\alpha| \le m - 1$, then*

$$\int_\Omega (Pu)v - u(P^t v)\, dx = 0.$$

The strategy is to use this identity in the lens-shaped region, bounded on one side by Σ and on the other by a smooth hypersurface, which is nearly parallel to and quite close to Σ (Figure 1.7.3). Since $\tilde{\Sigma}$ is nearly parallel to Σ, it is noncharacteristic for P and therefore noncharacteristic for P^t. If v is a C^m solution of

$$P^t v = g, \qquad \partial^\alpha v|_\Sigma = 0 \qquad \text{for} \quad |\alpha| \le m - 1, \tag{6}$$

and u satisfies (1), then

$$\int_\Omega ug\, dx = 0.$$

Figure 1.7.3

If we solve (6) for many g, this identity is sufficient to show $u = 0$ in Ω. For example, Weierstrass' Approximation Theorem shows that it is sufficient to solve for all polynomials g, since one could then choose g_n converging uniformly to $\bar{u}$ in $\bar{\Omega}$ so

$$\int_\Omega |u|^2\, dx = \lim \int_\Omega u g_n\, dx = \lim 0 = 0. \tag{7}$$

Theorem 3 (Holmgren Uniqueness Theorem). *Suppose that $P(x, D)$ is a linear partial differential operator with coefficients real analytic on a neighborhood of $\bar{x} \in \mathbb{R}^d$, and Σ is a C^ω embedded hypersurface noncharacteristic at $\bar{x}$. If u is a C^m solution of* (1) *on a neighborhood of $\bar{x}$, then u vanishes on a neighborhood of $\bar{x}$.*

Proof. The first step is to normalize Σ. Choose real analytic coordinates so that $\Sigma = \{x_1 = 0\}$ and $\bar{x} = (0, 0, \ldots, 0)$. Next let

$$t \equiv x_1 + (x_2^2 + \cdots + x_d^2),$$

$$y_2 \equiv x_2, \quad y_3 \equiv x_3, \quad \ldots \quad y_d \equiv x_d,$$

$$y \equiv (y_2, \ldots, y_d).$$

Then in t, y coordinates, $\Sigma = \{t = |y|^2\}$.

For $\varepsilon > 0$, let $\tilde{\Sigma}_\varepsilon$ be the surface $\{t = \varepsilon\}$ in t, y space. Let ω_ε be the region $|y|^2 < t < \varepsilon$ between Σ and $\tilde{\Sigma}_\varepsilon$ (Figure 1.7.4).

Since Σ is noncharacteristic

$$P = \sum_{j+|\beta|\le m} a_{j,\beta}\partial_t^j \partial_y^\beta \qquad \text{with} \quad a_{m,0}(0, 0) \neq 0.$$

Choose $r_1 > 0$ so that $a_{m,0}(t, y) \neq 0$ for $|t| + |y| \le r_1$ and the coefficients of P are real analytic on $|t| + |y| < 2r_1$. Divide P by $a_{m,0}$ so that

$$P = \partial_t^m + \cdots \tag{8}$$

in $|t| + |y| \le r_1$. Then there are constants C, B so that the coefficients a_β of P^t satisfy

$$|\partial_{t,y}^\alpha a_\beta| \le C(B)^{|\alpha|}\alpha! \qquad \text{for all } \alpha \quad \text{and} \quad |t| + |y| \le r_1. \tag{9}$$

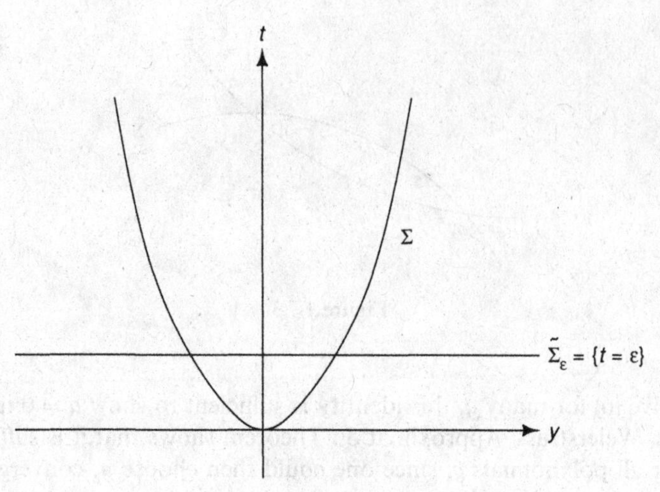

Figure 1.7.4

Choose $\varepsilon_0 > 0$ so that $\omega_{\varepsilon_0} \subset \{|t| + |y| \le r_1\}$. Consider the Cauchy problem with initial data at $t = \varepsilon \le \varepsilon_0$, that is,

$$P^t v = g(t, y), \qquad \partial^\alpha v|_{t=\varepsilon} = 0 \qquad \text{for all} \quad |\alpha| \le m - 1, \tag{10}$$

With g real analytic at $(\varepsilon, 0)$. Then the derivatives of g satisfy estimates analogous to (9)

$$\left| \frac{\partial^\alpha_{t,y} g(\varepsilon, 0)}{\alpha!} \right| \le \underline{C}(\underline{B})^{|\alpha|} \qquad \text{for all } \alpha. \tag{11}$$

For linear P satisfying (8) the method of majorants yields a solution of (10) with estimates for the derivatives $\partial^\alpha u(\varepsilon, 0)$ depending only on C, B, $\underline{C}$, $\underline{B}$. As P is linear, the domain of convergence of the power series solution to (10) does not change if g is multiplied by a constant, so one gets a uniform domain of convergence for $\varepsilon \in [0, \varepsilon_0]$ and g with $\underline{B}$ uniformly bounded. For g a polynomial, the estimate (11) holds for any $\underline{B} < \infty$. Thus, (10) has a solution in $|t - \varepsilon| + |y| \le \rho$ with ρ independent of the polynomial g and $\varepsilon \in [0, \varepsilon_0]$.

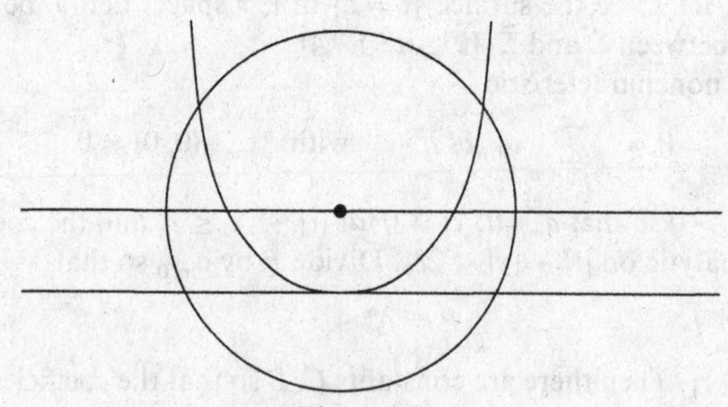

Figure 1.7.5

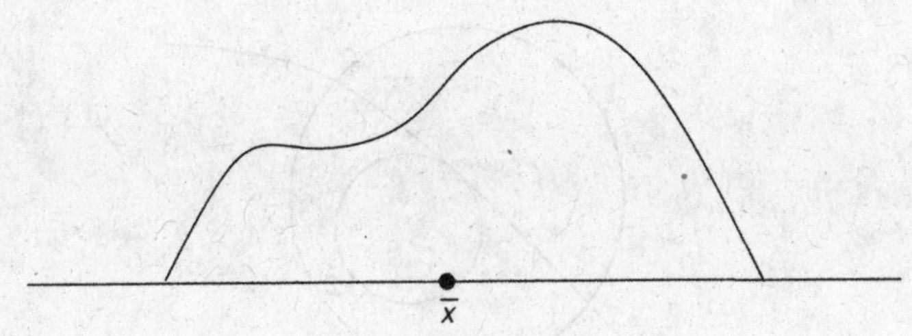

Figure 1.7.6

Choose $\varepsilon_1 \in]0, \varepsilon_0]$ so that (Figure 1.7.5)

$$\{|t - \varepsilon| + |y| \le \rho\} \supset\supset \omega_{\varepsilon_1}.$$

The argument in (7) shows that $u = 0$ in ω_{ε_1}. In the original coordinates, the set ω_{ε_1} is a one-sided neighborhood of $\bar{x}$ (Figure 1.7.6).

A similar argument works for the opposite side, and the proof is complete. □

Corollary 4. *The same result is true if Σ is supposed to be only $C^m \cap C^2$ instead of C^ω.*

PROOF. For r small, the ball $B_r(\bar{x})$ is divided cleanly in two by Σ, which is noncharacteristic at all points in B_{2r}. Thus $B_r = B^+ \cup B^- \cup (\Sigma \cap B_r)$. Define $\tilde{u}$ to be equal to u in B^+ and identically zero in B^- (Figure 1.7.7). Since Σ is noncharacteristic and $\partial^\alpha u|_\Sigma = 0$ for all $|\alpha| \le m - 1$, the same identity is true for $|\alpha| = m$ and therefore

$$\tilde{u} \in C^m, \qquad P\tilde{u} = 0.$$

For ρ small let $\tilde{\Sigma}$ be the sphere of radius ρ tangent to Σ at $\bar{x}$ and lying entirely in $B^- \cup \{\bar{x}\}$ (Figure 1.7.8). For ρ fixed, apply Holmgren's Theorem to $\tilde{u}$ with vanishing Cauchy data on $\tilde{\Sigma}$. Conclude that $\tilde{u} = 0$ on a neighborhood of $\bar{x}$, so that $u = 0$ on a neighborhood of $\bar{x}$ in $B^+ \cup \Sigma$. A similar argument works for B^-. □

Corollary 5 (Semiglobal Holmgren). *Suppose that $P(x, D)$ is an mth order linear partial differential operator with coefficients real analytic on a neighbor-*

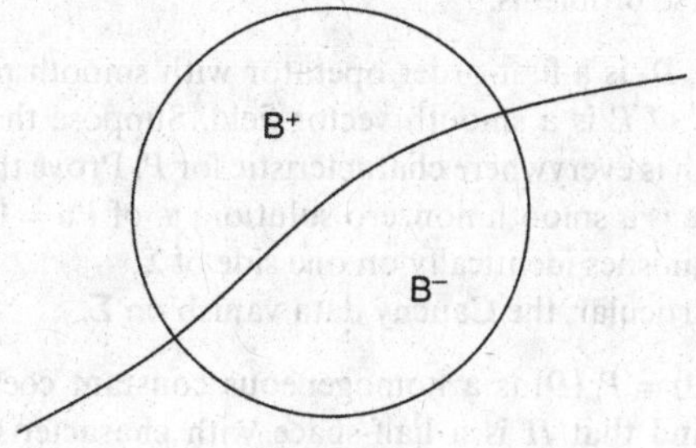

Figure 1.7.7

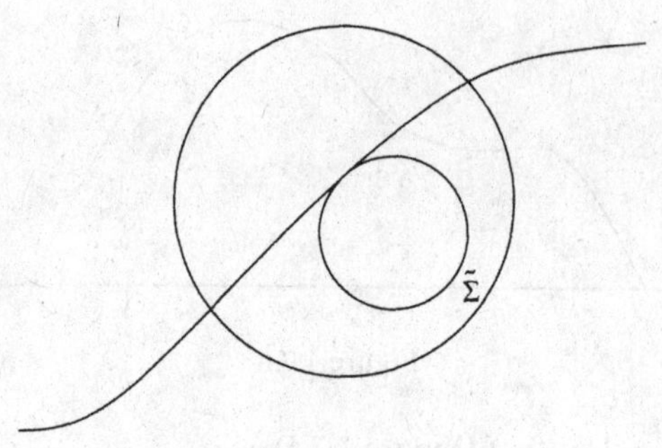

Figure 1.7.8

hood of Σ, a C^m noncharacteristic embedded hypersurface. If $u \in C^m$ on a neighborhood of Σ satisfies

$$P(x, D)u = 0, \qquad \partial^\alpha u|_\Sigma = 0 \qquad \text{for} \quad |\alpha| \leq m - 1,$$

then u vanishes on a neighborhood of Σ.

PROOF. For each $\bar{x} \in \Sigma$, Holmgren's Theorem asserts that there is an an open neighborhood, $\omega(\bar{x})$, of $\bar{x}$ with $u = 0$ on $\omega(\bar{x})$. Then u vanishes on the neighborhood $\bigcup_{\bar{x} \in \Sigma} \omega(\bar{x})$ of Σ. □

For linear equations with C^∞ instead of C^ω coefficients one may have nonuniqueness in the noncharacteristic Cauchy problem. The constructions of Plis and P. Cohen are deservedly famous (see [H2, Vol. 3]).

On the other hand, it is easy to find examples of nonuniqueness for the characteristic Cauchy problem. Consider, for instance, the operator $P = \partial_2$ and $\Sigma = \{x \in \mathbb{R}^2 : x_1 = 0\}$. Then $u = f(x_1)$ is a good example, provided $f \in C^\infty(\mathbb{R})$ has support in $x_1 \geq 0$ (see also Problems 1–3).

PROBLEMS

The next problems give further examples of nonuniqueness for the characteristic Cauchy problem. This topic is taken up again in §3.9 where it is seen to be related to some ill-posed "inverse problems."

1. Suppose that $P(x, D)$ is a first-order operator with smooth real coefficients. Then the principal part of P is a smooth vector field. Suppose that Σ is an embedded hypersurface which is everywhere characteristic for P. Prove that for any $x \in \Sigma$ with $P_1(x, D) \neq 0$, there is a smooth nonzero solution, u, of $Pu = 0$ on a neighborhood of x such that u vanishes identically on one side of Σ.
   DISCUSSION. In particular, the Cauchy data vanish on Σ.

2. Suppose that $P(D) = P_m(D)$ is a homogeneous constant coefficient partial differential operator, and that H is a half-space with characteristic boundary. Prove that there are smooth nontrivial solutions, u, to $Pu = 0$ which have support in H.
   Hint. Write the half-space as $\langle x, \xi \rangle \geq 0$ and try functions of the form $f(\langle x, \xi \rangle)$.

DISCUSSION. Such solutions are called *null solutions*. They exist without the homogeneity assumption but the construction is harder (see Problem 3 and §3.9).

3. In §4.2 the function

$$u(t, x) \equiv (4\pi t)^{-1/2} e^{-x^2/4t}$$

is shown to be a smooth solution of the heat equation $u_t = u_{xx}$ in $t > 0$. Verify this by direct computation. Extend u to vanish in $t \leq 0$. Prove that the resulting function is $C^\infty(\mathbb{R}^2 \backslash 0)$ and satisfies the heat equation on $\mathbb{R}^2 \backslash 0$. In addition, show that u does not vanish on a neighborhood of any point $(0, \underline{x})$ with $\underline{x} \neq 0$.
DISCUSSION. At such points $(0, \underline{x})$, u is a counterexample to local uniqueness for the characteristic Cauchy problem.

§1.8. Fritz John's Global Holmgren Theorem

Suppose that $Pu = 0$ and the Cauchy data of u vanish on a noncharacteristic surface Σ. It is natural to ask on how large a neighborhood of Σ must u vanish? The content of John's Global Holmgren Theorem is that u vanishes on any set swept out by deforming Σ through noncharacteristic surfaces whose ends stay in Σ (Figure 1.8.1). The precise description is somewhat long. The result was proved in 1948, though the methods were all available since the last century.

Suppose that $\Omega \subset \mathbb{R}^d$ is open and $P(x, D)$ is an mth order linear partial differential operator on Ω with coefficients in $C^\omega(\Omega)$. In addition, suppose that $\Sigma \subset \Omega$ is a noncharacteristic immersed C^m hypersurface.

We must define precisely what is meant by a continuous deformation through noncharacteristic surfaces Σ_λ whose ends lie in Σ. The surfaces Σ_λ will be images of a fixed set $\mathcal{O} \subset \mathbb{R}^d$ by a map σ depending on $\lambda \in [0, 1]$. We suppose that:

(i) $\mathcal{O} \subset\subset \mathbb{R}^{d-1}$ is open and σ: $[0, 1] \times \text{cl}(\mathcal{O}) \rightarrow \Omega \subset \mathbb{R}^d$ is continuous.
(ii) For each $\lambda \in [0, 1]$, $\sigma(\lambda, \cdot)$: $\mathcal{O} \rightarrow \mathbb{R}^d$ is a C^m immersion of a noncharacteristic hypersurface, Σ_λ.

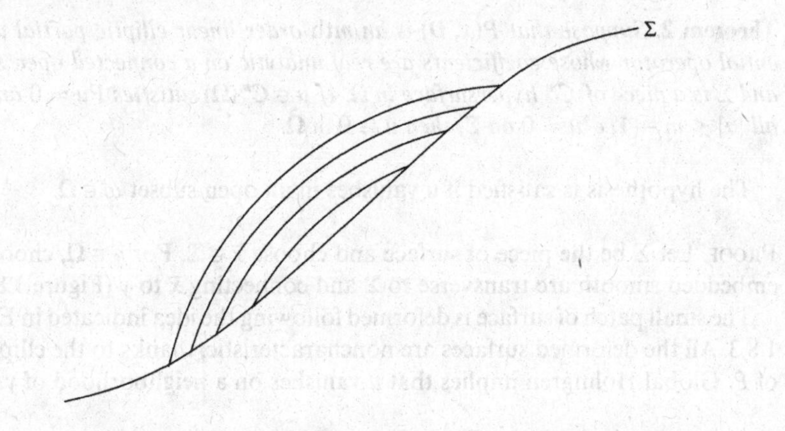

Figure 1.8.1

(iii) The initial surface, Σ_0, is a subset of Σ.
(iv) $\sigma([0, 1] \times \partial\mathcal{O}) \subset \Sigma$ which expresses the fact that the edge $\sigma(\lambda, \partial\mathcal{O})$ of Σ_λ lies in Σ.

Theorem 1 (John's Global Holmgren Theorem). *If $u \in C^m(\Omega)$, $Pu = 0$ in Ω and $\partial^\alpha u|_\Sigma = 0$ for $|\alpha| \le m - 1$, then for $|\alpha| \le m - 1$, $\partial^\alpha u = 0$ on $\sigma([0, 1] \times \mathrm{cl}(\mathcal{O}))$.*

PROOF. Let $A \subset [0, 1]$ be defined as

$$\{\lambda \in [0, 1]: \text{for all } |\alpha| \le m - 1, \partial^\alpha u = 0 \text{ on } \sigma([0, \lambda] \times \mathrm{cl}(\mathcal{O}))\}.$$

Since σ and $\partial^\alpha u$ for $|\alpha| \le m - 1$ are continuous, A is a closed subset of $[0, 1]$. By hypothesis, $0 \in A$.

By connectedness of $[0, 1]$, it suffices to show that A is open in $[0, 1]$.

If $\lambda \in A$, then $\partial^\alpha u|_{\Sigma_\lambda} = 0$ for $|\alpha| \le m - 1$, so by Corollary 1.7.5, u vanishes on a neighborhood of Σ_λ. Since $\sigma(\lambda, \partial\mathcal{O}) \subset \Sigma_0$, u vanishes identically on a neighborhood of $\sigma(\lambda, \partial\mathcal{O})$. Thus u vanishes on an open neighborhood $\mathcal{N}$ of the compact set $\sigma(\{\lambda\} \times \mathrm{cl}(\mathcal{O}))$.

Since σ is continuous we may choose, for each $p \in \{\lambda\} \times \mathrm{cl}(\mathcal{O})$, a relatively open neighborhood $\omega_p \subset [0, 1] \times \mathrm{cl}(\mathcal{O})$ of p such that $\sigma(\omega_p) \subset \mathcal{N}$. Then $\omega \equiv \bigcup \omega_p$ is a relatively open neighborhood of $\{\lambda\} \times \mathrm{cl}(\mathcal{O})$ whose image lies in $\mathcal{N}$.

For $0 < \varepsilon < 1$, let $\eta_\varepsilon \equiv ([\lambda - \varepsilon, \lambda + \varepsilon] \cap [0, 1]) \times \mathrm{cl}(\mathcal{O})$. Denote by $\sim$ complement in $[0, 1] \times \mathrm{cl}(\mathcal{O})$. The decreasing family of compact sets $(\sim\omega) \cap \eta_\varepsilon$ have empty intersection. It follows that $(\sim\omega) \cap \eta_\varepsilon$ is empty for all ε sufficiently small.

Thus for ε small, $\eta_\varepsilon \subset \omega$, so $\sigma(\eta_\varepsilon) \subset \mathcal{N}$ an open set on which u vanishes. Thus, for $|\alpha| \le m - 1$, $\partial^\alpha u$ vanishes on $\sigma(\eta_\varepsilon)$ and therefore $[\lambda - \varepsilon, \lambda + \varepsilon] \cap [0, 1] \subset A$, proving that A is open. □

Application 1. The unique continuation principle for real analytic elliptic partial differential equations.

Theorem 2. *Suppose that $P(x, D)$ is an* mth *order linear elliptic partial differential operator whose coefficients are real analytic on a connected open set Ω, and Σ is a piece of C^m hypersurface in Ω. If $u \in C^m(\Omega)$ satisfies $Pu = 0$ and for all $|\alpha| \le m - 1$, $\partial^\alpha u = 0$ on Σ, then $u \equiv 0$ in Ω.*

The hypothesis is satisfied if u vanishes in an open subset $\omega \in \Omega$.

PROOF. Let Σ be the piece of surface and choose $\bar{x} \in \Sigma$. For $y \in \Omega$, choose an embedded smooth arc transverse to Σ and connecting $\bar{x}$ to y (Figure 1.8.2).

The small patch of surface is deformed following the idea indicated in Figure 1.8.3. All the deformed surfaces are noncharacteristic, thanks to the ellipticity of P. Global Holmgren implies that u vanishes on a neighborhood of y.

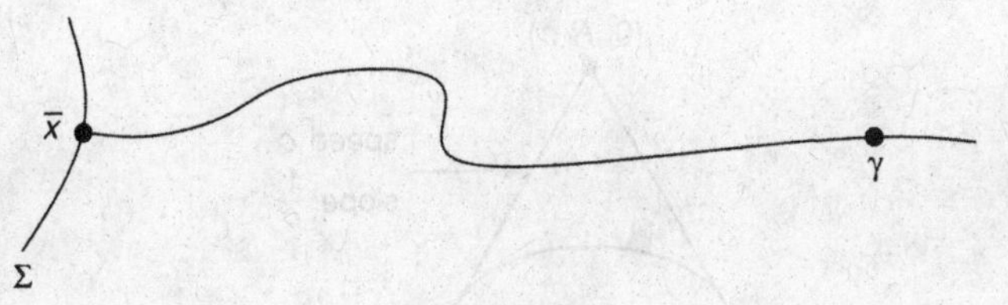

Figure 1.8.2

To describe precisely such a deformation, the first step is to change coordinates in a tubular neighborhood of the curve γ connecting $\bar{x}$ to y so that in the new coordinates $z = (z_1, \ldots, z_d) \equiv (z_1, z')$,

$$\gamma = \{(s, 0, \ldots, 0)\colon 0 \le s \le 1\} \text{ and } \Sigma \supset \{(0, z')\colon |z'| < 2r\}.$$

Then take $\mathcal{O} = \{z'\colon |z'| < r\}$ and $\sigma(\lambda, z') \equiv (a\lambda \cos^2 |z'|\pi/2r, z')$ with $a > 1$.

□

Remarks. 1. It is true but not easy to prove that, for elliptic $P(x, D)$ with real analytic coefficients, all solutions of $Pu \in C^\omega$ are real analytic. This gives a second proof of unique continuation since u vanishes to infinite order on Σ since Σ is noncharacteristic.

2. Sketched deformations are easier to understand than the precise formulas. For later examples we will give the sketch and leave the construction of precise σ's to the reader.

Application 2. Domains of influence and determinacy for the wave operator and D'Alembert's formula.

In the next sequence of results the Global Holmgren Theorem is applied to the partial differential operator $P(\partial_t, \partial_x) = \partial_t^2 - c^2\Delta \equiv \square$. P is called the *wave operator* or the *D'Alembertian.* The equation $Pu = 0$ is called the *wave equation.*

Theorem 3. *If $u \in C^2(\mathbb{R}_t \times \mathbb{R}_x^d)$ satisfies $\square u = 0$ and $u|_{t=0} = u_t|_{t=0} = 0$ on $|x| < R$, then*

$$u = 0 \quad in \quad \{(t, x)\colon |x| < R - c|t|\}.$$

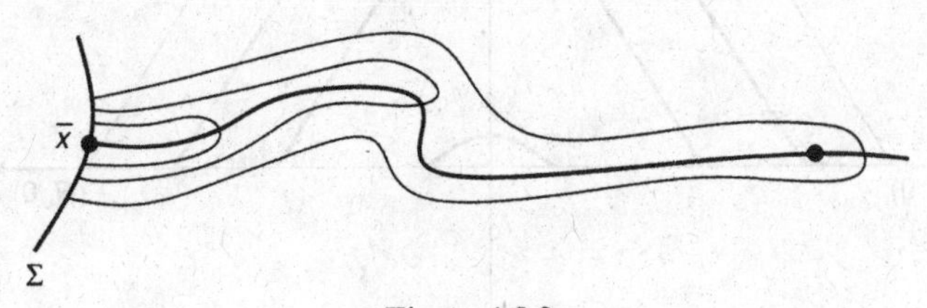

Figure 1.8.3

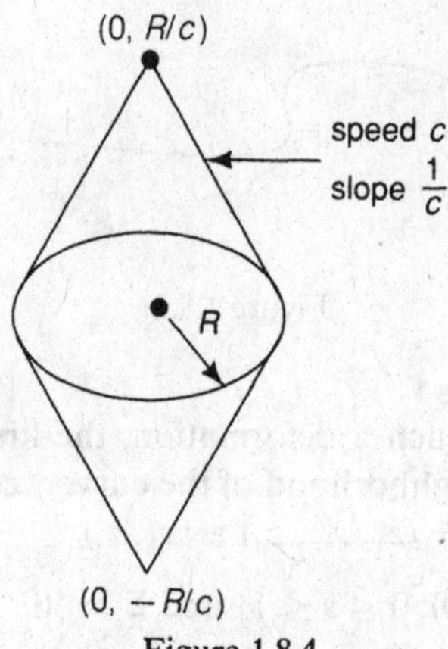

Figure 1.8.4

Proof. The domain is a double cone of revolution (Figure 1.8.4). To treat the upper half, the noncharacteristic initial disc is deformed through the surfaces of revolution with section sketched in Figure 1.8.5. The deformed surfaces are noncharacteristic provided their conormals satisfy $|\tau| > c|\xi|$. Thus the curves must have shallower slopes than the side of the triangle. This can be achieved without difficulty. □

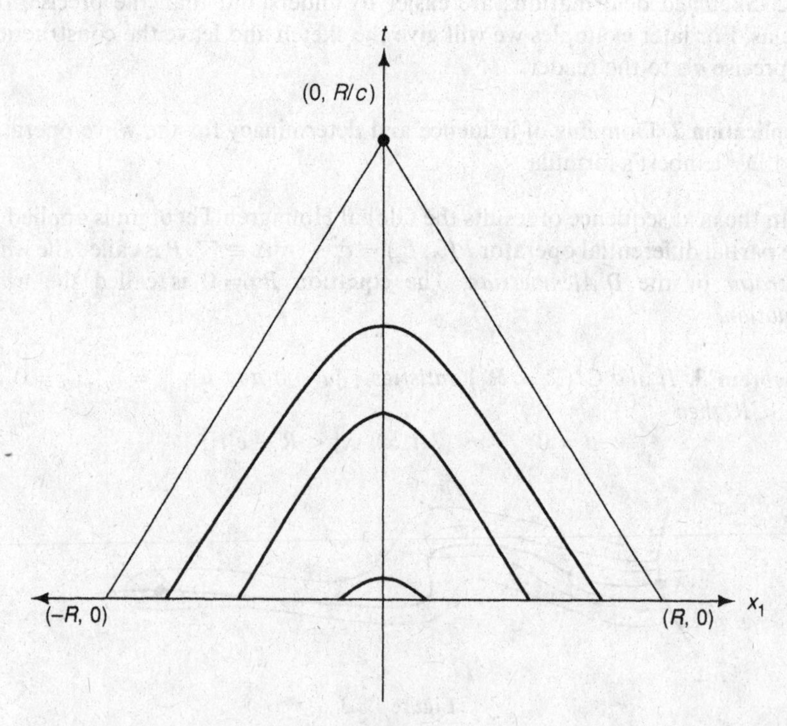

Figure 1.8.5

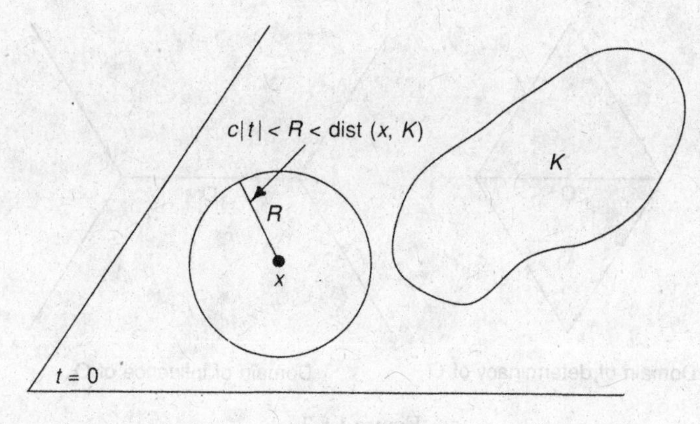

Figure 1.8.6

Corollary 4. *Suppose that* $u \in C^2(\mathbb{R}^{1+d})$ *is a solution of the wave equation,* $\Box u = 0$, *and that* K *is the support of the initial data*

$$K \equiv \text{supp } u_t|_{t=0} \cup \text{supp } u|_{t=0} \subset \mathbb{R}^d_x.$$

Then

$$\text{supp } u \subset \{(t, x) \colon \text{dist}(x, K) \le c|t|\}. \tag{1}$$

PROOF. If $\bar{t}$, $\bar{x}$ satisfies $\text{dist}(\bar{x}, K) > c|\bar{t}|$, choose $R \in]c|\bar{t}|, \text{dist}(\bar{x}, K)[$ so the initial data of u vanish on the ball of radius R with center $\bar{x}$ (Figure 1.8.6). Then Theorem 3 suitably translated shows that $u = 0$ on the double light cone with center at $(0, \bar{x})$ and radius R. The point $\bar{t}, \bar{x}$ lies in the interior of this set. $\Box$

This corollary asserts that waves propagate at speeds less than or equal to c. The set on the right-hand side of (1) is called the *domain of influence* of K. It is the set of points in space–time influenced by the Cauchy data in K. It consists exactly of those points which can be reached by curves starting in K at time $t = 0$ and never exceeding the speed c.

Restating the corollary we have:

If $u \in C^2(\mathbb{R}_t \times \mathbb{R}^d_x)$ *satisfies* $\Box u = 0$ *and* $u|_{t=0} = u_t|_{t=0} = 0$ *on the open subset* $\mathcal{O} \subset \mathbb{R}^d$, *then* $u = 0$ *on* $\{(t, x) \colon \text{dist}(x, \mathbb{R}^d \backslash \mathcal{O}) > c|t|\}$.

Thus, if the Cauchy data of two solutions u and v agree on the set $\mathcal{O}$, then u and v agree on $\{(t, x) \colon \text{dist}(x, \mathbb{R}^d \backslash \mathcal{O}) > c|t|\}$. This set is called the *domain of determinacy* of $\mathcal{O}$, since the values of solutions are determined in this set by the values of Cauchy data in $\mathcal{O}$. It consists of those points in space–time which cannot be reached by curves starting at $t = 0$ in the complement of $\mathcal{O}$ and never exceeding the speed c (Figure 1.8.7).

The boundaries of the domain of determinacy move inward at speed c. The boundaries of the domain of influence move outward at speed c.

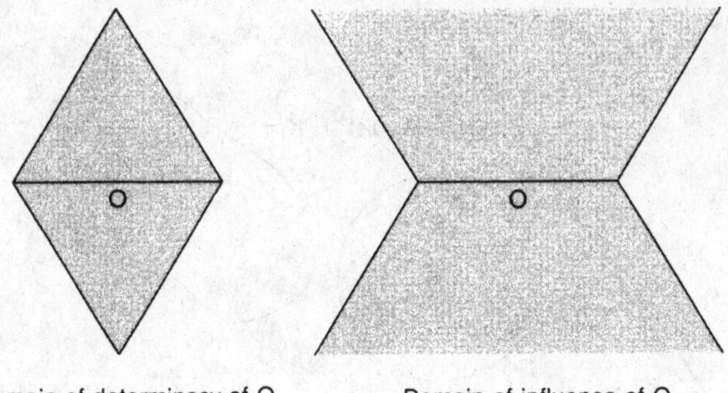

Figure 1.8.7

These last results are sharp. In the case of $x \in \mathbb{R}^1$, this is an immediate consequence of *D'Alembert's formula* for the solution of the Cauchy problem

$$u_{tt} - c^2 u_{xx} = 0, \qquad u(0, \cdot) = f(\cdot), \qquad u_t(0, \cdot) = g. \tag{2}$$

The derivation of the formula has two ingredients. First, one produces a solution of the Cauchy problem by a clever computation. That it is the only solution follows from the Global Holmgren Theorem.

The construction of a solution begins with the observation that

$$\partial_t^2 - c^2\partial_x^2 = (\partial_t + c\partial_x)(\partial_t - c\partial_x) = (\partial_t - c\partial_t)(\partial_t + c\partial_x),$$

so that if $(\partial_t \pm c\partial_x)v = 0$, then v satisfies the $1 - d$ wave equation. Theorem 1.1.1 shows that the general solution of $(\partial_t \pm c\partial_x)v = 0$ is a function of $x \mp ct$. Thus, for any $\varphi, \psi \in C^2(\mathbb{R})$,

$$u \equiv \varphi(x + ct) + \psi(x - ct) \tag{3}$$

is a solution of $\Box_{1+1} u = 0$.

Our strategy is to find φ, ψ so that u, given by (3), solves (1). The Cauchy data of such a u are given by

$$u(0, x) = \varphi(x) + \psi(x),$$
$$u_t(0, x) = c\varphi'(x) - c\psi'(x).$$

Differentiating the first of these equations gives the system

$$\varphi' + \psi' = f', \qquad \varphi' - \psi' = \frac{g}{c}.$$

Thus

$$\varphi' = \frac{f' + g/c}{2}, \qquad \psi' = \frac{f' - g/c}{2}.$$

Choose G with $G' = g$. Then with constants a and b

$$\varphi = \frac{f + G/c}{2} + a,$$

$$\psi = \frac{f - G/c}{2} + b.$$

The equation $\varphi + \psi = f$ forces $a + b = 0$. Note that adding a to φ, while subtracting a from ψ, does not affect the value of u defined by equation (3). This leads to the formula for u

$$u = \frac{f(x + ct) + f(x - ct)}{2} + \frac{1}{2c}(G(x + ct) - G(x - ct)). \tag{4}$$

To get an expression in terms of the data, note that

$$G(x + ct) - G(x - ct) = \int_{x-ct}^{x+ct} g(s)\, ds.$$

Theorem 5 (D'Alembert's Formula). *If $f \in C^2(\mathbb{R})$ and $g \in C^1(\mathbb{R})$, there is exactly one solution $u \in C^2(\mathbb{R}_t \times \mathbb{R}_x)$ of the Cauchy problem* (2). *The solution is given by the formula*

$$u(t, x) = \frac{f(x + ct) + f(x - ct)}{2} + \frac{1}{2c}\int_{x-ct}^{x+ct} g(s)\, ds. \tag{5}$$

Proof. Since this u has the form (4) it satisfies the wave equation. Formula (4) was derived exactly so that the initial conditions are satisfied. This establishes existence and the formula. Uniqueness is a consequence of the Global Holmgren Theorem. □

Corollary 6. *If $2 \le k \in \mathbb{N}$, $f \in C^k(\mathbb{R})$, and $g \in C^{k-1}(\mathbb{R})$, then $u \in C^k(\mathbb{R}^{1+1})$.*

Corollary 7. *If $u \in C^k(\mathbb{R}_t^1 \times \mathbb{R}_x^1)$, $k \ge 2$, satisfies $\Box u = 0$, then there exist φ, $\psi \in C^k(\mathbb{R})$ such that u is given by equation* (3).

Proof. Let $u|_{t=0} = f$, $u_t|_{t=0} = g \in C^{k-1}$, and choose $G \in C^k$ with $dG/dx = g$. Then

$$\varphi = \frac{f}{2} + \frac{G}{2c}, \qquad \psi = \frac{f}{2} - \frac{G}{2c},$$

do the trick. □

Examining (5) shows that our estimates for the domain of dependence and domain of influence are exact. The values of u at t, x depend on the values of f at the end points of the interval $[x - c|t|, x + c|t|]$ and on the integral of g

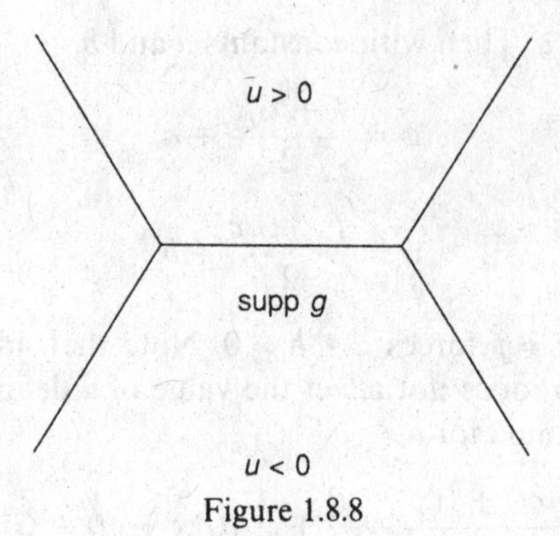

Figure 1.8.8

over the whole interval. Thus the domain of dependence of t, x is exactly the interval predicted by Holmgren's Theorem.

Similarly, if we consider the case $f = 0$ and $g \geq 0$, we find that supp u is exactly equal to the set on the right-hand side of (1). The data influence all the points which they could possibly influence. The case where supp g is an interval is sketched in Figure 1.8.8.

Application 3. The walls have ears.

Consider a wave propagating in $x \geq 0$, and an observer at $x = 0$ who measures $u(t, 0)$ and $u_x(t, 0)$ as functions of time. If he observes for $0 \leq t \leq T$, what part of the wave field can he determine from his measurement? The formulation in equations is

$$u_{tt} - c^2 u_{xx} = 0 \quad \text{in} \quad x \geq 0, t \in \mathbb{R},$$

$$u(t, 0) = f \quad \text{and} \quad u_x(t, 0) = g \qquad \text{known for} \quad 0 \leq t \leq T.$$

Thanks to the linearity of the wave equation, this is equivalent to asking at what points must u vanish if f and g vanish for $0 \leq t \leq T$. Thus, given

$$\square u = 0 \qquad \text{in} \quad \mathbb{R}_t \times \{x \geq 0\} \tag{6}$$

and

$$u(t, 0) = u_x(t, 0) = 0 \qquad \text{for} \quad 0 \leq t \leq T, \tag{7}$$

it suffices to determine where u is forced to be zero.

Solution. The Cauchy data of u vanish on the segment $\{x = 0 \text{ and } 0 < t < T\}$. Apply Fritz John's Global Holmgren Theorem with surfaces sketched in Figure 1.8.9. Conclude that u vanishes in the triangle swept out, namely $0 \leq x \leq c(T/2 - |t - T/2|)$. □

There are several multi-dimensional generalizations. Here is one. For $x \in \mathbb{R}^d$, an observer in the $d - 1$ plane $\{x_1 = 0\}$ measures $u(t, 0, x')$ and $u_x(t, 0, x')$ for

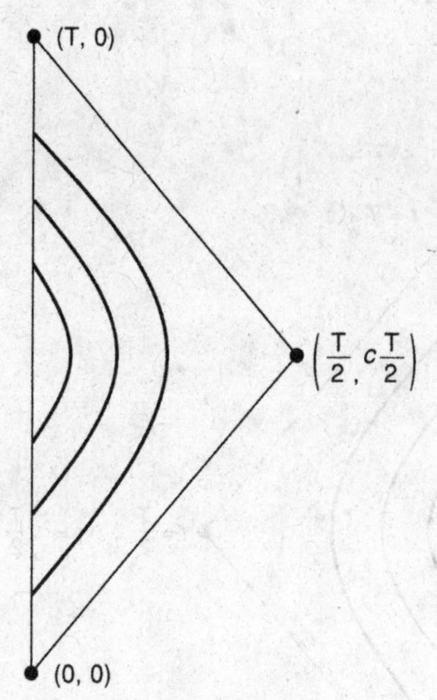

Figure 1.8.9

all $x' \equiv (x_2, \ldots, x_d)$ and $0 \le t \le T$. The analysis is similar. Take surfaces as above, and almost independent of $x_2, \ldots, x_d$. Precisely, if the deformation above is given by $\Sigma_\lambda = \{x_1 = F(\lambda, t)\}$, then for $\varepsilon > 0$ define multi-dimensional deformed surfaces by

$$\Sigma_\lambda \equiv \{x_1 = F(\lambda, t) - \varepsilon\lambda(x_2^2 + \cdots + x_{d-1}^2)\}.$$

Then let $\varepsilon \to 0$. One finds that u must vanish in the cylinder which is the product of the triangle from the one space dimension problem and $\mathbb{R}_{x'}^{d-1}$.

Corollary 8. *If* $u \in C^2(\mathbb{R}_t \times [0, \infty[\times \mathbb{R}_{x'}^{d-1})$, $\Box_{1+d}u = 0$ *in* $x_1 > 0$, *and*

$$u|_{x_1=0} = u_{x_1}|_{x_1=0} \quad \textit{for} \quad 0 \le t \le T,$$

then

$$u = 0 \quad \textit{in} \quad \left\{0 \le x_1 \le c\left(\frac{T}{2} - \left|t - \frac{T}{2}\right|\right)\right\}.$$

Application 4. What can a single snoop hear?

To observe wave motion along the entire strip $[0, T] \times \mathbb{R}_{x'}^{d-1}$ requires many observers when $d > 1$. We next ask what can be observed from a neighborhood of a single point.

Suppose $u \in C^2(\mathbb{R}_t \times \mathbb{R}_x^d)$, $\Box_{1+d}u = 0$, and $u = 0$ on a neighborhood of $[0, T] \times \{x = 0\}$. Where must u vanish?

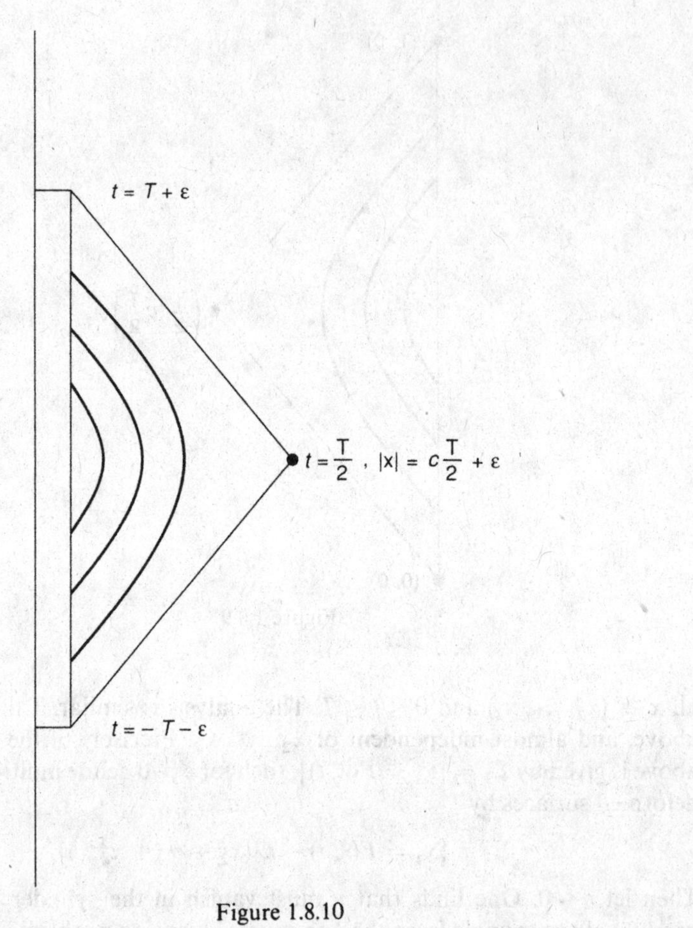

Figure 1.8.10

Solution. Choose $\varepsilon > 0$ so that $u(t, x)$ vanishes if $\text{dist}((t, x), [0, T] \times \{x = 0\}) < 3\varepsilon$. Apply the Global Holmgren Theorem with noncharacteristic surface Σ equal to $]-\varepsilon, T + \varepsilon[\times \{|x| = \varepsilon\}$. The deformed surfaces are surfaces of revolution about the t-axis with cross section sketched in Figure 1.8.10. $\square$

Corollary 9. *Suppose* $u \in C^2(\mathbb{R}_t \times \mathbb{R}^d_x)$, $\square_{1+d} u = 0$, *and* $u = 0$ *on a neighborhood of* $[0, T] \times \{x = 0\}$, *then* $u = 0$ *on* $|x| \le c(T/2 - |t - T/2|)$.

This result determines precisely the time of arrival at a point $y \notin K$ of a wave which begins at $t = 0$ in K. Toward this end consider

$$u \in C^2(\mathbb{R}_t \times \mathbb{R}^d_x), \qquad \square_{1+d} u = 0,$$

$$K \equiv \text{supp}\, u|_{t=0} \cup \text{supp}\, u_t|_{t=0}.$$

For $y \notin K$ let $\delta \equiv \text{dist}(y, K)$. Then Corollary 4 shows that $]-\delta/c, \delta/c[\times \{y\}$ is disjoint from supp u. This result is sharp.

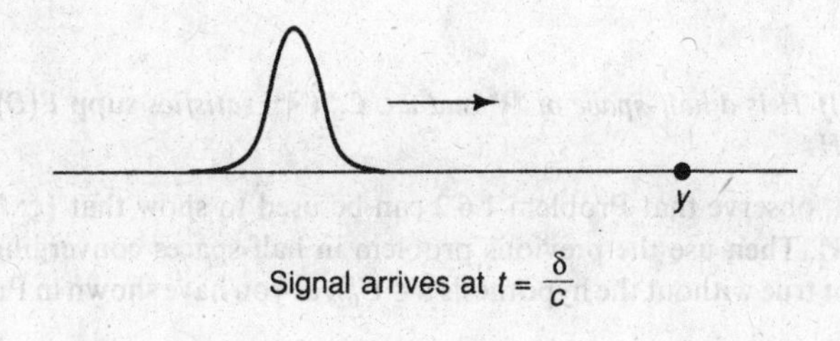

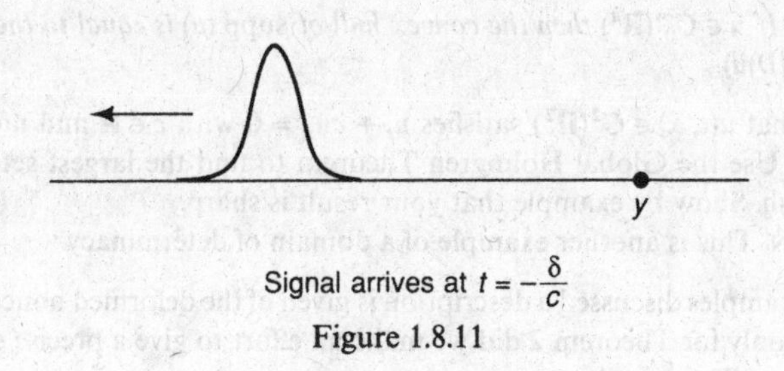

Figure 1.8.11

Corollary 10. *With the above notation, one of the two points* $(\pm\delta/c, y)$ *lies in* supp u.

Remark. A signal cannot reach y in time less than δ/c, but it reaches in time *exactly* δ/c in either forward or backward direction of time (or both). A wave moving away from you will arrive in the past.

EXAMPLE. With $d = 1$, consider a forward-moving blip $u = \psi(x - ct)$ which at $t = 0$ has support in $\{x \leq 0\}$ and $\{x = 0\}$ lies in the support. Then the signal arrives at $t = \delta/c$. A backward-moving blip, $u = \psi(x + ct)$, arrives at $\delta = -t/c$ (Figure 1.8.11).

PROOF OF COROLLARY 10. If $(\pm\delta/c, y) \notin \operatorname{supp} u$, then $[-\delta/c, \delta/c] \times \{y\}$ is disjoint from supp u. Thus, for some $\varepsilon > 0$,

$$\left[\frac{-\delta - \varepsilon}{c}, \frac{\delta + \varepsilon}{c}\right] \times \{|x - y| \leq \varepsilon\}$$

is disjoint from supp u. Apply Corollary 9 to conclude that at $t = 0$, u vanishes on the ball of radius $\delta + \varepsilon$ with center y. This contradicts the definition of δ. □

PROBLEMS

1. Prove

   Theorem. *If* $u \in C^m(\mathbb{R}^d)$ *satisfies* $P(D)u = 0$ *on* $x \cdot \xi \leq T_2$ *and* $u = 0$ *on* $x \cdot \xi \leq T_1$ *where* $P_m(\xi) \neq 0$, *then* $u = 0$ *on* $x \cdot \xi \leq T_2$.

   DISCUSSION. This result is false if $P_m(\xi) = 0$ as you showed in Problem 1.7.2.

2. (i) Prove

Theorem. *If H is a half-space of $\mathbb{R}^d$ and $u \in C_0^\infty(\mathbb{R}^d)$ satisfies* supp $P(D)u \subset H$, *then* supp $u \subset H$.

Hint. First, observe that Problem 1.6.2 can be used to show that $\{\xi: P_m(\xi) \neq 0\}$ is dense in $\mathbb{R}^d_\xi$. Then use the previous problem in half-spaces converging to H. This result is not true without the hypothesis $u \in C_0^\infty$, as you have shown in Problem 1.7.2.
(ii) Prove

Corollary. *If $u \in C_0^\infty(\mathbb{R}^d)$ then the convex hull of* supp (u) *is equal to the convex hull of* supp$(P(D)u)$.

3. Suppose that $u(t, x) \in C^2(\mathbb{R}^2)$ satisfies $u_t + cu_x = 0$ with $c \in \mathbb{R}$ and $u(0, x) = 0$ for $x \in [a, b]$. Use the Global Holmgren Theorem to find the largest set on which u must vanish. Show by example that your result is sharp.
DISCUSSION. This is another example of a domain of determinacy.

In all the examples discussed a description is given of the deformed noncharacteristic surfaces, but only for Theorem 2 did we make an effort to give a precise construction of a map σ as in Theorem 1.

4. For Theorem 3 or Corollary 8 give a precise explicit description of a mapping σ satisfying the conditions of the Global Holmgren Theorem and sweeping out the desired region.

5. Suppose that $P = \prod(\partial_t + c_i\partial_x)$, with real constants $c_1 < c_2 < \cdots < c_m$.
(i) Prove that if $g_j \in C^{m-j}(\mathbb{R})$ for $0 \leq j \leq m - 1$, then the Cauchy problem
$$Pu = 0, \qquad \partial_t^j u(0, \cdot) = g_j(\cdot), \qquad j \leq m - 1,$$
has a $C^m(\mathbb{R}^2)$ solution of the form $\sum \varphi_i(x - c_i t)$.
(ii) Show that u is C^k if the g_j are C^{k-j} with $k \geq m$.
(iii) Prove that the general C^k solutions of $Pu = 0$ is a sum of this form with φ_i in C^k.
(iv) Describe the domain of influence of an interval $[a, b]$ in $\{t = 0\}$.
Hint. Imitate the proof of D'Alembert's Theorem.

§1.9. Characteristics and Singular Solutions

The explicit solution formulas of §1.1, D'Alembert's formula, and the perturbation theory of §1.4 all involve characteristic curves. In the last section, characteristic curves and surfaces (for $d > 2$) played an important role in describing the propagation of zeros. In this section, we examine their role in the construction of simple singular ($\equiv$ not infinitely differentiable) solutions.

Consider first the equation $u_t + cu_x = 0$ with $c \in \mathbb{R}$. The general C^1 solution is $u = g(x - ct)$ with $g \in C^1(\mathbb{R})$. If $g \in C^1(\mathbb{R})$ is piecewise smooth on $\mathbb{R}$, with jumps at points $a_1 < a_2 < \cdots < a_M$ with jumps in derivatives $k_1, \ldots, k_M$, then u will be piecewise smooth on $\mathbb{R}^2$ with jumps in derivatives of order $k_j \geq 2$ along the curves $x - ct = a_j$.

More generally, consider the variable coefficient equation

$$u_t + c(t, x)u_x + d(t, x)u = 0, \qquad c, d \in C^\infty(\mathbb{R}^2 : \mathbb{R}). \tag{1}$$

This is an ordinary differential equation for u along the integral curves of the vector field $\partial_t + c(t, x)\partial_x$. These integral curves are the characteristic curves for this operator. So long as the integral curves do not escape to infinity in finite time one has an explicit representation of the solution. The escape time to infinity will be infinite provided c does not grow too fast as $|x|$ tends to infinity, for example, if

$$(\forall T > 0)(\exists K) \qquad (|t| \le T \Rightarrow |c(t, x)| \le K(1 + |x|)).$$

If g is piecewise smooth as in the previous paragraph, the solution u will be piecewise smooth with jumps in derivatives of order k_j across the characteristic curves Γ_j passing through the points $(0, a_j)$. These two examples suggest that characteristics are the carriers of singularities of piecewise smooth solutions. This is true is great generality.

Consider an mth order linear operator $P(x, \partial)$ with coefficients in $C^\infty(\mathbb{R}^d)$. Suppose that Σ is an infinitely differentiable embedded hypersurface in $\mathbb{R}^d$. Piecewise smooth functions singular across Σ are defined as follows. Since Σ is an embedded hypersurface, for each $x \in \Sigma$ there is small open ball B centered at $\underline{x}$ so that B is diffeomorphic to $\{|x| < 1\}$ by a diffeomorhism which carries $B \cap \Sigma$ to $\{x_1 = 0\} \cap \{|x| < 1\}$ (Figure 1.9.1).

Definition. A function u defined on a neighborhood of Σ is piecewise smooth if, for each $\underline{x} \in \Sigma$, there is a ball B as above such that u is C^∞ on both components, $B_\pm$, of $B \backslash \Sigma$, and the restriction of each derivative $\partial^\alpha u$ to $B_\pm$ extends to a continuous function on $\bar{B}_\pm$.

Theorem 1. *If P and Σ are as above and there is a piecewise smooth u defined on a neighborhood of Σ satisfying $u \in C^m$, $Pu \in C^\infty$, and u is not C^∞ on a neighborhood of $\underline{x}$, then Σ must be characteristic at $\underline{x}$.*

PROOF. If Σ is noncharacteristic at $\underline{x}$, choose B as in Figure 1.9.1 such that Σ is noncharacteristic at all points of $\Sigma \cap \bar{B}$.

Write the equation $Pu \in C^\infty(B)$ in coordinates with $\Sigma = \{x_1 = 0\}$. The coefficient of $(\partial/\partial x_1)^m$ is nonzero on Σ since Σ is noncharacteristic. Shrinking B if necessary, we may suppose that the coefficient is nonzero on $\bar{B}$. Dividing by this coefficient yields a relation

$$\partial_1^m u = \sum_{1 \le j \le m} A_j(x, \partial_2, \ldots, \partial_d)\partial_1^{m-j} u + f, \tag{2}$$

where A_j is a linear differential operator of degree j. The coefficients of A_j and the function f belong to $C^\infty(B)$.

Let $\partial' \equiv (\partial_2, \ldots, \partial_d)$ and $x' \equiv (x_2, \ldots, x_d)$. By hypothesis, $u \in C^m(B)$ and u is C^∞ on $\mathrm{cl}(B_+)$ and $\mathrm{cl}(B_-)$. Thus $\partial^\alpha u(0+, x') = \partial^\alpha u(0-, x')$ for all α with $|\alpha| \le m$.

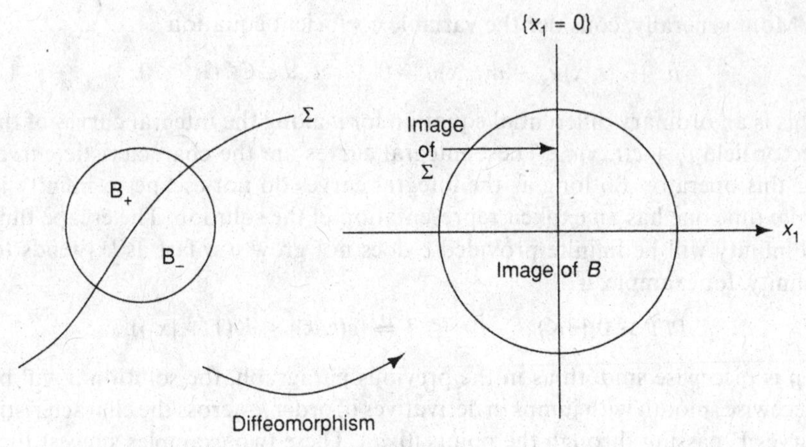

Figure 1.9.1

The theorem is proved by showing that $u \in C^\infty(B)$. This is done by proving, for all $k \geq m$, that $\partial^\alpha u(0+, x') = \partial^\alpha u(0-, x')$ for all $\alpha \in \mathbb{N}^d$ with $\alpha_1 \leq k$.

Applying $(\partial')^\beta$ to the identity $\partial_1^j u(0+, x') = \partial_1^j u(0-, x')$ for $j \leq m$, shows that the result is true for $k = m$.

Differentiating (2) yields

$$\partial_1^{k+1} u = \partial_1^{k+1-m} \partial_1^m u = \partial_1^{k+1-m}\left(\sum_{1 \leq j \leq m} A_j(x, \partial_2, \ldots, \partial_d) \partial_1^{m-j} u + f \right).$$

By the inductive hypothesis the limits of the right-hand side on the two sides of $\Sigma \cap B$ are equal. Therefore $\partial_1^{k+1} u(0+, x') = \partial_1^{k+1} u(0-, x')$. Applying $(\partial')^\beta$ to this identity proves the assertion for $k + 1$. □

We conclude that the only possible carriers of singularities of piecewise smooth solutions are the characteristic surfaces.

In particular, elliptic equations which have no characteristic surfaces can have no such singular solutions. In fact, solutions of $Pu \in C^\infty$ with P elliptic must themselves be C^∞. For P of order 2 this Interior Elliptic Regularity Theorem is proved in §5.9.

D'Alembert's formula (1.8.3) shows that all the characteristic curves, $x \pm ct =$ constant for the $1 - d$ wave operator $\square_{1+1}$, are carriers of such singular solutions. In Problem 1 you extend this result to all homogeneous constant coefficient operators. It is not difficult to show that if Σ is simply characteristic, in the sense that $\nabla_\xi P_m(x, \xi) \neq 0$ for all $x, \xi \in N^*(\Sigma)$, then there are many piecewise smooth solutions with jumps along Σ (see John's chapter in Bers, John, and Schecter [BJS]).

We next turn to a different phenomenon involving singularities and for which the characteristics play a crucial role, namely, *the formation of shock waves.*

For equation (1) in regions of space–time, where c is an increasing function of x, the characteristic curves spread apart as t increases. The domain of influence of an interval $[a, b]$ in $\{t = 0\}$ is bounded by the characteristic curves through a and b and is therefore an expanding region where c is increasing. These are expanding waves. Conversely, where c is decreasing the characteristics approach each other as t increases and one has compressive waves. Theorem 1.2.4 implies that integral curves of $\partial_t + c(t, x)\partial_x$ cannot cross, so that even though they may grow closer together they will never meet. The expansion and compression just described is caused by the fact that the speed of propagation c depends on t, x.

Consider next the inviscid Burgers equation

$$u_t + uu_x = 0. \tag{3}$$

Writing the second term as $(u^2/2)_x$ expresses the equation as a *conservation law*, that is, an expression of the form

$$f(x, u)_t + g(t, x, u)_x = 0.$$

The name is explained for the Burgers equation as follows. Consider the integral of u from a to b as the amount of u in $[a, b]$. The rate of change of this quantity is given by

$$\begin{aligned}\frac{d}{dt}\int_a^b u(t, x)\,dx &= \int_a^b u_t(t, x)\,dx \\ &= \int_a^b \left(\frac{u^2(t, x)}{2}\right)_x dx = \frac{u^2(b) - u^2(a)}{2}.\end{aligned}$$

The last expression involves only the values of u at the boundary of $[a, b]$, and is called the *flux* of u. The function $u^2/2$ is called the *flux density*. The quantity $u^2(b)/2$ (resp. $u^2(a)/2$) is interpreted as the amount of u flowing in at b (resp. out at a). If u is equal to a constant $u(\infty)$ outside a (possibly time-dependent) compact set, one sees that $\int u - u(\infty)\,dx$ is independent of t. Thus the amount of u is conserved. For the general form, it is $\int f(x, u) - f(x, u(\infty))\,dx$ which is conserved. Most of the laws of continuum mechanics are conservation laws for physical quantities like energy, mass, and momentum.

Consider next the initial value problem for the Burgers equation (3) with data

$$u(0, x) = g(x). \tag{4}$$

Suppose that a C^1 solution u exists on an as-yet unspecified region of space–time. In §1.6 we showed that u is constant on the integral curves of the vector field $\partial_t + u\partial_x$ which are straight lines. These lines are precisely the characteristic hypersurfaces (dimension $d - 1 = 1$ in this case). This gives a simple formula for the solution. At the point $(0, a)$ on the initial line we know the value of $u(0, a) = g(a)$. The characteristic through this point is then the line $(t, a + tg(a))$, so long as it lies within the region where u is C^1. The solution must have value $g(a)$ on this line.

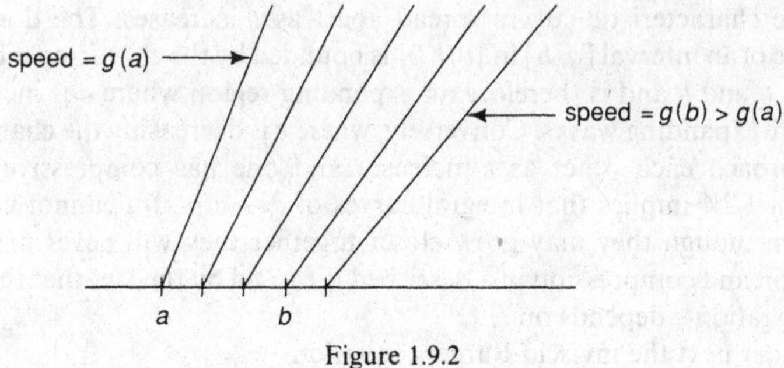

Figure 1.9.2

Consider the behavior of the characteristic lines through $(0, a)$ and $(0, b)$ with $a < b$. If $g(a) < g(b)$, the lines diverge from each other as t increases. If g is monotone increasing in the interval $[a, b]$, then characteristics starting in $\{0\} \times [a, b]$ fan apart and simply cover an expanding wedge-shaped region (Figure 1.9.2). Such a solution is called an *expansion wave* or a *rarefaction wave* or a *fan*. The amount of u in the fan at time t is given by

$$\int_{a+tg(a)}^{b+tg(b)} u(t, x)\, dx.$$

Its rate of change with time is equal to

$$\int_{a+tg(a)}^{b+tg(b)} u_t(t, x)\, dx + g(b)u(t, b + tg(b)) - g(a)u(t, a + tg(a)).$$

Using the differential equation converts the first term to a boundary term which exactly cancels the last two. Thus the amount of u in the fan is conserved and is spread over an ever-widening interval. This explains the origin of the name rarefaction wave.

A more striking phenomenon is produced if $g(a) > g(b)$. The two characteristic lines approach and cross (Figure 1.9.3). The fact that u has different values on the two lines is contradictory at the crossing point. The lines cross at $t = -(b - a)/(g(b) - g(a))$. This proves the following estimate on the time of existence of C^1 solutions.

Theorem 2. *If the initial value problem* (3), (4) *has a solution in* $C^1([0, T] \times \mathbb{R})$ *and* g *is not monotone increasing, then*

$$T \le \inf\left\{ -\frac{b-a}{g(b)-g(a)} : a < b \text{ and } g(a) > g(b) \right\}$$

$$= -1/\inf\{g'(x) : x \in \mathbb{R}\}.$$

The last equality is a simple consequence of the Mean Value Theorem. An alternate proof is presented in Problem 2. The estimate of Theorem 2 is sharp.

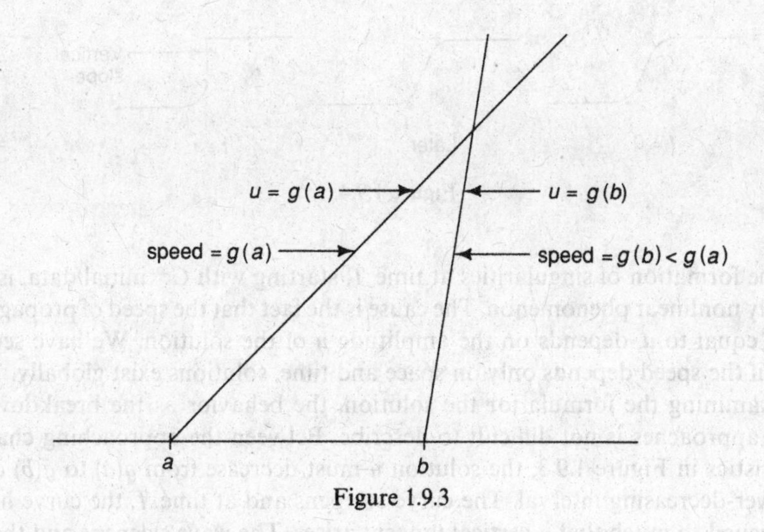

Figure 1.9.3

Theorem 3. *If $g \in C^1(\mathbb{R})$ and* inf g' *is finite, then with*

$$T \equiv \infty \quad \text{if } (\inf g') \geq 0 \quad \text{and} \quad T \equiv -(\inf g')^{-1} \quad \text{otherwise,}$$

there is a unique solution $u \in C^1([0, T[\times \mathbb{R})$ of the initial value problem (3), (4) *given implicitly by the formula*

$$u(t, x) - g(x - t(u(t, x))) = 0. \tag{5}$$

If $k \geq 2$ and g is C^k on a neighborhood of $\underline{x}$, then the solution u is C^k on a neighborhood of the characteristic $(t, \underline{x} + tg(\underline{x}))$, $0 \leq t < T$.

PROOF. If $u \in C^1([0, T[\times \mathbb{R}^2)$ is a solution, then u is constant on characteristics, so the value of u at $(\underline{t}, \underline{x})$ is equal to the value at all points $(\underline{t}, \underline{x}) + (t - \underline{t})(1, u(\underline{t}, \underline{x}))$, the characteristic line through $\underline{t}, \underline{x}$. Setting $t = 0$ yields (5).

The Implicit Function Theorem shows that this equation is uniquely solvable for $u(t, x)$ while $0 \leq t < T$. The key estimate is that if $G(t, x, u) \equiv u - g(x - tu)$, then $\partial G/\partial u > 0$ as long as $t \in [0, T[$. This proves both existence and uniqueness.

The final regularity assertion follows from the regularity part, as opposed to the existence part, of the Implicit Function Theorem. □

As an example, consider data, $g \in C^1$, which is piecewise smooth with singularities at $a_1 < a_2 < \cdots < a_M$. Then the construction of the solution u using characteristics shows that the solution is piecewise smooth with singularities propagating along characteristics. These singularities travel with the local speed of propagation u. Such signals are called *sound waves.* Note that the same name is used for the solutions in first-order perturbation theory in Figure 1.4.2. The common feature is the speed of propagation.

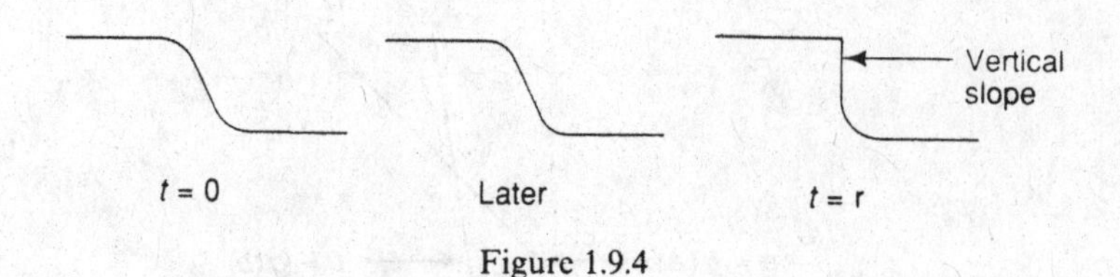

Figure 1.9.4

The formation of singularities at time T, starting with C^∞ initial data, is a purely nonlinear phenomenon. The cause is the fact that the speed of propagation, equal to u, depends on the amplitude u of the solution. We have seen that if the speed depends only on space and time, solutions exist globally.

Examining the formula for the solution, the behavior as the breakdown time approaches is not difficult to describe. Between the approaching characteristics in Figure 1.9.3, the solution u must decrease from $g(a)$ to $g(b)$ on an ever-decreasing interval. The curve steepens and at time T, the curve has steepened so much that a vertical tangent arises. The wave steepens and then breaks (Figure 1.9.4).

There is an even more explicit example of breakdown. Its construction is based on the observation that solutions of (3) are mapped to solutions by the scaling law

$$u(t, x) \mapsto u(\lambda t, \lambda x).$$

We seek solutions which are invariant under this transformation law (called *self-similar* solutions). A function is self-similar in this context if and only if it is positive homogeneous of degree 0 in t, x, that is, $u(t, x) = u(1, x/t) \equiv \psi(x/t)$ for $t > 0$. Then

$$u_t = \psi'\left(\frac{x}{t}\right)\left(\frac{-x}{t^2}\right), \qquad u_x = \psi'\left(\frac{x}{t}\right)\left(\frac{1}{t}\right),$$

$$u_t + uu_x = \psi'\left(\frac{x}{t}\right)\left(\psi\left(\frac{x}{t}\right) - \frac{x}{t}\right)\left(\frac{1}{t}\right).$$

Thus u satisfies the Burgers equation in $t > 0$ if and only if $\psi(s)$ satisfies

$$\psi'(s)(\psi(s) - s) = 0.$$

Thus the graph of ψ consists of two horizontal plateaus, where $\psi' = 0$, connected by a segment of the line $\psi = s$ (Figure 1.9.5). The graph of ψ is also the graph of $u|_{t=1}$, so u is not C^1 but Lipshitz continuous uniformly in each set $t \geq \varepsilon > 0$. As t decreases to zero the graph of $u(t, x) = \psi(x/t)$ is ψ simply scaled, so the connecting segment steepens to slope $1/t$.

The function $v(t, x) \equiv u(-t + 1, -x)$ is a solution which is Lipshitz continuous in $t < 1$ and which steepens as t increases toward $t = 1$.

The last paragraph is only informative to the extent that one is willing to accept a solution v which is not C^1. A similar situation arises if one accepts $g(x - ct)$ as solution of $u_t + cu_x = 0$ when g is only Lipshitz continuous. The

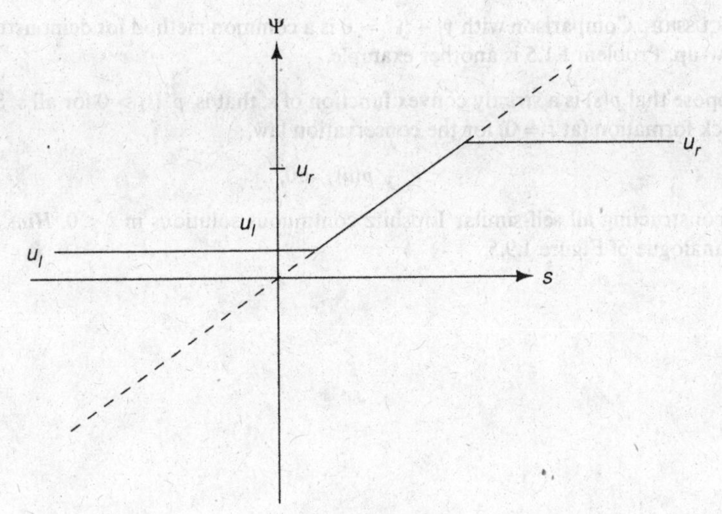

Figure 1.9.5

Theory of Distributions systematizes such so-called weak solutions. One good reason to accept generalized solutions is if they are limits of genuine classical solutions. For example, if g_n is a sequence of smooth functions converging uniformly to g, then the solutions $g_n(x - ct)$ converge to $g(x - ct)$. Similarly, if ψ_n is a sequence of smooth increasing functions converging uniformly to ψ and satisfying $\psi_n' < 1$, then the solutions of the Burgers equation with $u_n(0, x) = \psi_n(-x)$ will exist for $0 \le t \le 1$, and for any $\varepsilon > 0$ converge uniformly on $0 \le t \le 1 - \varepsilon$ to v. Thus, v is a reasonable generalized solution.

Surprisingly, solutions of the Burgers equation can be extended past the blow-up at $t = T$ to be discontinuous solutions which are still important for the physical interpretation. The discontinuities lie along *shock waves*, which are important in the modeling of combustion and supersonic flight (see [La], [Sm]). We have just been studying *spontaneous shock formation*.

Problems

1. Suppose that $P(\partial) = P_m(\partial)$ is a homogeneous constant coefficient partial differential operator and that Σ is a characteristic hyperplane. Find a piecewise smooth solution u to $Pu = 0$ which is singular at every point of Σ. *Hint*. Use the hint of Problem 1.7.2.

2. The following steps provide an alternate proof of Theorem 2:
   (i) Derive a partial differential equation satisfied by $u_x(t, x)$. *Hint*. Apply the result of Problem 1.4.2 to $u(t, x + \sigma)$. Alternatively, differentiate (3) with respect to x.
   (ii) Let $\gamma(t) \equiv (t, a + g(a)t)$ be the characteristic through $(0, a)$ and $y(t) \equiv u_x(\gamma(t))$. Show that $y' + y^2 = 0$ and $y(0) = g'(a)$.
   (iii) Show that y blows up at $t = -1/g'(a)$. Prove Theorem 2 by choosing a appropriately.

DISCUSSION. Comparison with $y' + y^2 = 0$ is a common method for demonstrating blow-up. Problem 1.1.5 is another example.

3. Suppose that $p(s)$ is a strictly convex function of s, that is, $p''(s) > 0$ for all s. Study shock formation (at $t = 0$) for the conservation law,

$$u_t + p(u)_x = 0,$$

by constructing all self-similar Lipshitz continuous solutions in $t < 0$. *Hint*. Find an analogue of Figure 1.9.5.

CHAPTER 2

Some Harmonic Analysis

§2.1. The Schwartz Space $\mathscr{S}(\mathbb{R}^d)$

The space $\mathscr{S}$ consists of smooth functions which together with all their derivatives decay rapidly to zero as $x \to \infty$. It is very useful in Fourier analysis, as it forms an easily manipulated family of functions which is mapped isomorphically onto itself by the Fourier transform. From this starting point, the classical theorems of Plancherel and Hausdorf–Young follow by straightforward completion arguments. This convenient formulation was exploited by S. Bochner. It was the inspired idea of L. Schwartz that an extension by duality, rather than continuity, gives a far-reaching generalization of the Fourier transform which has been crucial in modern analysis ever since. It is the goal of this chapter to give a brief description of these ideas. At the same time we review the basic techniques of the Theory of Distributions. It is assumed that the reader has a modest familiarity with the elementary Theory of Distributions. A brief introduction is presented in Appendix A.

Definition.

$$\mathscr{S}(\mathbb{R}^d) = \{u \in C^\infty(\mathbb{R}^d) : \forall \alpha, \beta \in \mathbb{N}^d, \sup_{x \in \mathbb{R}^d} |x^\alpha \partial^\beta u(x)| < \infty\}.$$

$\mathscr{S}(\mathbb{R}^d)$ is a vector space. Membership is tested by the countable family of seminorms

$$\|f\|_{\alpha,\beta} \equiv \sup_x |x^\alpha \partial^\beta f(x)|. \tag{1}$$

Functions in $\mathscr{S}$ are smooth and all derivatives tend to zero faster than any power of $|x|$ as $x \to \infty$.

EXAMPLES. 1. $C_0^\infty(\mathbb{R}^d) \subset \mathscr{S}(\mathbb{R}^d)$.

2. $e^{-|x|^2} \in \mathscr{S}(\mathbb{R}^d)$. Similarly, $e^{-(1+|x|^2)^{1/2}}$, and for $\varepsilon > 0$, $e^{-(1+|x|^2)^\varepsilon}$ belong to $\mathscr{S}(\mathbb{R}^d)$.

3. No matter how large N is, $(1 + |x|^2)^{-N}$ is not in $\mathscr{S}(\mathbb{R}^d)$.

4. Not all members of $\mathscr{S}(\mathbb{R}^d)$ decay exponentially (Problem 1).

Definition. A sequence $g_n \in \mathscr{S}$ is convergent to g in $\mathscr{S}$ if and only if $(\forall \alpha, \beta)$ $(\|g_n - g\|_{\alpha,\beta} \to 0$ as $n \to \infty)$.

Convergence, like membership in $\mathscr{S}$, is tested by a countable set of seminorms. It is easy to show that g_n converges to g in $\mathscr{S}$ if and only if $\rho(g_n, g)$ tends to zero where ρ is the metric

$$\rho(g, f) \equiv \sum 2^{-|\alpha|-|\beta|} \frac{\|g - f\|_{\alpha,\beta}}{1 + \|g - f\|_{\alpha,\beta}}. \tag{2}$$

This metric endows $\mathscr{S}(\mathbb{R}^d)$ with the structure of a metric space.

Proposition 1. *If $g \in \mathscr{S}(\mathbb{R}^d)$, $g(0) = 1$, then for any $f \in \mathscr{S}(\mathbb{R}^d)$,*

$$\mathscr{S}\text{-}\lim g(\varepsilon x) f = f \qquad \text{as} \quad \varepsilon \to 0.$$

The reason for this is that for ε small, $g(\varepsilon x)$ is a long flat plateau of height very close to 1 (Figure 2.1.1). Multiplying by g is close to multiplication by 1. The proof of Proposition 1 is Problem 2.

Corollary 2. *$C_0^\infty(\mathbb{R}^d)$ is dense in $\mathscr{S}(\mathbb{R}^d)$.*

PROOF. Choose $g \in C_0^\infty(\mathbb{R}^d)$ with $g(0) = 1$. Then $C_0^\infty(\mathbb{R}^d) \ni g(x/n) f \to f$, by Proposition 1. □

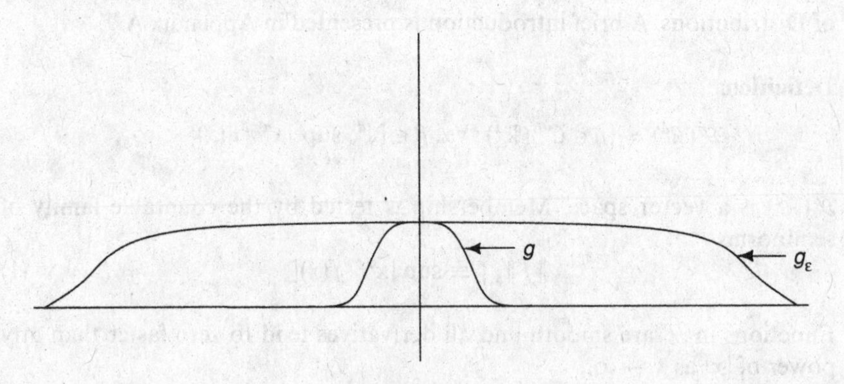

Figure 2.1.1

Proposition 3. *$\mathscr{S}(\mathbb{R}^d)$, ρ is a complete metric space.*

PROOF. If g_n is a Cauchy sequence, it follows that for any compact set $K \subset \mathbb{R}^d$ and any α, $\partial^\alpha g_n$ is uniformly bounded on K. The Arzela–Ascoli Theorem implies that the g_n converge to a limit $g \in C^\infty(\mathbb{R}^d)$, the convergence of all derivatives being uniform on compact sets.

For any compact K and α, β

$$\sup_K |x^\alpha \partial^\beta g| = \lim_{n\to\infty} \left(\sup_K |x^\alpha \partial^\beta g_n| \right) \leq \lim \sup \|g_n\|_{\alpha,\beta}.$$

As the right-hand side is independent of K we find that $\|g\|_{\alpha,\beta}$ is bounded by the right-hand side, so g belongs on $\mathscr{S}$.

Finally,

$$\sup_K |x^\alpha \partial^\beta (g_m - g)| = \lim_{n\to\infty} \left(\sup_K |x^\alpha \partial^\beta (g_m - g_n)| \right) \leq \lim_{n\to\infty} \|g_m - g_n\|_{\alpha,\beta}.$$

Given $\varepsilon > 0$, choose N so that if $n, m > N$, $\|g_m - g_n\|_{\alpha,\beta} < \varepsilon$. It follows that for $m > N$, $\|g_m - g\|_{\alpha,\beta} < \varepsilon$, proving convergence in $\mathscr{S}$. □

A vector space like $\mathscr{S}(\mathbb{R}^d)$, which is a complete metric space whose topology is defined by a countable family of seminorms, is called a *Fréchet space*. The basic principles of functional analysis; the Closed Graph Theorem, the Uniform Boundedness Principle, and the Hahn–Banach Theorem are valid in Fréchet spaces. We will not need these results.

Proposition 4. *If $F \in C^\infty(\mathbb{C}^k : \mathbb{C})$ with $F(0) = 0$ and $f_j \in \mathscr{S}(\mathbb{R}^d)$ for $1 \leq j \leq k$, then $F(f_1(x), \ldots, f_k(x)) \in \mathscr{S}(\mathbb{R}^d)$. The map from $\mathscr{S}(\mathbb{R}^d)^k$ to $\mathscr{S}(\mathbb{R}^d)$ so defined is continuous.*

Warning. The hypothesis on F does not mean that F is holomorphic. It means that $(\partial_x, \partial_y)^\alpha F(x + iy)$ exists and is continuous for any α. Holomorphic functions are those with $(\partial_x + i\partial_y)F = 0$.

EXAMPLES. 1. $\sin(\bar{f}(x)) \in \mathscr{S}$ whenever $f \in \mathscr{S}$. Here $\bar{f}$ is the complex conjugate of f.

2. The map $\varphi, \psi \mapsto \varphi\psi$ is a continous bilinear map of $\mathscr{S} \times \mathscr{S}$ to $\mathscr{S}$.

PROBLEMS

1. Construct a $u \in \mathscr{S}(\mathbb{R})$ which is not exponentially small at infinity. That is, for all $a > 0$, $e^{a|x|^a} u \notin L^\infty(\mathbb{R})$.

2. Prove Proposition 1.

3. Prove

Proposition.
(i) *If* $M \in C^\infty(\mathbb{R}^d)$ *and*

$$(\forall\alpha)(\exists N, c) \qquad (|\partial^\alpha M| \leq c(1 + |x|)^N),$$

then the map $f \mapsto Mf$ *is a continuous linear transformation of* $\mathscr{S}(\mathbb{R}^d)$ *into itself.*
(ii) *If in addition*

$$(\exists n, c > 0) \qquad (|M(x)| \geq c(1 + |x|)^{-n}),$$

then the mapping is one-to-one and onto with continuous inverse.

4. Prove Proposition 4.

§2.2. The Fourier Transform on $\mathscr{S}(\mathbb{R}^d)$

The Fourier transform and its inverse serve to express a function u as a superposition of oscillatory exponential functions $e^{-i\langle x,\xi\rangle}$, $\xi \in \mathbb{R}^d$, according to the formula

$$u(x) = (2\pi)^{-d/2} \int e^{i\langle x,\xi\rangle}\hat{u}(\xi)\, d\xi. \tag{1}$$

The goal of this section is to present the basic properties of the Fourier transform and, in particular, to prove the inversion formula (1).

Definition. For $u \in \mathscr{S}(\mathbb{R}^d)$, the Fourier transform of u is the function defined by

$$\mathscr{F}u(\xi) \equiv \hat{u}(\xi) \equiv (2\pi)^{-d/2} \int e^{-i\langle x,\xi\rangle} u(x)\, dx. \tag{2}$$

Here $\langle x, \xi\rangle \equiv \sum x_i\xi_i$. We will often abbreviate $\langle x, \xi\rangle$ as $x\xi$.

EXAMPLES. 1. For $u = e^{-x^2/2} \in \mathscr{S}(\mathbb{R})$

$$\hat{u}(\xi) = (2\pi)^{-1/2} \int_{-\infty}^{\infty} e^{-ix\xi}e^{-x^2/2}\, dx.$$

Complete the square in the exponent to find

$$\frac{x^2}{2} + ix\xi = \tfrac{1}{2}(x^2 + 2ix\xi) = \tfrac{1}{2}((x + i\xi)^2 + \xi^2).$$

Thus

$$\sqrt{2\pi}\,\hat{u} = e^{-\xi^2/2} \int_{-\infty}^{\infty} e^{-(x+i\xi)^2/2}\, dx$$

$$= e^{-\xi^2/2} \int_\Gamma e^{-z^2/2}\, dz,$$

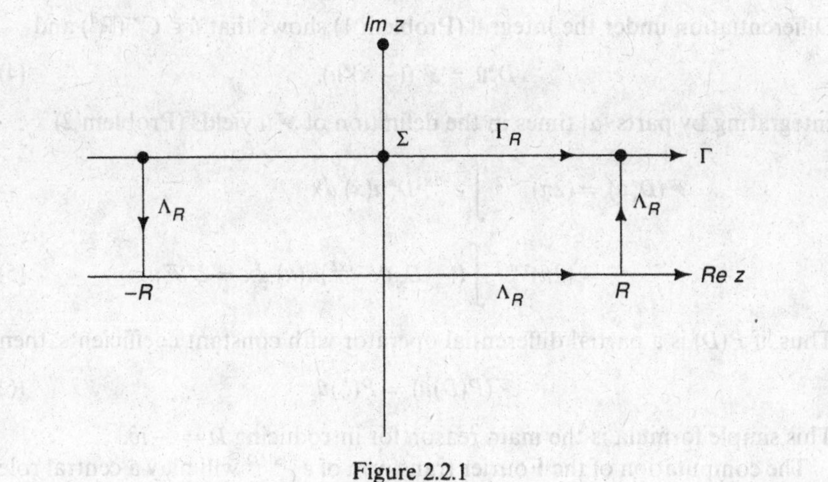

Figure 2.2.1

where Γ is the contour $\operatorname{Im} z = \xi$ traversed from left to right. Let Γ_R be the segment on Γ running from $-R + i\xi$ to $R + i\xi$ (Figure 2.2.1). Since the integrand decays exponentially fast as $|z|$ goes to infinity along Γ, the integral is equal to the limit

$$= \lim_{R\to\infty} \int_{\Gamma_R} e^{-z^2/2}\, dz.$$

By Cauchy's Theorem, the integral over Γ_R is equal to the integral over Λ_R, where $\Gamma_R - \Lambda_R$ is equal to the boundary of the rectangle whose side opposite Γ_R is the interval $[-R, R]$ on the x-axis. Thus our Fourier transform is

$$= \lim_{R\to\infty} \int_{\Lambda_R} e^{-z^2/2}\, dz.$$

As $R \to \infty$, the integrals over the vertical sides of the rectangle tend to zero exponentially rapidly and we find

$$\hat{u}(\xi) = e^{-\xi^2/2} \frac{1}{(2\pi)^{1/2}} \int_{-\infty}^{\infty} e^{-x^2/2}\, dx = e^{-\xi^2/2}.$$

2. To compute the Fourier transform of the multi-dimensional analogue, $u = e^{-|x|^2/2} \in \mathscr{S}(\mathbb{R}^d)$, note that $u = \prod_j e^{-x_j^2/2}$. Fubini's Theorem yields

$$\hat{u}(\xi) = (2\pi)^{-d/2} \int \prod e^{-ix_j\xi_j} e^{-x_j^2/2}\, d\xi_1 \cdots d\xi_d$$

$$= \prod (2\pi)^{-1/2} \int e^{-ix_j\xi_j} e^{-x_j^2/2}\, d\xi_j = \prod_j e^{-\xi_j^2/2} = e^{-|\xi|^2/2}.$$

For any $u \in \mathscr{S}(\mathbb{R}^d)$, $\hat{u}$ is bounded, and

$$\|\hat{u}\|_{L^\infty(\mathbb{R}^d)} \le (2\pi)^{-d/2} \|u\|_{L^1(\mathbb{R}^d)} \le c\|(1 + |x|)^{d+1} u\|_{L^\infty(\mathbb{R}^d)}. \tag{3}$$

Differentiation under the integral (Problem 1) shows that $\hat{u} \in C^\infty(\mathbb{R}^d)$ and

$$D_\xi^\alpha \hat{u} = \mathscr{F}((-x)^\alpha u). \tag{4}$$

Integrating by parts $|\alpha|$ times in the definition of $\mathscr{F}u$ yields (Problem 2)

$$\begin{aligned}\mathscr{F}(D_x^\alpha u) &= (2\pi)^{-d/2} \int e^{-ix\xi} D_x^\alpha u(x)\, dx \\ &= (2\pi)^{-d/2} \int ((-D_x)^\alpha e^{-ix\xi}) u(x)\, dx = \xi^\alpha \mathscr{F}u. \end{aligned} \tag{5}$$

Thus, if $P(D)$ is a partial differential operator with constant coefficients, then

$$\mathscr{F}(P(D)u) = P(\xi)\hat{u}. \tag{6}$$

This simple formula is the main reason for introducing $D = -i\partial$.

The computation of the Fourier transform of $e^{-|x|^2/2}$ will play a central role in the analysis of $\mathscr{F}$. We next present a second derivation starting from the fact that the function $u = e^{-x^2/2} \in \mathscr{S}(\mathbb{R})$ satisfies the ordinary differential equation $u' = -xu$. Since this equation is homogeneous, linear, and first order, the general solution is a constant multiple of u.

Take the Fourier transform of the differential equation to find

$$i\xi\hat{u} = \frac{1}{i}(\partial_\xi \hat{u}), \qquad \text{hence} \quad (\partial_\xi + \xi)\hat{u} = 0.$$

Thus the transform satisfies the same equation as u, so

$$\hat{u} = ce^{-\xi^2/2}.$$

The constant is evaluated using

$$c = \hat{u}(0) = \int e^{-x^2/2}\, dx \frac{1}{\sqrt{2\pi}} = 1.$$

Proposition 1. *For any $u \in \mathscr{S}(\mathbb{R}^d)$, $\hat{u} \in \mathscr{S}(\mathbb{R}^d)$, and the map $u \mapsto \hat{u}$ is a continuous transformation of $\mathscr{S}$ to itself.*

PROOF. Estimate the $\mathscr{S}(\mathbb{R}_\xi^d)$ seminorms of $\hat{u}$ as follows:

$$\begin{aligned}\|\hat{u}\|_{\alpha,\beta} &= \|\xi^\alpha \partial_\xi^\beta \hat{u}\|_{L^\infty(\mathbb{R}_\xi^d)} = \|\mathscr{F}(\partial_x^\alpha(x^\beta u))\|_{L^\infty(\mathbb{R}_\xi^d)} \\ &\le (2\pi)^{-d/2} \|\partial_x^\alpha(x^\beta u)\|_{L^1(\mathbb{R}_x^d)} \le c\|(1+|x|)^{d+1}\partial_x^\alpha(x^\beta u)\|_{L^\infty(\mathbb{R}_x^d)} < \infty. \end{aligned}$$

If $u_n \to u$ in $\mathscr{S}$, apply this estimate to $u_n - u$ to show that $\|\hat{u}_n - \hat{u}\|_{\alpha,\beta}$ converges to zero. □

The Fourier transform behaves well with respect to dilations and translations

$$(\sigma_\lambda u)(x) \equiv u(\lambda x), \quad \text{is called the dilation by } \lambda \in \mathbb{R},$$

$$(\tau_h u)(x) = u(x - h), \quad \text{is the translation by } h \in \mathbb{R}^d.$$

Proposition 2. *If $u \in \mathscr{S}(\mathbb{R}^d)$, then for any $h \in \mathbb{R}^d$, $\lambda \in \mathbb{R}\backslash 0$:*

(i) $\mathscr{F}(\tau_h u) \equiv \mathscr{F}(u(\cdot - h)) = e^{-ih\xi}\mathscr{F}(u) = e^{-ih\xi}\hat{u}(\xi)$.
(ii) $\mathscr{F}(e^{ihx}u) = \tau_h \mathscr{F}(u) \equiv \hat{u}(\cdot - h)$.
(iii) $\mathscr{F}(\sigma_\lambda u) = |\lambda|^{-d}(\sigma_{1/\lambda}\mathscr{F}(u))$, *that is,* $\mathscr{F}(\sigma_\lambda u)(\xi) = |\lambda|^{-d}(\mathscr{F}u)(\xi/\lambda)$.

PROOF. (i) In $\int e^{-i\langle x,\xi\rangle}u(x-h)\,dx$ make the change of variable $y \equiv x - h$ to find $\int e^{-i\langle y+h,\xi\rangle}u(y)\,dy$. Assertion (i) follows.

(ii) Left to the reader.

(iii) In $\int e^{-i\langle x,\xi\rangle}u(\lambda x)\,dx$ make the change of variable $y \equiv \lambda x$ with Jacobian $|\det dy/dx| = |\lambda|^d$. This yields $\int e^{-i\langle y/\lambda,\xi\rangle}u(y)|\lambda|^{-d}\,dy$, and (iii) follows. □

Finally, we require a simple duality identity. Denote by $\langle\cdot,\cdot\rangle$ the pairing $f, g \mapsto \int fg\,dx$ from $\mathscr{S} \times \mathscr{S}$ to $\mathbb{C}$. Then

$$\text{for all } \varphi, \psi \in \mathscr{S}, \qquad \langle\mathscr{F}\varphi, \psi\rangle = \langle\varphi, \mathscr{F}\psi\rangle. \tag{7}$$

PROOF. $K(\xi, x) \equiv (2\pi)^{-d/2}e^{-ix\xi}$ is the kernel of the integral operator $\mathscr{F}$. Note that K is symmetric under interchange of x and ξ. Then

$$\langle\mathscr{F}\varphi, \psi\rangle = \int \psi(\xi)\left(\int K(\xi, x)\varphi(x)\,dx\right)d\xi.$$

Since $K(\xi, x)\psi(\xi)\varphi(x) \in L^1(\mathbb{R}^{2d}, dx\,d\xi)$ Fubini's Theorem justifies an interchange of order of integration giving

$$= \iint \varphi(x)K(\xi, x)\psi(\xi)\,d\xi\,dx = \int \varphi(x)\left(\int K(x, \xi)\psi(\xi)\,d\xi\right)dx$$
$$= \langle\varphi, \mathscr{F}\psi\rangle. \qquad \square$$

Virtually the same proof yields an identity involving the scalar product in $L^2(\mathbb{R}^d : \mathbb{C})$

$$(\varphi, \psi)_{L^2(\mathbb{R}^d)} \equiv \int_{\mathbb{R}^d} \varphi(x)\overline{\psi(x)}\,dx,$$

which is linear in the first slot. One finds

$$(\mathscr{F}\varphi, \psi) = (\varphi, \mathscr{F}^*\psi), \qquad \forall\varphi, \psi \in \mathscr{S}(\mathbb{R}^d), \tag{8}$$

where $\mathscr{F}^*$, called the *Inverse Fourier Transform*, is the integral operator with kernel $(2\pi)^{-d/2}e^{ix\xi}$. The difference between $\mathscr{F}^*$ and $\mathscr{F}$ is the minus sign in the exponent.

In the proof of the Fourier Inversion Formula, we need a simple result about approximate δ functions, $\varepsilon^{-d}j(x/\varepsilon)$ (Figure 2.2.2).

The next proposition is more than sufficient.

Proposition 3. *If $j \in L^1(\mathbb{R}^d)$, $\int j(x)\,dx = 1$, and $u \in L^\infty(\mathbb{R}^d)$ is continuous at* 0, *then*

$$\lim_{\varepsilon\to 0+} \int u(x)\varepsilon^{-d}j(x/\varepsilon)\,dx = u(0).$$

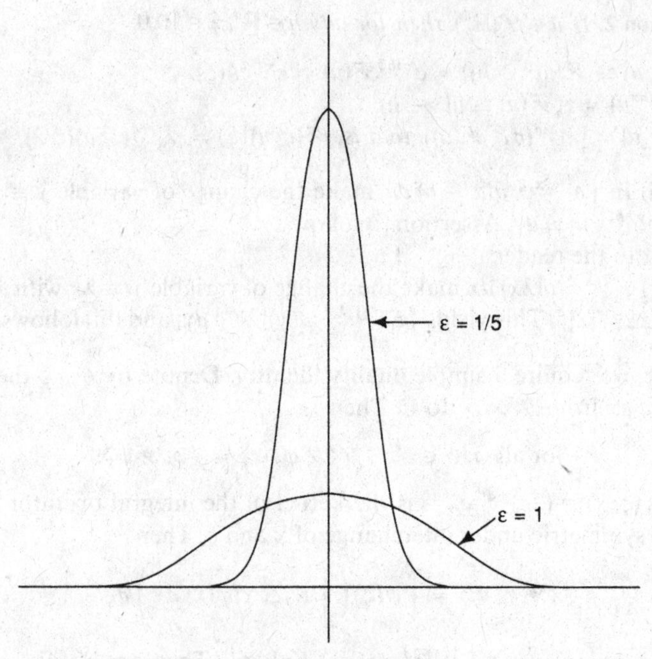

Figure 2.2.2

PROOF. Replacing u by $u - u(0)$, it is sufficient to consider the case $u(0) = 0$. Given a challenge number, $\eta > 0$, choose $R > 0$ so that $|u(x)| < \eta/\|j\|_{L^1}$ for $|x| \leq R$. Then

$$\left| \int_{|x| \leq R} u\varepsilon^{-d} j\left(\frac{x}{\varepsilon}\right) dx \right| < \frac{\eta}{\|j\|_{L^1}} \int \left| \varepsilon^{-d} j\left(\frac{x}{\varepsilon}\right) \right| dx = \eta.$$

For the rest

$$\left| \int_{|x| > R} u\varepsilon^{-d} j\left(\frac{x}{\varepsilon}\right) dx \right| \leq \|u\|_{L^\infty} \int_{|x| > R} \left| \varepsilon^{-d} j\left(\frac{x}{\varepsilon}\right) \right| dx$$

$$= \|u\|_{L^\infty} \int_{|x| > R/\varepsilon} |j(x)|\, dx = o(1).$$

Thus

$$\overline{\lim} \left| \int u\varepsilon^{-d} j\left(\frac{x}{\varepsilon}\right) dx \right| < \eta. \qquad \square$$

Theorem 4 (The Fourier Inversion Formula). *For all* $u \in \mathscr{S}(\mathbb{R}^d)$

$$u(x) = (2\pi)^{-d/2} \int e^{ix\xi} \hat{u}(\xi)\, d\xi = \mathscr{F}^* \mathscr{F} u.$$

PROOF. First consider the case $x = 0$. Let $v(x) = e^{-|x|^2/2}$. Since $\int v\,dx = (2\pi)^{d/2}$ and $\mathscr{F} v = v$, Proposition 3 shows that

$$\begin{aligned}(2\pi)^{+d/2}u(0) &= \lim_{\varepsilon\to 0} \langle u, \varepsilon^{-d} v(x/\varepsilon)\rangle = \lim_{\varepsilon\to 0} \langle u, \varepsilon^{-d}\sigma_{1/\varepsilon} v\rangle \\ &= \lim \langle u, \varepsilon^{-d}\sigma_{1/\varepsilon}(\mathscr{F} v)\rangle = \lim \langle u, \mathscr{F}(\sigma_\varepsilon v)\rangle \\ &= \lim \langle \mathscr{F} u, \sigma_\varepsilon v\rangle.\end{aligned}$$

Now, $\sigma_\varepsilon v(\xi) = v(\varepsilon\xi) \to v(0) = 1$ pointwise, and $|\sigma_\varepsilon v| \le 1$. Furthermore, $\mathscr{F} u \in \mathscr{S}(\mathbb{R}^d) \subset L^1(\mathbb{R}^d)$. Lebesgue's Dominated Convergence Theorem shows that the right-hand side converges to $\int (\mathscr{F} u)(\xi)\,d\xi$. The proof of the special case, $x = 0$, is complete.

For the general case, note that

$$\begin{aligned}u(x) = (\tau_{-x}u)(0) &= (2\pi)^{-d/2} \int \mathscr{F}(\tau_{-x}u)(\xi)\,d\xi \\ &= (2\pi)^{-d/2} \int e^{ix\xi}\hat{u}(\xi)\,d\xi \quad \text{(by Proposition 2 (i)).} \qquad \square\end{aligned}$$

This proof rests on little more than the identity $\mathscr{F} v = v$. The explanation why this is sufficient is that the elementary properties of $\mathscr{F}$ and $\mathscr{F}^*$ imply that $\{u \in \mathscr{S}(\mathbb{R}^d)\colon \mathscr{F}^*\mathscr{F} u = u\}$ is a closed linear subspace which is invariant under translation and dilation. It is not hard to show that the span of the translates and dilates of our v form a dense subset of $\mathscr{S}(\mathbb{R}^d)$ (Problem 4).

That $\mathscr{F}^*\mathscr{F} = \mathrm{Id}$ on $\mathscr{S}$ implies that $\mathscr{F}$ is one-to-one and $\mathscr{F}^*$ is onto. However, $\mathscr{F}^*$ differs from $\mathscr{F}$ by a simple sign change $x \mapsto -x$ in the kernel, so we have $\mathscr{F}^* = \mathscr{R}\mathscr{F}$ where $\mathscr{R}$ is the reflection operator $(\mathscr{R}u)(x) \equiv u(-x)$. Then $\mathscr{F} = \mathscr{R}^2\mathscr{F} = \mathscr{R}\mathscr{F}^*$ is also onto. Summarizing, we have proved the following corollary.

Corollary 5. *The Fourier transform, $\mathscr{F}$, is a linear bijection of $\mathscr{S}(\mathbb{R}^d)$ to itself with inverse equal to $\mathscr{F}^*$.*

There is a parallel theory of Fourier series of smooth 2π multiply periodic functions whose Fourier expansion is

$$u(\theta) = (2\pi)^{-d/2} \sum_{n\in\mathbb{Z}^d} u_n e^{i\langle n,\theta\rangle},$$

$$u_n = (2\pi)^{-d/2} \int_{[0,2\pi]^d} u(\theta) e^{-i\langle n,\theta\rangle}\,d\theta.$$

The transform $u \mapsto \{u_n\}$ is a bijection of smooth multiply periodic u to the space $\mathfrak{s}$ of rapidly decreasing complex sequences. That is, sequences such that $|n|^k u_n \in l^\infty$ for all k. The analogue of Plancherel's identity (2.3.2) is called

Bessel's identity

$$\sum |u_n|^2 = \int_{[0,2\pi]^d} |u(\theta)|^2 \, d\theta. \tag{9}$$

An elegant derivation of these results, using the framework of §2.5, can be found in Donoghue [Don].

PROBLEMS

1. Carefully justify the differentiations leading to the formula $D_j \mathscr{F}(u) = \mathscr{F}(-x_j u)$. Then prove that $D^\alpha \mathscr{F}(u) = \mathscr{F}((-x)^\alpha u)$. *Hint*. Induction on $|\alpha|$.
2. Carefully justify the integration by parts leading to the formula $\mathscr{F}(D_j u) = \xi_j \mathscr{F} u$. Then prove $\mathscr{F}(D^\alpha u) = \xi^\alpha \mathscr{F} u$. *Hint*. Induction.
3. If A is an invertible linear map of $\mathbb{R}^d$ to itself and $u \in \mathscr{S}(\mathbb{R}^d)$, define u_A by $u_A(x) \equiv u(A^{-1}x)$. Compute a formula for the Fourier transform of u_A. Prove that $\mathscr{F}(u_A) = (\mathscr{F}u)_A$ for all u if and only if A is an orthogonal transformation.
   DISCUSSION. The special case of A equal to a reflection was used in the discussion of the Fourier Inversion Formula. A simple consequence of this special case is that the Fourier transform of an odd (resp. even) function is odd (resp. even).
4. Prove that the linear span of the translates and dilates of $\exp(-|x|^2/2)$ are dense in $\mathscr{S}(\mathbb{R}^d)$, thereby giving an alternate derivation of the Fourier Inversion Formula. *Hint*. For $u \in C_0^\infty(\mathbb{R}^d)$ consider
   $$\varepsilon^{-d} \int u(y) \exp\left(-\left| \frac{x-y}{\varepsilon} \right|^2 /2 \right) dy.$$
5. Show that if $u \in \mathscr{S}$ satisfies $\mathscr{F}(u) = i^n u$, then $v \equiv (\partial_j + x_j)u$ satisfies $\mathscr{F}(v) = i^{n+1} v$.
   DISCUSSION. Thus $w \equiv (\partial + x)^\alpha (\exp(-|x|^2/2)$ satisfies $\mathscr{F}(v) = v$ provided $|\alpha|$ is a multiple of four. The Gaussian is by no means the only function which is its own Fourier transform.
   The function w is an eigenfunction of $\mathscr{F}$ for all α. These eigenfunctions form an orthogonal basis for $L^2(\mathbb{R}^d)$. This gives the spectral decomposition of the unitary operator $\mathscr{F}$ from $L^2(\mathbb{R}^d)$ to itself (Theorem 2.4.4). These eigenfunctions were well known in the nineteenth century and give yet another approach to the Fourier Inversion Formula. In fact, this is the derivation in Weiner's text on the Fourier transform.

§2.3. The Fourier Transform on $L^p(\mathbb{R}^d)$: $1 \le p \le 2$

The Fourier transform, defined on $\mathscr{S}(\mathbb{R}^d)$ can be extended to more general classes of functions and distributions. Linear maps are commonly extended using one of two algorithms. The first is extension by continuity.

Proposition 1. *Suppose that X is a normed linear space, $E \subset X$ is a dense linear space, and Y is a Banach space. If $T: E \to Y$ is a continuous linear map, that is,*

$$(\exists c > 0)(\forall e \in E), \qquad \|Te\|_Y \le c\|e\|_X,$$

then there is one and only one continuous linear map $T_{\text{ext}}: X \to Y$ *with* $T_{\text{ext}}|_E = T$.

As an example, we extend the Fourier transform as a continuous linear map of $L^1(\mathbb{R}^d)$ to $L^\infty(\mathbb{R}^d)$ and from $L^2(\mathbb{R}^d)$ to itself. Toward this end, we need a simple approximation theorem.

Proposition 2. *$C_0^\infty(\mathbb{R}^d)$ is dense in $L^p(\mathbb{R}^d)$ for $1 \le p < \infty$.*

PROOF. The set of finite linear combinations of characteristic functions of bounded measurable sets is dense. Thus we need only prove that if A is bounded and measurable, then for any $\varepsilon > 0$, there is a $\varphi \in C_0^\infty$ with $\|\varphi - \chi_A\|_p < \varepsilon$.

Given $\varepsilon > 0$, choose compact K and open $\mathcal{O}$, with $K \subset A \subset \mathcal{O}$ and $\mu(\mathcal{O}\backslash K) < \varepsilon^p$. Choose $\varphi \in C_0^\infty(\mathbb{R}^n)$ with $0 \le \varphi \le 1$, $\text{supp}(\varphi) \subset \mathcal{O}$, and $\varphi|_K = 1$ (Problem 1). Then

$$\|\varphi - \chi_A\|_p^p = \int |\varphi - \chi_A|^p \le \int_{\mathcal{O}\backslash K} 1 < \varepsilon^p. \qquad \square$$

Basic Estimates. If $f \in \mathcal{S}(\mathbb{R}^d)$, then

$$\|\mathcal{F}f\|_{L^\infty} \le (2\pi)^{-d/2}\|f\|_{L^1}, \tag{1}$$

$$\|\mathcal{F}f\|_{L^2} = \|f\|_{L^2} = \|\mathcal{F}^*f\|_{L^2}. \tag{2}$$

PROOF. The first estimate is immediate and was already observed in (2.2.3). For the second compute

$$\|\mathcal{F}f\|_{L^2}^2 = (\mathcal{F}f, \mathcal{F}f) = (f, \mathcal{F}^*\mathcal{F}f) = (f, f) = \|f\|_{L^2}^2.$$

The estimate for $\mathcal{F}^*f$ follows in the same way from the identity $\mathcal{F}\mathcal{F}^* = \text{Id}$ (Corollary 2.2.6). $\square$

Definition. $\dot{C}(\mathbb{R}^d)$ denotes the set of $u \in C(\mathbb{R}^d)$ such that $\lim_{x\to\infty} u(x) = 0$.

Such functions are said to vanish at infinity. $\dot{C}(\mathbb{R}^d)$ is a closed subspace of $L^\infty(\mathbb{R}^d)$, so is a Banach space in the supremum norm. $\dot{C}$ is the closure of $\mathcal{S}(\mathbb{R}^d)$ in $L^\infty(\mathbb{R}^d)$.

Theorem 3 (Riemann–Lebesgue Lemma). *$\mathcal{F}: \mathcal{S}(\mathbb{R}^d) \to \mathcal{S}(\mathbb{R}^d)$ extends uniquely to a continuous linear map $L^1(\mathbb{R}^d) \to \dot{C}(\mathbb{R}^d)$ with norm equal to $(2\pi)^{-d/2}$. For $u \in L^1(\mathbb{R}^d)$ the value of $(\mathcal{F}u)(\xi)$ is equal to the absolutely convergent integral $(2\pi)^{-d/2}\int e^{-ix\xi}u(x)\,dx$.*

PROOF. The existence follows from the density of $\mathcal{S}$ in L^1 together with basic estimate (1). The upper bound for the norm is achieved at $\hat{u}(0)$ for any positive $u \in \mathcal{S}$. To prove the formula, choose $u_n \in \mathcal{S}$, $u_n \to u$ in L^1. Then

$\mathcal{F}u_n \to \mathcal{F}u$ in $\dot{C}$. In particular, $\mathcal{F}u_n(\xi) \to \mathcal{F}u(\xi)$. Now

$$(2\pi)^{d/2}\mathcal{F}u_n(\xi) = \int e^{-ix\xi}u_n(x)\,dx \to \int e^{-ix\xi}u(x)\,dx,$$

since the difference of the two integrals is dominated by $\|u_n - u\|_{L^1}$. □

Theorem 4 (Plancherel). *The Fourier transforms $\mathcal{F}$ and $\mathcal{F}^*$ on $\mathcal{S}$ extend uniquely to unitary maps of L^2 to itself satisfying $\mathcal{F}^*\mathcal{F} = \mathcal{F}\mathcal{F}^* = I$.*

PROOF. The existence of isometric extensions for $\mathcal{F}$ and $\mathcal{F}^*$ follows from the basic estimates (2).

That $\mathcal{F}\mathcal{F}^* = \mathcal{F}^*\mathcal{F} = \mathrm{I}$ follows since both sides are continuous on L^2 and they are equal on the dense subset, $\mathcal{S}$. Unitarity follows. □

A typical u in L^2 is not in L^1 (e.g. $(1 + |x|)^{-1} \in L^2(\mathbb{R})$) so $\int e^{-ix\xi}u(x)\,dx$ is not absolutely convergent. Thus, the Fourier transform of a typical element of L^2 is not given by the usual integral formula.

For $u \in L^1 \cap L^2$ we have given two meanings to $\mathcal{F}u$. Thinking of u as an element of L^1 (resp. L^2), $\mathcal{F}u$ is defined as an element of $\dot{C}$ (resp. L^2). It is important to know that the two results are equal. Equality is interpreted as either equality in the sense of distributions or equality almost everywhere. To prove the latter, fix $R > 0$ and $B \equiv \{|x| < R\}$. Choose $u_n \in \mathcal{S}$ with $u_n \to u$ in both L^1 and L^2. Simple plateau cutoff, followed by convolution with an approximate delta, achieves this goal. Then

$$L^2\text{-lim}\ \mathcal{F}u_n = \mathcal{F}_{L^2}u \qquad \text{and} \qquad \dot{C}\text{-lim}\ \mathcal{F}u_n = \mathcal{F}_{L^1}u.$$

In particular, $\mathcal{F}u_n$ converges in $L^2(B)$ to both $(\mathcal{F}_{L^2}u)|_B$ and to $(\mathcal{F}_{L^1}u)|_B$. Thus $\mathcal{F}_{L^2}u = \mathcal{F}_{L^1}u$ almost everywhere on B.

If u belongs to L^2 we obtain classical formulas for the Fourier transform by choosing u_n with $u_n \to u$ in L^2. Then $\mathcal{F}u = L^2\text{-lim}_{n\to\infty}\mathcal{F}u_n$. For example, if one takes $u_n \equiv \chi_{\{|x|<R\}}u$ or $u_n \equiv e^{-\varepsilon|x|}u$, one finds

$$\begin{aligned}\mathcal{F}u &= L^2\text{-}\lim_{R\to\infty}\ (2\pi)^{-d/2}\int_{|x|<R} e^{-ix\xi}u(x)\,dx \\ &= L^2\text{-}\lim_{\varepsilon\to 0}\ (2\pi)^{-d/2}\int e^{-\varepsilon|x|}e^{-ix\xi}u(x)\,dx.\end{aligned}$$

Note that these are L^2 limits, not limits almost everywhere.

$\mathcal{F}$ maps L^1 to L^∞ with norm $(2\pi)^{-d/2}$ and L^2 to L^2 with norm 1. It follows, by interpolation, that $\mathcal{F}$ maps the L^p spaces "between" L^1 and L^2 to those "between" L^∞ and L^2. The precise result is given by the Riesz–Thorin Convexity Theorem. This result asserts that if K is bounded from L^{r_0} to L^{s_0} and bounded from L^{r_1} to L^{s_1}, then K is bounded from L^{r_θ} to L^{s_θ} for $0 \le \theta \le 1$, where r_θ, s_θ are defined by

$$\frac{1}{r_\theta} \equiv \theta\frac{1}{r_0} + (1-\theta)\frac{1}{r_1}, \qquad \frac{1}{s_\theta} \equiv \theta\frac{1}{s_0} + (1-\theta)\frac{1}{s_1}. \tag{3}$$

Two conventions are in force here. First, r and s are between 1 and infinity, and second, in (3) we take $1/\infty = 0$, $1/0 = \infty$.

Definition. A linear map $K: \mathcal{S}(\mathbb{R}^d) \to \mathcal{S}(\mathbb{R}^d)$ is of type (r, s) if and only if

$$(\exists c \ge 0)\ (\forall \varphi \in \mathcal{S}(\mathbb{R}^d)), \qquad \|K\varphi\|_{L^s(\mathbb{R}^d)} \le c\|\varphi\|_{L^r(\mathbb{R}^d)}.$$

If $r < \infty$, then $\mathcal{S}$ is dense in L^r and K has a unique extension to a bounded linear map of L^r to L^s. For $r = \infty$, the extension maps $\dot{C}(\mathbb{R}^d)$ to L^s.

Theorem 5 (Riesz–Thorin Convexity Theorem). *Suppose that for $i = 0, 1$, $1 \le r_i$, $s_i \le \infty$, and K is a linear map which is of type (r_i, s_i). Then for all $\theta \in [0, 1]$, K is of type (r_θ, s_θ) where r_θ and s_θ are defined in (3). Furthermore, if*

$$(\forall u \in L^{r_i}), \qquad \|Ku\|_{s_i} \le B_i\|u\|_{r_i}, \qquad i = 0, 1,$$

the for all $\theta \in [0, 1]$, $u \in L^{r_\theta}$,

$$\|Ku\|_{s_\theta} \le B_0^\theta B_1^{1-\theta}\|u\|_{r_\theta}.$$

An elegant proof using Hadamard's Three Circle Theorem from complex analysis can be found in many functional analysis texts.

The above theorem shows that the set of points $(1/r, 1/s) \in [0, 1] \times [0, 1]$, such that K is of type (r, s) in a convex set, and the norm of K from L^r to L^s is a convex function on that set.

Corollary 6 (Hausdorff–Young Inequality). *The Fourier transform $\mathcal{F}: \mathcal{S} \to \mathcal{S}$ extends uniquely to a bounded linear map from L^p to L^q for $1 \le p \le 2$, $1/q + 1/p = 1$. Furthermore, for such p and any $f \in L^p$,*

$$\|\mathcal{F}u\|_{L^q} \le (2\pi)^{-d(2-p)/2p}\|u\|_{L^p}.$$

PROOF. Apply the Riesz–Thorin Theorem with $r_0 = 1$, $s_0 = \infty$, $r_1 = 2$, $s_1 = 2$. □

The type of $\mathcal{F}$ thus contains the segment sketched in Figure 2.3.1.

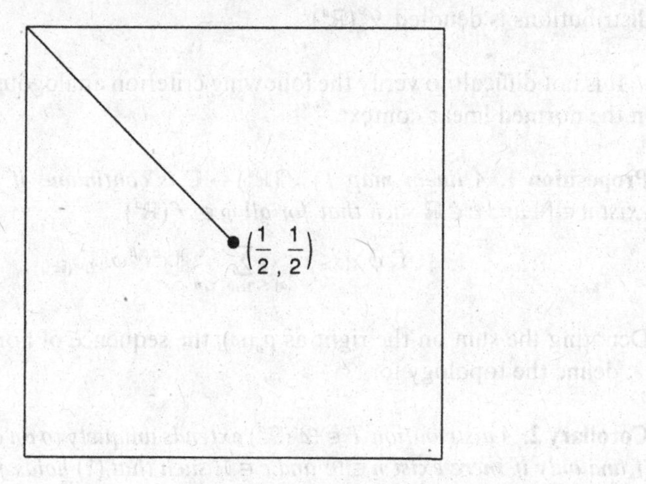

Figure 2.3.1

PROBLEMS

1. If $\mathcal{K}$ is a compact subset of the open set $\mathcal{O}$ in $\mathbb{R}^d$, construct a $\varphi \in C_0^\infty(\mathcal{O})$ with $0 \le \varphi \le 1$ and φ equal to 1 on a neighborhood of $\mathcal{K}$.

2. Let $u \equiv e^{-x}\chi_{]0,\infty[} \in L^1(\mathbb{R})$. Compute the Fourier transform of:
   (a) u; (b) the reflection $\mathscr{R}u$; (c) $\tau_h u$;
   (d) $e^{ixh}u$; (e) $(\sin x)u$.

3. For $c \in \mathbb{C}\backslash\mathbb{R}$ and $k = 2, 3, \ldots$, use the calculus of residues to compute the Fourier transform of $1/(x-c)^k \in L^1(\mathbb{R})$.

4. For the functions $v \equiv e^{-|x|}$ and $u \equiv (1+x^2)^{-1} \in L^1(\mathbb{R})$ verify the Fourier Inversion Formula by computing the Fourier transforms and then applying directly $\mathscr{F}^*$.
   DISCUSSION. This special case can be used as the keystone of the proof of the inversion formula in the same way that we used the Gaussian, $\exp(-x^2/2)$.

§2.4. Tempered Distributions

The three spaces in which $\mathscr{F}$ acts most naturally are $\mathscr{S}(\mathbb{R}^d)$, $L^2(\mathbb{R}^d)$, and $\mathscr{S}'(\mathbb{R}^d)$. The last, the dual of $\mathscr{S}(\mathbb{R}^d)$, is called the space of *tempered distributions.* The extension of $\mathscr{F}$ from $\mathscr{S}$ to $\mathscr{S}'$ is by a duality argument quite different from the extension process in §2.3. This section is an introduction to $\mathscr{S}'$ including a discussion of the extension.

Recall that the distributions $T \in \mathscr{D}'(\mathbb{R}^d)$ are linear functionals on $\mathscr{D}(\mathbb{R}^d) = C_0^\infty(\mathbb{R}^d)$ (see Appendix A). The tempered distributions are those distributions which extend to continuous linear maps from $\mathscr{S}(\mathbb{R}^d)$ to $\mathbb{C}$. Since $\mathscr{D}(\mathbb{R}^d)$ is dense in $\mathscr{S}(\mathbb{R}^d)$ the extension is uniquely determined.

Definition. A *tempered distribution* is a continuous linear functional on $\mathscr{S}(\mathbb{R}^d)$, that is, a continuous linear map from $\mathscr{S}(\mathbb{R}^d)$ to $\mathbb{C}$. The space of tempered distributions is denoted $\mathscr{S}'(\mathbb{R}^d)$.

It is not difficult to verify the following criterion analogous to boundedness in the normed linear context.

Proposition 1. *A linear map $T\colon \mathscr{S}(\mathbb{R}^d) \to \mathbb{C}$ is continuous if and only if there exist $n \in \mathbb{N}$ and $c \in \mathbb{R}$ such that for all $\varphi \in \mathscr{S}(\mathbb{R}^d)$*

$$|\langle T, \varphi \rangle| \le c \sum_{|\alpha| \le n,\, |\beta| \le n} \|x^\alpha \partial^\beta \varphi\|_{L^\infty(\mathbb{R}^d)}. \tag{1}$$

Denoting the sum on the right as $p_n(\varphi)$, the sequence of norms $p_0 \le p_1 \le p_2 \ldots$ define the topology for $\mathscr{S}$.

Corollary 2. *A distribution $T \in \mathscr{D}'(\mathbb{R}^d)$ extends uniquely to an element of $\mathscr{S}'(\mathbb{R}^d)$ if and only if there exist $n \in \mathbb{N}$ and $c \in \mathbb{R}$ such that* (1) *holds for all $\varphi \in \mathscr{D}(\mathbb{R}^d)$.*

In particular, we have

$$\mathscr{D}(\mathbb{R}^d) \subset \mathscr{E}'(\mathbb{R}^d) \subset \mathscr{S}'(\mathbb{R}^d) \subset \mathscr{D}'(\mathbb{R}^d).$$

EXAMPLES. 1. If f is a Lebesgue measurable function on $\mathbb{R}^d$ such that for some $M, (1 + |x|^2)^{-M} f \in L^1(\mathbb{R}^d)$, then the distribution defined by f is tempered since

$$\begin{aligned}\langle f, \varphi \rangle &= \langle (1 + |x|^2)^{-M} f, (1 + |x|^2)^M \varphi \rangle \\ &\leq \|(1 + |x|^2)^{-M} f\|_{L^1} \|(1 + |x|^2)^M \varphi\|_{L^\infty} \leq c_f p_{2M}(\varphi).\end{aligned}$$

In particular, $\mathscr{S}(\mathbb{R}^d) \subset \mathscr{S}'(\mathbb{R}^d)$.

2. If μ is a Borel measure such that for some M, $(1 + |x|^2)^{-M} \mu$ is a finite measure, then the distribution defined by μ is tempered. Reason as in Example 1 to show that

$$\langle \mu, \varphi \rangle \leq \|(1 + |x|^2)^{-M} \mu\|_{\text{Tot. Var.}} \|(1 + |x|^2)^M \varphi\|_{L^\infty(\mathbb{R}^d)} \leq c_\mu p_{2M}(\varphi).$$

3. If $f \in L^p(\mathbb{R}^d)$, $1 \leq p \leq \infty$, then $f \in \mathscr{S}'$ since these functions satisfy the condition of Example 1, if one chooses M so large that $(1 + |x|^2)^{-M} \in L^q(\mathbb{R}^d)$ and then uses Hölder's inequality.

Definition. A sequence $T_n \in \mathscr{S}'(\mathbb{R}^d)$ converges to $T \in \mathscr{S}'(\mathbb{R}^d)$ if and only if for all $u \in \mathscr{S}(\mathbb{R}^d)$

$$\langle T_n, \varphi \rangle \to \langle T, \varphi \rangle \qquad \text{as} \quad n \to \infty.$$

We write $T_n \rightharpoonup T$ or $\mathscr{S}'$-lim $T_n = T$.

EXAMPLE. If $\varphi \in \mathscr{D}(\mathbb{R}^d)$, $\varphi(0) = 1$, then

$$\mathscr{S}'\text{-}\lim_{n \to \infty} \varphi\left(\frac{x}{n}\right) T = T. \tag{2}$$

Note that $\mathscr{S}'(\mathbb{R}^d) \subset \mathscr{D}'(\mathbb{R}^d)$ so if $T \in \mathscr{S}'$, then φT is a well-defined element of $\mathscr{D}'(\mathbb{R}^d)$ for $\varphi \in C^\infty(\mathbb{R}^d)$. If $\varphi \in C_0^\infty$, then $\varphi T \in \mathscr{E}' \subset \mathscr{S}'$, so the assertion makes sense.

To prove (2), note that for $\psi \in \mathscr{S}$, $\langle \varphi(x/n) T, \psi \rangle \equiv \langle T, \varphi(x/n) \psi \rangle$ and $\varphi(x/n)\psi \to \psi$ in $\mathscr{S}$. Thus $\langle T, \varphi(x/n)\psi \rangle \to \langle T, \psi \rangle$, thanks to the continuity of T. □

The topology in $\mathscr{S}'$ associated with this convergence is called the weak-star topology defined by the (uncountable) family of seminorms

$$|T|_\psi \equiv |\langle T, \psi \rangle|, \qquad \psi \in \mathscr{S}.$$

We will have no need for topological subtleties but note in passing that this topology in $\mathscr{S}'$ is not metrizable.

Our next, and principal, concern will be to extend to $\mathscr{S}'$ the basic linear operators of analysis, for example, ∂^α and $\mathscr{F}$. The extensions will be proved to be sequentially continuous and it is important to know that such extensions are uniquely determined by their values on $\mathscr{S}$. This follows from the density of $\mathscr{S}$ in $\mathscr{S}'$.

Proposition 3. *$C_0^\infty(\mathbb{R}^d)$ is sequentially dense in $\mathscr{S}'(\mathbb{R}^d)$.*

PROOF. Choose $\varphi \in \mathscr{D}(\mathbb{R}^d)$ with $\varphi(0) = 1$. Then for $n = 1, 2, \ldots, \varphi(x/n)T \in \mathscr{E}'(\mathbb{R}^d) \subset \mathscr{S}'(\mathbb{R}^d)$.

Choose $j \in \mathscr{D}(\mathbb{R}^d)$ with $\int j(x)\,dx = 1$ and $j(-x) = j(x)$. Let $j_n(x) \equiv n^d j(nx)$, so that $j_n \rightharpoonup \delta$. Then $\varphi(x/n)T$ has compact support so Proposition 4 of the Appendix implies that

$$T_n \equiv j_n * \left(\varphi\left(\frac{x}{n}\right)T\right) \in \mathscr{D}(\mathbb{R}^d) \subset \mathscr{S}(\mathbb{R}^d).$$

It remains to show that for all $\psi \in \mathscr{S}$, $\langle T_n, \psi\rangle \to \langle T, \psi\rangle$. Now $\langle T_n, \psi\rangle = \langle T, (\varphi(x/n)(j_n * \psi))\rangle$. Thus it suffices to show that $\varphi(x/n)(j_n * \psi) \to \psi$ in $\mathscr{S}$. Toward that end, compute

$$\varphi\left(\frac{x}{n}\right)(j_n * \psi) - \psi = \int \varphi\left(\frac{x}{n}\right)\left\{\psi\left(x + \frac{y}{n}\right) - \psi(x)\right\} j(y)\,dy.$$

For $\alpha \in \mathbb{N}^d, \beta \in \mathbb{N}^d$, apply $x^\alpha \partial_x^\beta$ to this difference to obtain a finite sum of terms of the form

$$c(\alpha, \beta, \gamma) \int \frac{x^\alpha}{n^{|\gamma|}} \partial_x^\gamma \varphi\left(\frac{x}{n}\right)\left\{\partial_x^{\beta-\gamma}\psi\left(x + \frac{y}{n}\right) - \partial_x^{\beta-\gamma}\psi(x)\right\} j(y)\,dy,$$

the sum over those $\gamma \in \mathbb{N}^d$ with $0 \le \gamma_i \le \beta_i$ for $i = 1, \ldots, d$. The difference of ψ's is estimated using the mean value theorem and the fact that the derivatives of ψ decrease faster than any power

$$\left|\partial_x^{\beta-\gamma}\psi\left(x + \frac{y}{n}\right) - \partial_x^{\beta-\gamma}\psi(x)\right| \le \left|\frac{y}{n}\right| c_M (1 + |x|)^{-M}.$$

Performing the y integral yields the estimate

$$c\,\frac{|x|^\alpha (1 + |x|)^{-M}}{n^{|\gamma|+1}}.$$

Choosing $M > |\alpha|$ the result follows. □

Given a continuous linear operator $L\colon \mathscr{S} \to \mathscr{S}$, the *transpose* L' maps $\mathscr{S}' \to \mathscr{S}'$. For $T \in \mathscr{S}'(\mathbb{R}^d)$, $L'T \in \mathscr{S}'$ is defined by

$$\langle L'T, \varphi\rangle \equiv \langle T, L\varphi\rangle \qquad \text{for all} \quad \varphi \in \mathscr{S}. \tag{3}$$

The next proposition shows that the identity can sometimes be used to extend L.

Proposition 4. *Suppose that $L: \mathcal{S}(\mathbb{R}^d) \to \mathcal{S}(\mathbb{R}^d)$ is a continuous linear map and that the restriction of the transposed operator to $\mathcal{S}$, $L'|_{\mathcal{S}}$, is a continuous map of $\mathcal{S}$ to itself. Then L has a unique sequentially continuous extension to a linear map $L: \mathcal{S}'(\mathbb{R}^d) \to \mathcal{S}'(\mathbb{R}^d)$ defined by*

$$\langle LT, \varphi \rangle \equiv \langle T, L'\varphi \rangle, \qquad \text{for all} \quad T \in \mathcal{S}', \quad \varphi \in \mathcal{S}.$$

PROOF. If $\tilde{L}$ is such an extension, $T \in \mathcal{S}'$, and $\varphi \in \mathcal{S}$, choose $T_n \in \mathcal{S}$, converging to T in $\mathcal{S}'(\mathbb{R}^d)$. Then

$$\langle LT_n, \varphi \rangle = \langle T_n, L'\varphi \rangle.$$

Passing to the limit $n \to \infty$ yields

$$\langle \tilde{L}T, \varphi \rangle = \langle T, L'\varphi \rangle. \tag{4}$$

Conversely, defining $\tilde{L}$ by this identity yields a sequentially continuous linear extension. □

This proposition identifies when passing the operator to the test function yields a good extension.

The extension is continuous for the weak-star topology on $\mathcal{S}'$, a fact we will not need.

For general L, one will not even have $L'\varphi \in \mathcal{S}$ for $\varphi \in \mathcal{S}$. The hypothesis on L' is very restrictive. However, the following list shows that many operators are included. The translation and dilation operators were defined before Proposition 2.2.2 and the multiplication operators M were defined in Problem 2.1.3.

L	$L'\vert_{\mathcal{S}}$
∂^α	$(-\partial)^\alpha$
τ_h	τ_{-h}
σ_k	$k^{-d}\sigma_{1/k}$
$\mathcal{F}$	$\mathcal{F}$
$M(x)$	$M(x)$

As the demonstrations of the assertions in the table follow a single pattern we consider only the formulas for $(\partial^\alpha)'|_{\mathcal{S}}$ and $\mathcal{F}'|_{\mathcal{S}}$.

For $T \in \mathcal{S}'$ and $\varphi \in \mathcal{S}$, we have

$$\langle (\partial^\alpha)'T, \varphi \rangle \equiv \langle T, (\partial^\alpha)\varphi \rangle.$$

If $T \in \mathcal{S}$, the right-hand side is equal to

$$\int T(x)\partial_x^\alpha \varphi(x)\, dx = \int (-\partial_x)^\alpha T(x)\varphi(x)\, dx = \langle (-\partial)^\alpha T, \varphi \rangle.$$

Thus, for such T, $(\partial^\alpha)'T = (-\partial)^\alpha T$.

Similarly, for $T \in \mathscr{S}'$ and $\varphi \in \mathscr{S}$,

$$\langle \mathscr{F}'T, \varphi \rangle \equiv \langle T, \mathscr{F}\varphi \rangle.$$

For $T \in \mathscr{S}$ formula (2.2.7) shows that this is equal to $\langle \mathscr{F}T, \varphi \rangle$, whence $\mathscr{F}'|_{\mathscr{S}} = \mathscr{F}$.

Theorem 5. *Each of the operators $M, \partial^\alpha, \tau_h, \sigma_k, \mathscr{F}$ from $\mathscr{S}$ to itself has a unique extension to a sequentially continuous map of $\mathscr{S}'$ to $\mathscr{S}'$. The extensions are defined for $T \in \mathscr{S}'$, $\varphi \in \mathscr{S}$, by*

$$\langle MT, \varphi \rangle \equiv \langle T, M\varphi \rangle,$$

$$\langle \partial^\alpha T, \phi \rangle \equiv \langle T, (-\partial^\alpha \varphi \rangle,$$

$$\langle \tau_n T, \varphi \rangle \equiv \langle T, \tau_{-h}\varphi \rangle,$$

$$\langle \sigma_h T, \varphi \rangle \equiv \langle T, k^{-d}\sigma_{1/k}\varphi \rangle,$$

$$\langle \mathscr{F}T, \varphi \rangle \equiv \langle T, \mathscr{F}\varphi \rangle.$$

All the operators except $\mathscr{F}$ are well defined as maps of $\mathscr{D}'(\mathbb{R}^d)$ to itself. Thus, for $T \in \mathscr{S}' \subset \mathscr{D}'$, the action on T yields a well-defined element of $\mathscr{D}'$. The expressions above show that the two definitions agree for test functions in $\mathscr{D}'$. What is asserted is that the expressions are still meaningful and continuous for test functions in $\mathscr{S}$.

EXAMPLE. The smooth function $u = \sin(e^x)$ is uniformly bounded and so defines a tempered distribution. The distribution derivative of u must be equal to its classical derivative, namely, $v = e^x \cos(e^x)$, which grows exponentially as x tends to infinity. Thus $\int v\varphi \, dx$ is not an absolutely convergent integral for all $\varphi \in \mathscr{S}(\mathbb{R}^d)$. Nevertheless, v is a tempered distribution since it is the derivative of a tempered distribution. This seemingly paradoxical result is resolved by observing that the map $\mathscr{D} \ni \varphi \mapsto \int v\varphi$ does extend to a tempered distribution as one sees immediately from the identity, valid for $\varphi \in \mathscr{D}$,

$$\int \varphi v \, dx = \int \varphi \partial u \, dx = -\int u \partial\varphi \, dx.$$

Thus $|\int v\varphi| \le \|\partial\varphi\|_{L^1} \le c p_{d+2}(\varphi)$.

Most basic identities involving these operators in $\mathscr{S}$ extend to $\mathscr{S}'$, since $\mathscr{S}$ is sequentially dense in $\mathscr{S}'$. For example, the Fourier transform on $\mathscr{S}$ satisfies

$$\mathscr{F}D^\alpha T = \xi^\alpha \mathscr{F}T, \qquad \mathscr{F}^*\mathscr{F}T = T, \qquad \mathscr{F}(\tau_h T) = e^{ih\xi}\mathscr{F}T. \tag{5}$$

To prove the first note that $\mathscr{F}D^\alpha$ and $\xi^\alpha \mathscr{F}$ are sequentially continuous on $\mathscr{S}'$ and they agree in $\mathscr{S}$, thus for any $T \in \mathscr{S}'$ choose $T_n \in \mathscr{S}$ with $T_n \to T$ in $\mathscr{S}'$, then

$$\mathscr{F}D^\alpha T = \lim \mathscr{F}D^\alpha T_n = \lim \xi^\alpha \mathscr{F}T_n = \xi^\alpha \mathscr{F}T.$$

The other identities in (5) are proved in the same way. These identities allow for elegant manipulation of $\mathscr{F}$ as the following examples illustrate.

EXAMPLE. Compute the Fourier transform of 1.

Using the definition of $\mathscr{F}$ on $\mathscr{S}'$ the Fourier Inversion Formula yields

$$\langle \mathscr{F}1, \varphi \rangle = \langle 1, \mathscr{F}\varphi \rangle = \int \mathscr{F}\varphi(\xi)\, d\xi = (2\pi)^{d/2}\varphi(0) = \langle (2\pi)^{d/2}\delta, \varphi \rangle.$$

Thus

$$\mathscr{F}1 = (2\pi)^{d/2}\delta. \tag{6}$$

Formula (6) is equivalent to the identity $(2\pi)^{-d/2}\int \mathscr{F}\varphi(\xi)\, d\xi = \varphi(0)$ which implies the Fourier Inversion Formula.

We next give an alternate derivation of (6) without using the Fourier Inversion Formula, thereby giving Laurent Schwartz's elegant proof of Fourier's Theorem.

For $\alpha \neq 0$, $D^\alpha 1 = 0$. Thus, for $\alpha \neq 0$,

$$\xi^\alpha \mathscr{F}1 = \mathscr{F}D^\alpha 1 = 0.$$

In particular, $(\sum \xi_i^2)\mathscr{F}1 = 0$. Thus, $\operatorname{supp} \mathscr{F}1 \subset \{0\}$, so there are constants $c_\beta \in \mathbb{C}$ such that

$$\mathscr{F}1 = \sum_{|\beta| < N} c_\beta \partial_\xi^\beta \delta.$$

To evaluate the c_α for $\alpha \neq 0$, use the fact that $\xi^\alpha \mathscr{F}1 = 0$. Choose $\varphi \in \mathscr{S}(\mathbb{R}^d)$, $\varphi(\xi) = 1$ for $|\xi| \leq 1$. Then for $\alpha \neq 0$, $\langle \mathscr{F}1, \xi^\alpha\varphi \rangle = 0$, but

$$0 = \langle \mathscr{F}1, \xi^\alpha \varphi \rangle = \sum \langle c_\beta \partial^\beta \delta, \xi^\alpha \varphi \rangle = \sum c_\beta (-\partial)^\beta (\xi^\alpha \varphi)|_{\xi=0}.$$

Since φ is constant on a neighborhood of the origin the summands vanish at $\xi = 0$ unless $\beta = \alpha$. In that case, the sum is equal to $\alpha!\, c_\alpha$, so c_α vanishes unless $\alpha = 0$.

Thus $\mathscr{F}1 = c\delta$. To evaluate the constant c apply $c\delta$ to $e^{-\xi^2/2}$ to find

$$c = \langle c\delta, e^{-\xi^2/2} \rangle = \langle \mathscr{F}1, e^{-\xi^2/2} \rangle$$

$$= \langle 1, \mathscr{F}e^{-\xi^2/2} \rangle = \int e^{-\xi^2/2}\, d\xi = (2\pi)^{d/2}.$$

A third computation of $\mathscr{F}1$ starts with $1 = \mathscr{S}'\text{-}\lim e^{-|x/n|^2/2}$. Applying $\mathscr{F}$ yields

$$\mathscr{F}1 = \lim \mathscr{F}(e^{-|x/n|^2/2}) = \lim (2\pi)^{-d/2} n^d e^{-|nx|^2/2} = (2\pi)^{-d/2}\delta.$$

This argument is only a slight variant of the proof of Theorem 2.2.5.

EXAMPLE. Compute $\mathscr{F}\delta$.

The definition of $\mathscr{F}$ yields

$$\langle \mathscr{F}\delta, \varphi \rangle \equiv \langle \delta, \mathscr{F}\varphi \rangle = (\mathscr{F}\varphi)(0) = (2\pi)^{-d/2} \int \varphi(x)\, dx = \langle (2\pi)^{-d/2}, \varphi \rangle.$$

Thus $\mathscr{F}\delta = (2\pi)^{-d/2}$.

The next example illustrates how analytic continuation can be used in computing Fourier transforms. The result is needed in our study of initial value problems.

EXAMPLE. For $\mathrm{Re}(a) \geq 0$, $a \neq 0$, compute $\mathscr{F}(e^{-ax^2/2})$.

For any $\psi \in \mathscr{S}$, $\langle e^{-ax^2/2}, \psi \rangle$ is holomorphic in $\mathrm{Re}(a) > 0$ and continuous in $\mathrm{Re}(a) \geq 0$. Thus

$$\langle \mathscr{F} e^{-ax^2/2}, \psi \rangle = \langle e^{-ax^2/2}, \mathscr{F}\psi \rangle$$

is holomorphic and continuous in the same half-space.

For $a \in \mathbb{R}_+$ the Fourier transform of $\exp(-ax^2/2)$ is computed by writing it as $\exp(-(a^{1/2}x)^2/2)$ and using the dilation identity in Proposition 2.2.2 (iii) with $u = \exp(-x^2/2)$ and $\lambda = \sqrt{a}$ to find

$$\mathscr{F} e^{-ax^2/2} = a^{-d/2} e^{-x^2/2a}.$$

Thus

$$\langle \mathscr{F} e^{-ax^2/2}, \psi \rangle = a^{-d/2} \int e^{-x^2/2a}\, \psi(x)\, dx \qquad \text{for} \quad a \in \mathbb{R}_+.$$

Furthermore, the two sides are holomorphic in $\mathrm{Re}(a) > 0$, continuous in $\{\mathrm{Re}\, a \geq 0 \text{ and } a \neq 0\}$, and equal on $\mathbb{R}_+$. The unique continuation principle for analytic functions implies that the two sides are equal for all $\mathrm{Re}(a) > 0$. By continuity this extends to $\mathrm{Re}(a) \geq 0$, $a \neq 0$. Thus

$$\mathscr{F} e^{-ax^2/2} = a^{-d/2} e^{-\xi^2/2a}, \qquad \mathrm{Re}\, a \geq 0 \quad \text{and} \quad a \neq 0, \tag{7}$$

where $a^{1/2}$ is defined as the branch with $\mathrm{Re}(a) > 0$. In particular, $1^{1/2} = 1$.

We give two simple but striking applications of the Fourier transform on $\mathscr{S}'$.

First consider the solvability of the equation

$$(1 - \Delta)u = f. \tag{8}$$

For u, f in $\mathscr{S}'$ this is equivalent to

$$(1 + |\xi|^2)\hat{u} = \mathscr{F} f, \tag{9}$$

hence

$$\hat{u} = (1 + |\xi|^2)^{-1} \mathscr{F} f. \tag{10}$$

Proposition 6. *For any $f \in \mathscr{S}'(\mathbb{R}^d)$ there is exactly one solution $u \in \mathscr{S}'(\mathbb{R}^d)$ to* (8). *The solution is given by formula* (10). *In particular, if $f \in \mathscr{S}$, then $u \in \mathscr{S}$. If $f \in L^2$, then for all $|\alpha| \leq 2$, $D^\alpha u \in L^2(\mathbb{R}^d)$.*

The conclusion means that the distribution derivatives $D^\alpha u$ of the tempered distribution u are equal to L^2 functions.

PROOF. Only the last assertion must be proved. Compute

$$\mathscr{F}(D^\alpha u) = \frac{\xi^\alpha}{1+|\xi|^2}\mathscr{F}f \in L^2(\mathbb{R}^d).$$

By Plancherel's Theorem, this proves that $D^\alpha u \in L^2$. □

We have seen that the equation $(\partial_t + \partial_x)u \in L^2$ does not imply that all first derivatives of u are in L^2_{loc}. The regularity in the above proposition is typical of elliptic equations. Note that the equation asserts that one linear combination of $\{D^\alpha u: |\alpha| \le 2\}$ lies in L^2 and the proposition shows that all of the $D^\alpha u$ with $|\alpha| \le 2$ are in L^2. In this sense the result is surprising. It is false (and not obviously so) that if $f \in C_0(\mathbb{R}^d)$, then $u \in C^2(\mathbb{R}^d)$ (Problem 5.9.3). The gain of two derivatives is correct if one measures differentiability using L^2 derivatives from the Theory of Distributions and is false for the classical partial derivatives of second-year calculus.

The second application is to a Liouville-type theorem.

Theorem 7 (Generalized Liouville Theorem). *Suppose that $P(D)$ is a constant coefficient partial differential operator such that $P(\xi) \ne 0$ for $\xi \ne 0$. If $u \in \mathscr{S}'(\mathbb{R}^d)$ satisfies $Pu = 0$, then u is a polynomial in x.*

EXAMPLES. 1. $P = P_m$ is a homogeneous elliptic operator. In particular, if P is the Cauchy–Riemann operator $\partial_x + i\partial_y$ or the Laplace operator Δ.

2. $P = \partial_t - \Delta$, the heat operator.

3. The wave operators $\partial_t - \partial_x$ and $\partial_t^2 - \Delta$ do not satisfy the hypothesis.

4. In Problem 1.3.3 many such polynomial solutions were constructed for the heat equation. The real and imaginary parts of $(x + iy)^n$ are polynomials on $\mathbb{R}^2$ which satisfy Laplace's equation.

PROOF. Take the Fourier transform of the equation to obtain

$$\mathscr{F}(P(D))u = P(\xi)\hat{u} = 0.$$

Since $P(\xi) \ne 0$ if $\xi \ne 0$ it follows that $\operatorname{supp} \hat{u} \subset \{0\}$.

Thus $\mathscr{F}u$ must be a finite linear combination of derivatives of the delta function

$$\hat{u} = \sum c_\alpha D^\alpha \delta.$$

Apply the inverse Fourier transform to obtain

$$u = \sum c_\alpha \mathscr{F}^* D^\alpha \delta = \sum c_\alpha (-x)^\alpha \mathscr{F}^*\delta = \sum c_\alpha (-x)^\alpha (2\pi)^{-d/2},$$

a polynomial in x. □

Corollary 8. *The only bounded harmonic (resp. holomorphic) functions on $\mathbb{R}^n$ (resp. $\mathbb{C}$) are the constants.*

Another classical example which is easily understood is the operator $(d/dx)^n$. Contrast this with the constant coefficient nonhomogeneous operator $d^2/dx^2 + 1$ which has $\sin x$ as a nonpolynomial bounded solution.

PROBLEMS

1. Prove Proposition 1 and its corollary.

2. Compute the Fourier transform of the following tempered distributions.
 (a) $1/(x - c)$, $c \in \mathbb{C}\backslash\mathbb{R}$. *Hint.* Use Problem 2.3.3.
 (b) The Heaviside function $h \equiv \chi_{]0,\infty[}$. *Hint.* Consider the derivative of h. Find all distribution solutions of $xT = 1$ (there are infinitely many).
 (c) $\partial^\alpha \delta \in \mathscr{S}'(\mathbb{R}^d)$.
 (d) $x^\alpha \in \mathscr{S}'(\mathbb{R}^d)$.

There are situations where the Fourier transform has been defined in more than one way as an element of $\mathscr{S}'$. For example, if $u \in L^p$ for all $p \in [1, 2]$, then $\mathscr{F}u$ is defined by a different extension process for each such p, and in addition $\mathscr{F}u$ is defined in yet another process as an element of $\mathscr{S}'$. To show that all the extensions agree, the simplest algorithm is to show that they all define the same element of $\mathscr{S}'$.

3. Prove.

 Proposition. *Suppose that X and Y are topological spaces with $\mathscr{S} \subset X \subset \mathscr{S}'$, $\mathscr{S} \subset Y \subset \mathscr{S}'$, where each inclusion is continuous and sequentially dense. If $K: \mathscr{S} \to \mathscr{S}$ is continuous and has sequentially continuous extensions $K_1: X \to Y$ and $K_2: \mathscr{S}' \to \mathscr{S}'$, then the restriction of K_2 to X is equal to K_1.*

EXAMPLES. 1. (Fourier Transform of L^p). Here $K_2 = \mathscr{F}$, $X = L^p$, with $p \in [1,2]$, and $Y = L^q$ and K_1 is the Fourier transform as defined in §2.3. The proposition shows that the two definitions of the Fourier transform of an element of L^p define the same tempered distribution.

2. (Fourier Transform of Measures). Let $\mathscr{M}(\mathbb{R}^d)$ denote the set of finite Borel measures on $\mathbb{R}^d$. $\mathscr{M}$ is a Banach space with norm given by the total variation. For $\mu \in \mathscr{M}$, there is an elementary definition of a Fourier transform

$$(K_1 \mu)(\xi) \equiv (2\pi)^{-d/2} \int e^{-ix\xi}\, d\mu(x). \tag{11}$$

Then $K_1: X \to Y \equiv L^\infty(\mathbb{R}^d) \cap C(\mathbb{R}^d)$ has norm $2\pi^{-d/2}$ (exercise). However, $\mathscr{S}(\mathbb{R}^d)$ is not dense in $\mathscr{M}$. On the other hand, $\mathscr{S}$ is sequentially dense in $\mathscr{M}$ if $\mathscr{M}(\mathbb{R}^d) = \dot{C}(\mathbb{R}^d)'$ is given the weak-star topology. The proposition applies with $K_2 = \mathscr{F}$ showing that the two definitions of the Fourier transform of a measure yield the same tempered distribution. That is, the Distribution Theory Fourier Transform of μ is equal to a continuous function of ξ given by the absolutely convergent integral (11).

It is also true that $\mathscr{S}$ is not dense in $L^\infty = (L^1)'$ but is dense in the weak-star topology. Here, as for measures, it is often useful to use this weaker topology.

4. (a) For $\operatorname{Re} a \geq 0$, $a \neq 0$, and $b \in \mathbb{C}$ compute the Fourier transform of $e^{-(ax^2+2bx)/2}$.
 (b) Discuss the limit as a converges to zero.

§2.5. Convolution in $\mathscr{S}(\mathbb{R}^d)$ and $\mathscr{S}'(\mathbb{R}^d)$

In this section the extension process is applied to the operator convolution by $\varphi \in \mathscr{S}(\mathbb{R}^d)$. For $\psi \in \mathscr{S}(\mathbb{R}^d)$

$$(\varphi * \psi)(x) \equiv \int \varphi(x-y)\psi(y)\,dy. \tag{1}$$

The integral is absolutely convergent and Lebesque's Dominated Convergence Theorem shows that $\varphi * \psi$ is continuous and vanishes at infinity, that is, belongs to $\dot{C}(\mathbb{R}^d)$. Differentiating under the integral sign (Problem 1) shows that $\varphi * \psi \in C^\infty(\mathbb{R}^d)$ and

$$D^\alpha(\varphi * \psi) = (D^\alpha \varphi) * \psi = \varphi * D^\alpha \psi. \tag{2}$$

If $1/p + 1/q = 1$, Hölder's inequality implies

$$\|\varphi * \psi\|_{L^\infty} \le \|\varphi\|_{L^q} \|\psi\|_{L^p}. \tag{3}$$

To estimate the L^1 norm of $\varphi * \psi$ one integrates to find

$$\|\varphi * \psi\|_{L^1} \le \int\left(\int |\varphi(x-y)\psi(y)|\,dy\right)dx = \int\left(\int |\varphi(x-y)\psi(y)|\,dx\right)dy,$$

The last equality using Fubini's Theorem. The integral on the right is exactly equal to the product of the L^1 norms of φ and ψ proving that

$$\|\varphi * \psi\|_{L^1} \le \|\varphi\|_{L^1} \|\psi\|_{L^1}. \tag{4}$$

Estimates (3) and (4) form the heart of the L^p extensions for $\varphi *$ discussed in Problems 2–4.

For the $\mathscr{S}$, $\mathscr{S}'$ theory we need to show that $\varphi *$ is continuous from $\mathscr{S}$ to itself. An elegant proof uses the Fourier transform of $\varphi * \psi$. Since $\varphi * \psi$ is in L^1 its transform is computed as the absolutely convergent integral

$$\mathscr{F}(\varphi * \psi) = (2\pi)^{-d/2} \int e^{-ix\xi}\left(\int \varphi(x-y)\psi(y)\,dy\right)dx.$$

Fubini's Theorem justifies reversing the order

$$= \int \varphi(y)\left((2\pi)^{-d/2} \int e^{-ix\xi}\varphi(x-y)\,dy\right)dx.$$

The inner integral is $\mathscr{F}(\tau_y \varphi) = e^{-iy\xi}\hat{\varphi}(\xi)$, so

$$\mathscr{F}(\varphi * \psi) = \int \hat{\varphi}(\xi) e^{-iy\xi}\psi(y)\,dy = (2\pi)^{d/2}\hat{\varphi}\hat{\psi}. \tag{5}$$

Thus

$$\varphi * \psi = \mathscr{F}^*((2\pi)^{d/2}\hat{\varphi}\hat{\psi}). \tag{6}$$

Since $\mathscr{F}$ and $\mathscr{F}^*$ are continuous maps of $\mathscr{S}$ to itself and multiplication is a continuous map from $\mathscr{S} \times \mathscr{S}$ to $\mathscr{S}$ (see Example 2 at the end of §2.1), it follows that $\varphi, \psi \mapsto \varphi * \psi$ is a continuous bilinear map from $\mathscr{S} \times \mathscr{S}$ to $\mathscr{S}$.

Formula (6) also gives a short proof of the $\mathscr{S}$ convergence of mollification.

Proposition 1. *If $\varphi \in \mathscr{S}$ with $\int \varphi = 1$ and $\varphi_\varepsilon \equiv \varepsilon^{-d}\varphi(x/\varepsilon)$, then for all $u \in \mathscr{S}$*

$$\mathscr{S}\text{-}\lim_{\varepsilon \to 0} \varphi_\varepsilon * u = u.$$

PROOF.

$$\mathscr{F}(\varphi_\varepsilon * u) = (2\pi)^{d/2}\mathscr{F}\varphi_\varepsilon(\xi)\mathscr{F}u(\xi) = (2\pi)^{d/2}(\mathscr{F}\varphi)(\varepsilon\xi)\mathscr{F}u(\xi).$$

Since $\mathscr{F}\varphi(0) = (2\pi)^{-d/2}\int \varphi\, dx = (2\pi)^{-d/2}$, Proposition 2.1.1 shows that $\mathscr{S}\text{-}\lim \mathscr{F}(\varphi_\varepsilon * u) = \hat{u}(\xi)$. □

The map $\varphi\, *$ is a continuous linear map with transpose computed from

$$\langle (\varphi\, *)'T, \psi \rangle \equiv \langle T, \varphi * \psi \rangle.$$

For $T \in \mathscr{S}$ this is an integral

$$\iint T(x)\varphi(x - y)\psi(y)\, dy\, dx = \int\left(\int \varphi(x - y)T(x)\, dx\right)\psi(y)\, dy,$$

and the inner integral is the convolution of T with the reflection, $\mathscr{R}\varphi$, defined by $\mathscr{R}\varphi(x) \equiv \varphi(-x)$. Thus $(\varphi\, *)'|_{\mathscr{S}} = (\mathscr{R}\varphi)\, *$.

Proposition 2. *For $\varphi \in \mathscr{S}$, the operator $\varphi\, *$ extends uniquely to a sequentially continuous operator of $\mathscr{S}'$ to itself. This operator satisfies*

$$\langle \varphi * T, u \rangle = \langle T, (\mathscr{R}\varphi) * u \rangle,$$

$$\mathscr{F}(\varphi * T) = (2\pi)^{d/2}\hat{\varphi}\hat{T},$$

$$D^\alpha(\varphi * T) = (D^\alpha\varphi) * T = \varphi * D^\alpha T.$$

PROOF. All the identities continue from the sequentially dense subset of all $T \in \mathscr{S}(\mathbb{R}^d)$. □

For any distribution $T \in \mathscr{D}'$ the convolution $T * \varphi \in C^\infty(\mathbb{R}^d)$ is defined for all $\varphi \in \mathscr{D}(\mathbb{R}^d)$ by $\langle T, \tau_y\varphi \rangle$. Thus $T\, *$ maps $\mathscr{D}$ to C^∞. For $T \in \mathscr{S}'$, we extend the map $T\, *$ to a map of $\mathscr{S}$ to $\mathscr{S}'$.

Proposition 3. *For $T \in \mathscr{S}'(\mathbb{R}^d)$, the map $\mathscr{D}(\mathbb{R}^d) \ni \varphi \mapsto T * \varphi$ extends uniquely to a sequentially continuous map of $\mathscr{S}(\mathbb{R}^d)$ to $\mathscr{S}'(\mathbb{R}^d)$ which satisfies*

$$\langle T * \varphi, \psi \rangle = \langle T, \check{\varphi} * \psi \rangle. \tag{7}$$

PROOF. For φ and ψ in $\mathscr{D}$, the identity (7) is known. As we have observed, the map $\varphi, \psi \mapsto \varphi * \psi$ is a continuous bilinear map of $\mathscr{S} \times \mathscr{S}$ to $\mathscr{S}$. $T*$ can therefore be extended using formula (7). □

EXAMPLES

1. $\delta * = \mathrm{Id}$,

2. $(D^\alpha \delta) * = D^\alpha$.

3. $(P(D)\delta) * = P(D)$.

4. The unique tempered solution of $(1 - \Delta)u = f \in \mathscr{S}$ is given by $\mathscr{F}u = (1 + |\xi|^2)^{-1}\mathscr{F}f$. Define a tempered distribution K by

$$\mathscr{F}(K) \equiv (2\pi)^{-d/2}(1 + |\xi|^2)^{-1}. \tag{8}$$

Then, for $f \in \mathscr{S}$, the solution u is given by

$$u = K * f \text{ “=” } \int K(x - y) f(y)\, dy. \tag{9}$$

K is the unique $\mathscr{S}'(\mathbb{R}^d)$ solution of

$$(1 - \Delta)K = \delta. \tag{10}$$

Formulas (8)–(10) are typical of Green's functions. The identity (10) says that K is a *fundamental solution.* Equation (9) expresses the solution of the problem "find a tempered solution of $(1 - \Delta)u = f$" as an integral involving the data f. The kernel of the integral expression is called the Green's function. For $d = 1$, K is computed in Problem 2.3.4 (see also Problem A.3(vii)). The computation for $d = 3$ is outlined in Problem 5.

We next use convolution to prove that the Fourier transform of a distribution with compact support is a smooth function.

Theorem 4. *If $u \in \mathscr{E}'(\mathbb{R}^d)$ then the Fourier transform of u is the restriction to $\mathbb{R}^d$ of the entire holomorphic function*

$$\mathbb{C}^d \ni \zeta \mapsto 2\pi^{-d/2}\langle u, e^{-i\sum x_j \zeta_j}\rangle. \tag{11}$$

PROOF. Choose $j \in C_0^\infty(\mathbb{R}^d)$, $j(x) = j(-x)$, $\int j\, dx = 1$, and let $j_\varepsilon \equiv \varepsilon^{-d} j(x/\varepsilon)$, $u_\varepsilon \equiv j_\varepsilon * u \in C_0^\infty(\mathbb{R}^d)$. Then $u_\varepsilon \to u$ in $\mathscr{S}'(\mathbb{R}^d)$, so $\hat{u}_\varepsilon \rightharpoonup \hat{u}$ in $\mathscr{S}'$. Therefore

$$\hat{u}_\varepsilon \rightharpoonup \hat{u} \quad \text{in } \mathscr{D}'(\mathbb{R}^d). \tag{12}$$

Now, $\hat{u}_\varepsilon$ is entire analytic with

$$\begin{aligned}\hat{u}_\varepsilon(\zeta) &= \langle u_\varepsilon, (2\pi)^{-d/2} e^{-i\sum x_j \zeta_j}\rangle \\ &= \langle j_\varepsilon * u, (2\pi)^{-d/2} e^{-i\sum x_j \zeta_j}\rangle \\ &= \langle u, j_\varepsilon * (2\pi)^{-d/2} e^{-i\sum x_j \zeta_j}\rangle.\end{aligned}$$

As $\varepsilon \to 0$, $j_\varepsilon * (2\pi)^{-d/2} e^{-ix\zeta}$ converges to $(2\pi)^{-d/2} e^{-ix\zeta}$ in $C^\infty(\mathbb{R}_x^d)$, uniformly for ζ in compact subsets of $\mathbb{C}^n$. Thus, uniformly in compacts in $\mathbb{C}^n$, $\hat{u}_\varepsilon(\zeta) \to \langle u, (2\pi)^{-d/2} e^{-ix\zeta}\rangle$. In particular, the limit is entire analytic, and restricting to

$\mathbb{R}^d$,

$$\mathscr{D}'(\mathbb{R}^d) - \lim \mathscr{F} u_\varepsilon = \langle u, (2\pi)^{-d/2} e^{-ix\xi} \rangle. \tag{13}$$

Equating the limits in (12) and (13) yields

$$\hat{u}(\xi) = \langle u, (2\pi)^{-d/2} e^{-ix\xi} \rangle,$$

and the right-hand side is the restriction to $\mathbb{R}^d$ of the entire function (11). □

This result is typical of *Paley–Weiner Theorems* which establish connections between the support properties of distributions and analyticity properties of their Fourier transforms. As a second example, we mention that tempered distributions with support in $x_1 \geq 0$ have Fourier transforms which are analytic in $\text{Im}(\xi_1) < 0$. The construction in §3.9 relies on this idea.

PROBLEMS

1. Justify the differentiations leading to (2).

L^p Theory of Convolutions
The basic estimates (3), (4) show that $*$ extends uniquely from $\mathscr{S} \times \mathscr{S}$ to a continuous bilinear map $L^1(\mathbb{R}^d) \times L^1(\mathbb{R}^d) \to L^1(\mathbb{R}^d)$ and $L^p(\mathbb{R}^d) \times L^q(\mathbb{R}^d) \to L^\infty(\mathbb{R}^d)$. As in the proof of the Riemann–Lebesgue Lemma, the image of the first map is in $\dot{C}(\mathbb{R}^d)$.

2. Prove that for $p \in [1, \infty]$ the map $*$ defined on $\mathscr{S} \times \mathscr{S}$ extends uniquely to a continuous bilinear map from $L^1(\mathbb{R}^d) \times L^p(\mathbb{R}^d) \to L^p(\mathbb{R}^d)$. *Hint*: Interpolation, or use Hölder's inequality.

3. For $1/p + 1/q = 1$ and $r \in [1, q]$, find $s \in [1, \infty]$ so that $*$ extends uniquely to a continuous bilinear map $L^r(\mathbb{R}^d) \times L^p(\mathbb{R}^d) \to L^s(\mathbb{R}^d)$. Express the continuity in the form of an inequality. *Hint*: Interpolation in the first slot. This result is called the *Hausdorf–Young Inequality*.

4. For $j \in L^1(\mathbb{R}^d)$ with $\int j(x)\,dx = 1$, let $j_\varepsilon(x) \equiv \varepsilon^{-d} j(x/\varepsilon)$. Prove that for any $p \in [1, \infty[$ and $u \in L^p(\mathbb{R}^d)$, $j_\varepsilon * u$ converges to u in L^p as $\varepsilon \to 0$. For $p = \infty$, show that the convergence is valid in the weak-star topology for $L^\infty = (L^1)'$, but not in the norm topology.

Computation of the Green Kernel for $(a^2 - \Delta)^{-1}$, $d = 3$
For $a > 0$, the Green's function, K, for $a^2 - \Delta$ is the inverse Fourier transform of $(2\pi)^{-d/2}/(a^2 + |\xi|^2)$. Then $K \in \mathscr{S}'(\mathbb{R}^d)$ and $(a^2 - \Delta)K = \delta$. The next problem leads you through a computation of K for $d = 3$. The case $d = 1$ is contained in Problem 2.3.4. See also Problem A.4(vii).

5. (i) Show that $K \in L^2(\mathbb{R}^3)$ and is rotation invariant, that is, $K_A = K$ for any orthogonal transformation A (see Problem 2.2.3).
   (ii) Conclude that $K(x) = g(|x|)$ for a $g \in L^2(\mathbb{R}_+, r^2\,dr)$. *Warning*: K is in $L^2(\mathbb{R}^3)$ so evaluating at points is risky, be careful.
   (iii) Show that in $\mathscr{D}'(\mathbb{R}_+)$, $(a^2 - r^{-2}\partial_r r^2 \partial_r)g = 0$. *Hint*: Consider $\langle K, (a^2 - \Delta)\psi \rangle = 0$ for suitable radial test functions on $\mathbb{R}^3$.
   (iv) Conclude that $g = (Be^{-ar} + Ce^{+ar})/r$ with $B, C \in \mathbb{R}$.
   (v) Show that C must vanish. Thus $\text{supp}(K - Be^{-ar}/r) \subset \{0\}$, so $K - Be^{-ar}/r = \sum c_\alpha \partial^\alpha \delta$ a finite sum.

(vi) Show that $c_\alpha = 0$ for all α.
(vii) Determine B using $\mathscr{F}K = (2\pi)^{-d/2}/(a^2 + |\xi|^2)$. *Hint*: $\xi = 0$.
DISCUSSION. For even d, K involves higher transcendental functions (see Courant and Hilbert [CH, Vol. II]). Having kernels often allows one to find supplementary estimates as in the next problem.

6. (i) For which $p \in [1, \infty]$ is K in $L^p(\mathbb{R}^3)$.
   (ii) Use Problem 3 to find an interval $I \subset [1, \infty]$ such that if $f \in L^p(\mathbb{R}^3)$, then $(a^2 - \Delta)^{-1} f \in L^s(\mathbb{R}^3)$ for all $s \in I$. Is your I as large as possible?
   DISCUSSION. Such results can be proved directly from the Fourier transform formula, but not nearly so simply. It is also true that if $f \in L^p$, and $p \neq 1, \infty$ then, for all $|\alpha| \leq 2$, $D^\alpha(a^2 - \Delta)^{-1} f \in L^p$. The case $p = 2$ is Proposition 2.4.5. For $p \neq 2$, this is a special case of the *Calderon–Zygmund inequality* and lies considerably deeper.

§2.6. L^2 Derivatives and Sobolev Spaces

The regularity of a function is often clearly expressed by saying that its distribution derivatives up to some order lie in one of the classical spaces L^p. For example, if $I =]a, b[$ is an interval in $\mathbb{R}$, then for $f \in \mathscr{D}'(I)$:

(1) $f \in C^1(I)$ if and only if $Df \in C(I)$.
(2) $f \in \text{Lip}(I)$, if and only if $Df \in L^\infty(I)$. Here

$$\text{Lip}(I) \equiv \left\{ f \in C(I) \colon \sup_{x \neq y} \frac{|f(x) - f(y)|}{|x - y|} < \infty \right\}.$$

(3) f is absolutely continuous in I if and only if $Df \in L^1(I)$.
(4) f is of bounded variation in I if and only if $Df \in \mathscr{M}(I)$, the finite Borel measures on I.

Caution. Changing f on a set of measure zero does not effect the distribution defined by f, so the inclusion $f \in C^1(I)$ must be interpreted as meaning that f is equal almost everywhere to an element of $C^1(I)$. Similar interpretations apply for $f \in AC$, $f \in BV$, and $f \in \text{Lip}$.

There are many ways of defining the notion of a function with derivative in $L^2(I)$. Most are equivalent and useful. One which is *not* good is that there exists $g \in L^2$ such that

$$\frac{f(x + h) - f(x)}{h} \to g \quad \text{pointwise a.e.}$$

To see the inadequacy of this notion, note that the Cantor function satisfies the above with $g = 0$. However, the Cantor function is strictly monotone. Its distribution derivative is not equal to zero but is equal to a measure singular with respect to Lebesgue measure. The conclusion is clear, pointwise differentiation is *not* the correct notion.

A second example is $u = (1 - \Delta)^{-1} f$ for $f \in L^2(\mathbb{R}^d)$. Here u has distributional second derivatives in L^2, showing that such things arise for partial differential equations. Our first result shows that, with the exception of pointwise convergence of difference quotients, most natural ways of defining "$\partial f/\partial x_1 \in L^2(\mathbb{R}^d)$" are equivalent.

Proposition 1. *For $f \in L^2(\mathbb{R}^d)$ the following are equivalent:*

(1) (Schwartz). *The distribution derivative $\partial f/\partial x_1 \in L^2(\mathbb{R}^d)$.*
(2) (Fourier). $\xi_1 \hat{f} \in L^2(\mathbb{R}^d)$.
(3) (Newton). *As $h \to 0$, $(f(x_1 + h, x_2, \ldots, x_j) - f(x_j))/h$ converges in $L^2(\mathbb{R}^d)$.*
(4) (Friedrichs). *There exists a sequence $f_n \in \mathscr{S}(\mathbb{R}^d)$ such that $f_n \to f$ in $L^2(\mathbb{R}^d)$ and $\partial f_n/\partial x_1$ converges in $L^2(\mathbb{R}^d)$.*

Remark. Each of the conditions provides a natural candidate for $\partial f/\partial x_1$. For (2) it is $\mathscr{F}^*(i\xi_1 \hat{f})$ and for (3) and (4) it is the L^2 limit asserted to exist. The proof below establishes the equivalence and the equality of the four candidates for $\partial f/\partial x_1$.

PROOF. (1) $\Leftrightarrow$ (2) Since $\mathscr{F}(D_1 f) = \xi_1 f$, the equivalence is an immediate consequence of the Plancherel Theorem.

(2) $\Rightarrow$ (3) Introduce the notation

$$\delta_1^h = h^{-1}(\tau_{-h(1,0,\ldots,0)} - I)$$

for the forward difference operator converging to $\partial/\partial x_1$. We must show that $\delta_1^h f$ converges in $L^2(\mathbb{R}^d)$. By Plancherel's Theorem, it suffices to prove that $g_h \equiv \mathscr{F}(\delta_1^h f)$ converges in $L^2(\mathbb{R}^d)$.

Compute

$$g_h = [(e^{ih\xi_1} - 1)/h\xi_1]\xi_1 \hat{f}.$$

The factor in square brackets converges to $i = de^{i\theta}/d\theta|_{\theta=0}$ for all $\xi \in \mathbb{R}^d$. Since $|de^{i\theta}/d\theta| \le 1$, the Mean Value Theorem implies that the factor in square brackets has modulus less than or equal to one. Thus $g_h \to i\xi_1 \hat{f}$ pointwise and $|g_h| \le |\xi_1 f|$. Lebesgue's Dominated Convergence Theorem yields $g_h \to i\xi_1 \hat{f}$ in $L^2(\mathbb{R}^d)$.

(3) $\Rightarrow$ (1) For any $f \in \mathscr{S}'(\mathbb{R}^d)$, $\delta_1^h f \to \partial f/\partial x_1$ in $\mathscr{S}'(\mathbb{R}^d)$ (Problem 1). Let g be the L^2 limit of $\delta_1^h f$. Then $\delta_1^h f \to g$ in $\mathscr{S}'(\mathbb{R}^d)$. Equating the two $\mathscr{S}'$ limits yields $\partial f/\partial x_1 = g \in L^2(\mathbb{R}^d)$.

At this stage (1) $\Leftrightarrow$ (2) $\Leftrightarrow$ (3) is proved.

(4) $\Rightarrow$ (1). Since $f_n \to f$ in L^2 we have $f_n \to f$ in $\mathscr{S}'$, so $\partial f_n/\partial x_1 \to \partial f/\partial x_1$ in $\mathscr{S}'$ since $\partial/\partial x_1$ is sequentially continuous on $\mathscr{S}'$.

Let g be the $L^2(\mathbb{R}^d)$ limit of $\partial f_n/\partial x_1$. Then $\partial f_n/\partial x_1 \to g$ in $\mathscr{S}'$. Equating the two $\mathscr{S}'(\mathbb{R}^d)$ limits yields $\partial f/\partial x_1 = g \in L^2(\mathbb{R}^d)$.

(2) $\Rightarrow$ (4). Write $\hat{f} = h + k$, $h \equiv \hat{f}\chi_{|\xi_1| \le 1}$. Choose $h_n \in \mathscr{S}$, $h_n \to h$ in $L^2(\mathbb{R}^d)$, and

supp $h_n \subset \{|\xi_1| \le 2\}$. This can be done by mollifying by a compactly supported approximate delta function of sufficiently small width.

Similarly, choose a sequence $l_n \subset \mathscr{S}$ converging in $L^2(\mathbb{R}^d)$ to $\xi_1 k$ with supp $l_n \subset \{|\xi_1| \ge \frac{1}{2}\}$. Define f_n by $\mathscr{F} f_n \equiv h_n + l_n/\xi_1$. Then l_n/ξ_1 converges to k in $L^2(\mathbb{R}^d)$ since $|1/\xi_1| \le 2$ on supp l_n.

Thus $\mathscr{F} f_n \to h + k$ in L^2 and $\mathscr{F}(\partial f_n/\partial x_1) = i\xi_1 \mathscr{F} f_n = i\xi_1 h_n + i l_n \to i\xi_1 \hat{f}$ in $L^2(\mathbb{R}^d)$. □

For $s \in \mathbb{Z}_+$, the Sobolev space $H^s(\mathbb{R}^d)$ is defined to be the set of all $u \in \mathscr{S}'(\mathbb{R}^d)$ with the property that for all $|\alpha| \le s$, $D^\alpha u \in L^2(\mathbb{R}^d)$. The analogue of Proposition 1 requires difference operators converging to ∂^α. For $h \ne 0$ define the vector of difference operators $\delta^h \equiv (\delta_1^h, \ldots, \delta_d^h)$, where $\delta_j^h \equiv (\tau_{-he_j} - I)/h$ and $e_j \equiv (0, \ldots, 1, \ldots, 0)$ with 1 in the jth component. As with multi-index notation for derivatives, let

$$(\delta^h)^\alpha \equiv \prod (\delta_i^h)^{\alpha_i}.$$

Note that the operators δ_i^h commute so the order of factors in the above product does not matter.

Proposition 2. *For $u \in \mathscr{S}'(\mathbb{R}^d)$ and $s \in \mathbb{Z}_+$, the following are equivalent:*

(1) *For all $\alpha \in \mathbb{N}^d$ with $|\alpha| \le s$, $D^\alpha u \in L^2(\mathbb{R}^d)$.*
(2) *For any α with $|\alpha| \le s$, $\xi^\alpha \hat{u} \in L^2(\mathbb{R}^d)$.*
(3) *$u \in L^2$ and, for any α with $|\alpha| \le s$, $(\delta^h)^\alpha u$ converges in $L^2(\mathbb{R}^d)$ as $h \to 0$.*
(4) *There exists a sequence $u_n \in \mathscr{S}(\mathbb{R}^d)$ such that $u_n \to u$ in $L^2(\mathbb{R}^d)$ and, for all $|\alpha| \le s$, $D^\alpha u_n$ converges in $L^2(\mathbb{R}^d)$.*

The proof is Problem 2.

The characterization (2) can be rewritten as $w(\xi)|\hat{u}|^2 \in L^1$ where $w(\xi) \equiv \sum |\xi^\alpha|^2$, the sum over $|\alpha| \le s$. The condition also suggests introducing the norm

$$\left(\sum_{|\alpha| \le s} \|D^\alpha u\|_{L^2}^2 \right)^{1/2} = \left(\int w(\xi)|\hat{u}|^2 \, d\xi \right)^{1/2}.$$

With this norm, $\mathscr{F}$ establishes an isomorphism between $H^s(\mathbb{R}^d)$ and $L^2(\mathbb{R}^d_\xi, w(\xi)\, d\xi)$. In particular, H^s is complete. One defines the same set of distributions using any other weight w' such that $c_1 w \le w' \le c_2 w$ with positive constants c_j. A convenient and nearly universal choice is $w' = (1 + |\xi|^2)^s \equiv \langle \xi \rangle^{2s}$. Here $\langle \xi \rangle \equiv (1 + |\xi|^2)^{1/2}$ is a smooth strictly positive function which grows like $|\xi|$ as $\xi \to \infty$. One advantage of this choice of weight function is that it suggests a generalization of H^s to arbitrary real s.

Definition. For $s \in \mathbb{R}$, $H^s(\mathbb{R}^d) = \{u \in \mathscr{S}'(\mathbb{R}^d) : \langle \xi \rangle^s \hat{u} \in L^2(\mathbb{R}^d)\}$. The norm in H^s is defined by $\|u\|_{H^s} \equiv \|\langle \xi \rangle^s \hat{u}\|_{L^2(\mathbb{R}^d)}$.

EXAMPLES. 1. Since $\langle\xi\rangle^s\mathcal{F}\delta = (2\pi)^{-d/2}\langle\xi\rangle^s$ is square integrable if and only if $s < -d/2$, we see that $\delta \in H^s(\mathbb{R}^d)$ if and only if $s < -d/2$.

2. The spaces H^s consist of distributions which are in a sense square integrable at infinity. For example, the distribution 1 whose Fourier transform is $(2\pi)^{d/2}\delta$ belongs to no H^s.

The Fourier transform $\mathcal{F}$ is an isometric isomorphism ($\equiv$ unitary map) of H^s onto $L^2(\mathbb{R}^d, \langle\xi\rangle^{2s}\, d\xi)$. In particular, H^s is a separable Hilbert space.

Proposition 3. *For any* $-\infty < s < t < \infty$, $\mathcal{S} \subset H^t \subset H^s \subset \mathcal{S}'$, *each inclusion being (sequentially) continuous.*

PROOF. If u_n converges to zero in $\mathcal{S}$, then $(1 + |x|^2)^d(1 - \Delta)^M u_n$ converges uniformly to zero. Thus $(1 - \Delta)^M u_n$ converges to zero in L^2. Taking Fourier transform yields $\langle\xi\rangle^{2M}\hat{u}_n \to 0$ in L^2. Thus u_n converges to zero in all the H^s spaces. This proves the continuity of the first inclusion.

For the second note that the H^t norm is greater than or equal to the H^s norm.

For the final inclusion, note that for $u \in H^s$ and $\varphi \in \mathcal{S}$

$$|\langle u, \varphi\rangle| = |\langle\hat{u}, \hat{\varphi}\rangle \le \left|\int \langle\xi\rangle^s\hat{u}\langle\xi\rangle^{-s}\varphi d\xi\right| \le \|u\|_{H^s}\|\varphi\|_{H^{-s}}, \tag{1}$$

the last estimate following from the Schwartz inequality. Thus if u_n converges to zero in H^s, then $\mathcal{S}'$-lim $u_n = 0$. □

The converse Schwartz inequality applied in $L^2(\mathbb{R}^d_\xi)$ shows that

$$\|u\|_{H^s} = \sup\left\{\int u\varphi\, dx : \varphi \in \mathcal{S}, \|\varphi\|_{H^{-s}} = 1\right\}.$$

Inequality (1) shows that $\langle u, \varphi\rangle$ extends uniquely to a bilinear map $H^s \times H^{-s} \to \mathbb{C}$ satisfying (1) for all $u \in H^s$, $\varphi \in H^{-s}$. We continue to denote this pairing by $\langle\, ,\, \rangle$. Inequality (1) is then called the *generalized Schwartz inequality*. Note that for $u, \varphi \in H^s \times H^{-s}$, $\hat{u}\hat{\varphi} \in L^1(\mathbb{R}^d_\xi)$, and

$$\langle u, \varphi\rangle = \int \hat{u}(\xi)\hat{\varphi}(\xi)\, d\xi.$$

The pairing $\langle\, ,\, \rangle$ gives a representation of the dual of H^s called *Lax's Duality Theorem* (Problem 3.5.1). Note that $u(x)\varphi(x)$ is not necessarily locally integrable.

EXAMPLE. For $\varphi \in H^{d/2+1+\varepsilon}$ and $\psi = \partial\delta/\partial x_1 \in H^{-d/2-1-\varepsilon}$ we have

$$\langle\varphi, \psi\rangle = -\partial_1\varphi(0) = -(2\pi)^{-d/2}\int \hat{\varphi}(\xi)i\xi_1\, d\xi.$$

Here $\varphi(x)\psi(x) = \varphi\partial\delta/\partial x_1$ is not a locally integrable function even if $\varphi \in \mathscr{S}$. Nevertheless, many authors use the formal expression $\int \varphi\psi\, dx$ for $\langle\ ,\ \rangle$, just as one can write formally $\int T(x)\psi(x)\, dx$ for $\langle T, \psi\rangle$ when $T \in \mathscr{S}'$ and $\psi \in \mathscr{S}$.

If $P(D)$ is a constant coefficient partial differential operator of order m, then $|P(\xi)| \leq c\langle\xi\rangle^m$ and it follows that $P(D)$ maps H^s to H^{s-m} continuously. The operator $(1 - \Delta)^m$ is a unitary map of H^s to H^{s-2m}. More generally, if $\Lambda^m \equiv \mathscr{F}^*\langle\xi\rangle^m\mathscr{F}$, then Λ^m is unitary from H^s to H^{s-m}.

Proposition 4. *If $\varphi \in \mathscr{S}(\mathbb{R}^d)$ and $s \in \mathbb{R}$, then the map $u \mapsto \varphi u$ is a bounded linear map of $H^s(\mathbb{R}^d)$ to itself. Moreover,*

$$\|\varphi u\|_{H^s(\mathbb{R}^d)} \leq c(s, d)\|\langle\xi\rangle^{|s|}\hat{\varphi}(\xi)\|_{L^1(\mathbb{R}^d)}\|u\|_{H^s(\mathbb{R}^d)}.$$

PROOF. The Fourier transform of φu is given by

$$\mathscr{F}(\varphi u) = (2\pi)^{d/2}\int \hat{\varphi}(\xi - \eta)\hat{u}(\eta)\, d\eta.$$

Then

$$\langle\xi\rangle^s\mathscr{F}(\varphi u) = (2\pi)^{d/2}\int \frac{\langle\xi\rangle^s}{\langle\eta\rangle^s}\hat{\varphi}(\xi - \eta)\langle\eta\rangle^s\hat{u}(\eta)\, d\eta.$$

We need the following simple estimate for $\langle\cdot\rangle^\sigma$

$$\langle\xi\rangle^\sigma \leq 2^{|\sigma|/2}\langle\eta\rangle^\sigma\langle\xi - \eta\rangle^{|\sigma|}. \tag{18}$$

To prove (18) begin by noting that $\operatorname{grad}\langle\omega\rangle = \omega\langle\omega\rangle^{-1}$ is of length at most 1. Thus

$$\langle\xi\rangle \leq \langle\eta\rangle + |\xi - \eta| \leq \langle\eta\rangle(1 + |\xi - \eta|).$$

On the other hand, for any positive c,

$$(1 + c)^2 = 1 + 2c + c^2 \leq 2(1 + c^2).$$

Apply with $c = |\xi - \eta|$ to find $\langle\xi\rangle \leq 2^{1/2}\langle\eta\rangle\langle\xi - \eta\rangle$. Estimate (19) for $\sigma \geq 0$ follows.

The case of $\sigma < 0$ then follows from

$$\frac{\langle\xi\rangle^\sigma}{\langle\eta\rangle^\sigma} = \frac{\langle\eta\rangle^{|\sigma|}}{\langle\xi\rangle^{|\sigma|}} \leq 2^{|\sigma|/2}\langle\xi - \eta\rangle^{|\sigma|}.$$

Using (19) in (18) yields

$$|\langle\xi\rangle^s\mathscr{F}(\varphi u)| \leq c\int |\langle\xi - \eta\rangle^{|s|}\hat{\varphi}(\xi - \eta)|\,|\langle\eta\rangle^s\hat{u}(\eta)|\, d\eta.$$

Then Young's inequality implies

$$\|\langle\xi\rangle^s\mathscr{F}(\varphi u)\|_{L^2(\mathbb{R}^d)} \leq c\|\langle\xi\rangle^{|s|}\hat{\varphi}(\xi)\|_{L^1(\mathbb{R}^d)}\|\langle\eta\rangle^s\hat{u}(\eta)\|_{L^2(\mathbb{R}^d)},$$

which is the desired estimate. □

It is important to know that every compactly supported distribution lies in some H^s.

Proposition 5. *If* $T \in \mathscr{E}'(\mathbb{R}^d)$, *then there is an* $s \in \mathbb{R}$ *such that* $T \in H^s(\mathbb{R}^d)$.

PROOF. There are constants c, R and an integer N so that, for all $\varphi \in C^\infty(\mathbb{R}^d)$,

$$|\langle T, \varphi \rangle| \le c\left(\sum_{|\alpha| \le N} \sup_{|x| \le R} |D^\alpha \varphi(x)| \right).$$

Thus using Theorem 2.5.4 yields

$$|\hat{T}(\xi)| = \left| \frac{\langle T, e^{-ix\xi} \rangle}{(2\pi)^{d/2}} \right| \le c\langle \xi \rangle^N.$$

For s so negative that $s + N < -d/2$, it follows that $T \in H^s(\mathbb{R}^d)$. □

In dealing with function spaces the most usual method is to prove assertions on a conveniently chosen dense subset and then pass to the limit through approximations from that subspace. The limit is justified using appropriate inequalities. For example, in $L^p(\mathbb{R}^d)$, $p \ne \infty$, a convenient dense set is the simple functions $\sum c_i \chi_{E_i}$ with E_i bounded measurable. Equally important is the dense set $C_0^\infty(\mathbb{R}^d)$.

Proposition 6. *For any* $s \in \mathbb{R}$, $C_0^\infty(\mathbb{R}^d)$ *is dense in* $H^s(\mathbb{R}^d)$.

PROOF. Since $\mathscr{S}(\mathbb{R}^d_\xi)$ is dense in $L^2(\mathbb{R}^d, \langle \xi \rangle^{2s}\, d\xi)$, it follows that $\mathscr{S}(\mathbb{R}^d)$ is dense in $H^s(\mathbb{R}^d)$.

Choose $\chi \in C_0^\infty(\mathbb{R}^d)$ with $\chi(s) = 1$. Then for $f \in \mathscr{S}(\mathbb{R}^d)$, $\chi(x/n)f \to f$ in $H^s(\mathbb{R}^d)$, and $\chi(x/n)f \in C_0^\infty(\mathbb{R}^d)$. □

Since the integral is supported in $|\xi| \le 1 + R$, the integral in (2) is estimated by

$$\le (1 + (1 + R)^2)^s \int_{|\xi| \le 1 + R} |j_\varepsilon * \hat{u}_R - \hat{u}_R|^2\, d\xi.$$

Problem 2.5.4 implies that the right-hand side converges to zero, so we may choose n so that $\|v_n - u_R\|_s < \varepsilon/3$.

Choose $\chi \in C_0^\infty(\mathbb{R}^d)$ with $\chi(0) = 1$. Since $v_n \in \mathscr{S}$, $\chi(x/m)v_n \to v_n$ in $\mathscr{S}$ as m tends to infinity. This implies convergence to v_n in H^s. Thus we may choose

m so that $\|\chi(x/m)v_n - v_n\|_s < \varepsilon/3$. Then $\chi(x/m)v_n$ is within ε of u in s-norm and the proof is complete. □

The next result asserts that if $u \in H^s$ with s large, then u has classical derivatives. Roughly the number of classical derivatives is equal to the number of Sobolev derivatives minus one-half the dimension. There is a "loss" of one-half of a derivative per dimension.

Theorem 7 (Sobolev Embedding Theorem). *If $k \in \mathbb{N}$ and $u \in H^s(\mathbb{R}^d)$ with $s > d/2 + k$, then for all $\alpha \in \mathbb{N}^d$ with $|\alpha| \le k$, $D^\alpha u \in \dot{C}(\mathbb{R}^d)$. In addition, there is a constant $c = c(s, \alpha, d)$ so that for all $u \in H^s(\mathbb{R}^d)$*

$$\|D^\alpha u\|_{L^\infty(\mathbb{R}^d)} \le c\|u\|_{H^s(\mathbb{R}^d)}. \tag{3}$$

Proof. If suffices to show that (3) holds for all $u \in \mathcal{S}(\mathbb{R}^d)$, since given any $u \in H^s(\mathbb{R}^d)$ we may choose $u_n \in \mathcal{S}$, $u_n \to u$ in H^s. Then, for $|\alpha| \le k$,

$$\|D^\alpha u_n - D^\alpha u_m\|_{\dot{C}} = \le c\|u_n - u_m\|_{H^s},$$

so $D^\alpha u_n$ is a Cauchy sequence in $\dot{C}$ hence convergent in $\dot{C}$ to a function g_α. Then $D^\alpha u_n \rightharpoonup g_\alpha$ in $\mathcal{S}'(\mathbb{R}^d)$.

However, since $u_n \to u$ in $\mathcal{S}'$ we have $D^\alpha u_n \to D^\alpha u$ in $\mathcal{S}'$. Equating the two $\mathcal{S}'$ limits yields $D^\alpha u = g_\alpha \in \dot{C}$. Finally, passing to the limit $n \to \infty$ in

$$\|D^\alpha u_n\|_{L^\infty} \le c\|u_n\|_{H^s}$$

yields (3) for u.

To prove (3) for $u \in \mathcal{S}$ observe that

$$D^\alpha u(x) = (2\pi)^{-d/2} \int \frac{\xi^\alpha}{\langle\xi\rangle^s} \langle\xi\rangle^s \hat{u}\, d\xi.$$

Now $\xi^\alpha/\langle\xi\rangle^s \in L^2(\mathbb{R}^d_\xi)$ if (and only if) $s > d/2 + |\alpha|$. The Schwartz inequality yields (3) with $c = (2\pi)^{-d/2}\|\xi^\alpha\langle\xi\rangle^{-s}\|_{L^2(\mathbb{R}^d)}$. □

Problems

1. Prove that for any $f \in \mathcal{S}'(\mathbb{R}^d)$, $\delta_1^h f \rightharpoonup \partial f/\partial x_1$ in $\mathcal{S}'(\mathbb{R}^d)$.
2. Prove Proposition 2.
3. (i) For which values of s is $\chi_{[0,1]}$ in $H^s(\mathbb{R})$?
   (ii) For which values of s is $\chi_{[0,1]\times[0,1]}$ in $H^s(\mathbb{R}^2)$?
   (iii) If $K \in \mathcal{S}'(\mathbb{R}^d)$ is the tempered solution of $(1 - \Delta)K = \delta$, for which s is $K \in H^s(\mathbb{R}^d)$?

In the next two problems you will show, by explicit construction, that the Sobolev Embedding Theorem is sharp.

4. Using functions of the form $r^a(\ln r)^b$ near $x = 0$, construct a $u \in H^1(\mathbb{R}^2) \cap \mathcal{E}'(\mathbb{R}^2)$ which is not bounded near $(0, 0)$.
   Discussion. If u were in $H^{1+\varepsilon}$ for any $\varepsilon > 0$, then Sobolev's Theorem would imply that $u \in \dot{C}$.

5. Construct a $u \in H^{1/2}(\mathbb{R})$ such that $u \notin L^\infty([-1, 1])$. *Hint*. Choose $f \in L^2(\mathbb{R})$, $f \geq 0$, $\langle\xi\rangle^{-1/2} f \notin L^1(\mathbb{R})$. Define $u \equiv \mathcal{F}^{-1}(\langle\xi\rangle^{-1/2} f)$. Formally, $u(0) = (2\pi)^{-1/2} \int \langle\xi\rangle^{-1/2} f \, d\xi = \infty$. To show that u is not bounded on a neighborhood of 0, show (using (2.2.7)) that as $n \to \infty$, $\lim \int u(x) n e^{-(nx)^2/2} \, dx = \infty$. Why is this sufficient?
DISCUSSION. It is not hard to generalize this last construction to show that if $s \leq d/2$, then there is an unbounded element of $H^s(\mathbb{R}^d)$. Similarly, if $s \leq k + d/2$, $k \in \mathbb{N}$, and $\alpha \in \mathbb{N}^d$ with $|\alpha| = k$, then there is a $u \in H^s(\mathbb{R}^d)$ with $\partial^\alpha u$ unbounded.

The list of equivalent conditions in Proposition 1 is far from exhaustive. Two interesting additions are to include approximate derivatives $j_\varepsilon * \partial_1 f = \partial_1(j_\varepsilon * f)$ and to replace convergence in $L^2(\mathbb{R}^d)$ by boundedness.

6. Prove that the following conditions are equivalent to those in Proposition 1.
(5) As $h \to 0$, $\delta_1^h f$ is bounded in $L^2(\mathbb{R}^d)$.
(6) As $\varepsilon \to 0$, $\partial_1(j_\varepsilon * f)$ converges in $L^2(\mathbb{R}^d)$.
(7) As $\varepsilon \to 0$, $\partial_1(j_\varepsilon * f)$ is bounded in $L^2(\mathbb{R}^d)$.

CHAPTER 3

Solution of Initial Value Problems by Fourier Synthesis

§3.1. Introduction

This chapter describes a method for solving and/or analyzing partial differential equations using the Fourier transform. The key ingredient is that constant coefficient equations have explicit exponential solutions. Both the power series method and the Fourier analysis method have as point of departure explicit exact solutions. This is a severe limitation. Some more recent developments, for example, pseudodifferential and Fourier integral operator methods depend on explicit approximate solutions which exist in more general situations.

The goal of this chapter is to find formulas which are sufficiently informative to at least distinguish between good and bad initial value problems for constant coefficient linear equations such as $\partial_t + \partial_x$ and $\partial_t + i\partial_x$.

The idea has a predecessor for ordinary differential equations. If $P(D_t)$ is a constant coefficient ordinary differential operator, then

$$P(D_t)e^{i\tau t} = P(\tau)e^{i\tau t},$$

so that $e^{i\tau t}$ is a solution if and only if $P(\tau) = 0$. If this equation has simple roots $\tau_1, \tau_2, \ldots$, then the functions $\exp(i\tau_j t)$ span the solution set of the homogeneous equation $Pu = 0$. The recipe is only slightly more complicated if there are multiple roots.

For a constant coefficient linear partial differential operator, $P(D)$, we have

$$P(D)e^{ix\xi} = P(\xi)e^{ix\xi},$$

so there are solutions for any $\xi \in \mathbb{C}^n$ satisfying $P(\xi) = 0$, that is, any point in the complex characteristic variety. For initial value problems, one separates t and x,

$$P(D_t, D_x)e^{i\tau t + ix\xi} = P(\tau, \xi)e^{i\tau t + ix\xi}.$$

In many problems from applications the solutions tend to zero when $|x| \to \infty$ corresponding to the fact that the phenomena described take place near the observer. If ξ is real the exponential solutions are bounded as x tends to infinity while for $\xi \in \mathbb{C}\backslash\mathbb{R}$ they explode exponentially. When constructing solutions which decay at infinity, one takes a superposition of exponential solutions with $\xi \in \mathbb{R}^n$,

$$\int_{\mathbb{R}^n} a(\xi) e^{i\tau(\xi)t + ix\xi}\, d\xi.$$

We will be studying the consequences of this simple idea for a while. It is surprisingly rich.

Problems

1. For the Cauchy–Riemann operator, $\partial_y - i\partial_x$, show that the exponential solutions are exactly the functions e^{az} for $a \in \mathbb{C}$ and $z \equiv x + iy$.

2. Are there any operators $P(D)$ which have no exponential solutions?

§3.2. Schrödinger's Equation

The first example which we discuss in detail is the Schrödinger equation for a particle of mass equal to $\frac{1}{2}$,

$$u_t = i\Delta u, \qquad t, x \in \mathbb{R} \times \mathbb{R}^d. \tag{1}$$

Here the units of length and time have been chosen so that Planck's constant is equal to 1. The particle is moving in the absence of external forces. This is the quantum analogue of Galileo's particle moving in a straight line at constant speed. The text of Messiah [Me] is a standard introduction to quantum mechanics including motivation for the above equation.

In the traditional classification of partial differential operators, the Schrödinger equation is neither elliptic, parabolic, nor hyperbolic. The equation is an example where the precise existence and regularity results require Sobolev spaces. It is used here as a model problem to develop generally applicable techniques. The resulting formulas are somewhat simpler than those for the other natural candidate, the wave equation, which is presented in §3.7.

The physical interpretation, in quantum mechanics, is that the square of the modulus, $|u(t, x)|^2$, is the probability density for finding the particle at time t and place x. Precisely, for a Lebesgue measurable subset $E \subset \mathbb{R}^d$,

$$\text{Probability(particle is in } E \text{ at time } t) = \int_E |u(t, x)|^2\, dx. \tag{2}$$

The probability density for the particle's momentum is given by the Fourier transform of $u(t, \cdot)$

$$\text{Probability(momentum is in } E \text{ at time } t) = \int_E |\hat{u}(t, \xi)|^2\, d\xi. \tag{3}$$

This is definitely not obvious. For some motivation the reader is referred to texts of quantum mechanics (e.g. Messiah, [Me]). There is no simple, clear, and convincing derivations, just as one cannot *derive* Newton's laws. However, Examples 1 and 2 of §3.3, Theorems 4.4.4, 4.4.6, and Corollary 4.4.5 show that (3) is reasonable.

The probability interpretation requires that

$$\int |u(t, x)|^2 \, dx = 1 \qquad \text{for all} \quad t \geq 0, \tag{4}$$

for physically relevant solutions. One is immediately led to think that $L^2(\mathbb{R}^d)$ will play a distinguished role. Note that if (4) holds, then the Plancherel Theorem shows that $\mathscr{F}u$ also has L^2 norm equal to 1, so the interpretation of $|\mathscr{F}u|^2$ as a probability density is consistent.

To derive (4), note that if a solution u is small enough, as $x \to \infty$, to justify differentiation under the integral and neglecting terms at infinity from integrations by parts, we have

$$\begin{aligned} \partial_t \int |u|^2 \, dx = \partial_t \int u\bar{u} \, dx &= \int u\bar{u}_t + u_t\bar{u} \, dx \\ &= \sum_j \int u\overline{i\partial_j^2 u} + (i\partial_j^2 u)\bar{u} \, dx = \sum_j i \int \partial_j u \overline{\partial_j u} - \partial_j u \overline{\partial_j u} \, dx = 0. \end{aligned}$$

The integration by parts can be viewed another way. Writing the steps out shows that solutions of (1) satisfy a conservation law

$$\partial_t |u|^2 = i \sum \partial_j (u\overline{\partial_j u} - (\partial_j u)\bar{u}). \tag{5}$$

Integrating this identity over $\mathbb{R}^d$ yields $\partial_t \int |u|^2 \, dx = 0$. Alternatively, integrating over $[0, T] \times \mathbb{R}^d$ yields $\int |u(T, x)|^2 \, dx = \int |u(0, x)|^2 \, dx$. All roads lead to the conclusion that (4) is satisfied if $\int |u(0, x)|^2 \, dx = 1$.

As a final remark on the physical interpretation, we describe Heisenberg's uncertainty principle. Note that the probability interpretation (2) implies that the expected position x^{av} is given by

$$x^{\text{av}} = \int_{\mathbb{R}^d} x|u|^2 \, dx. \tag{6}$$

Similarly, (3) implies that the expected momentum, ξ^{av}, is equal to

$$\xi^{\text{av}} = \int_{\mathbb{R}^d} \xi |\hat{u}(t, \xi)|^2 \, d\xi. \tag{7}$$

The dispersions are

$$(\delta x_j)^2 = \int (x_j - x_j^{\text{av}})^2 |u|^2 \, dx, \qquad 1 \leq j \leq d, \tag{8}$$

$$(\delta \xi_j)^2 = \int (\xi_j - \xi_j^{\text{av}})^2 |\hat{u}|^2 \, d\xi, \qquad 1 \leq j \leq d. \tag{9}$$

These quantities are related by the Heisenberg uncertainty relation $(\delta x_j)^2 (\delta \xi_j)^2 \geq \frac{1}{4}$ (Problems 1, 2, 3) which, given the present perspective, is a theorem in Fourier analysis.

In spite of the probabilistic interpretation, the Schrödinger equation is a deterministic physical theory in the sense that, given the initial state, $u(0, \cdot) = f(\cdot)$, the solution u is determined for all t. The evolution is determined by solving the initial value problem

$$u_t = i\Delta u, \qquad u(0, \cdot) = f. \tag{10}$$

We already know a certain amount about this problem. Since $(t = 0)$ is characteristic, we do not expect the Taylor series for u to converge, even if f is real analytic (Problem 1.3.4). However, if f is polynomial all goes well (Problems 1.3.3 and 1.3.5). The physical requirement (4) renders such polynomial solutions uninteresting. We will see in §3.9 that since the initial plane is charateristic there is not uniqueness for the initial value problem in the category of all smooth solutions. There are nonzero $u \in C^\infty(\mathbb{R}_t \times \mathbb{R}^d_x)$ with

$$u_t = i\Delta u, \qquad u = 0 \qquad \text{for} \quad t < 0.$$

Though disconcerting at first glance, this is not terrible since the physical solutions must be square integrable in x. The null solutions above grow rapidly as $|x| \to \infty$.

Holmgren's Theorem shows that if $u = 0$ on a neighborhood of $]t_1, t_2[\times \{\bar{x}\}$, then u vanishes on $]t_1, t_2[\times \mathbb{R}^d$ (Problem 4). This result leads to the quantum mechanical way to catch a lion: If there exists a lion, then putting a cage anywhere, there is a strictly positive probability that the lion is in the cage. It also shows that it is not reasonable to look for solutions campactly supported in x.

The physical interpretation of $\int_E |u|^2 \, dx$, as the probability of finding the particle in E, suggests that we want solutions for which this quantity is continuous in t. We will look for solutions such that $t \mapsto u(t, \cdot)$ is continuous on $\mathbb{R}$ with values in $L^2(\mathbb{R}^d)$. This also forces related measurements of momenta to be continuous.

Problems

In these problems you will prove *Heisenberg's uncertainty inequality.*

1. For $u \in \mathcal{S}(\mathbb{R}^d)$ with $\int |u|^2 \, dx = 1$, let $v \equiv e^{-i\xi^{\text{av}} x} u(x + x^{\text{av}})$. Show that the average position and average momentum for v are both equal to zero. Show that the dispersions $(\delta x_j)^2$ and $(\delta \xi_j)^2$ are the same for v as for u.

2. With v as in Problem 1, let $Q \equiv \int x_j \bar{v} \partial_j v \, dx$.
   (i) Show that $|Q|^2 \leq (\delta x_j)^2 (\delta \xi_j)^2$.
   (ii) Perform an integration by parts in the definition of Q, to show that $\text{Re } Q = -\frac{1}{2}$.
   (iii) Conclude that $(\delta x_j)^2 (\delta \xi_j)^2 \geq \frac{1}{4}$.
   This is Heisenberg's Theorem.

Theorem 1. *If* $u \in \mathscr{S}(\mathbb{R}^d)$ *with* $\int |u|^2\,dx = 1$, *then* $(\delta x_j)^2(\delta\xi_j)^2 \geq \frac{1}{4}$ *where the dispersions are defined in equations* (6), (7), (8), *and* (9).

DISCUSSION. (1) The case of equality occurs when $\partial_j v = cx_j v$. These are exactly the Gaussians. (2) Crucial in the derivation is the Heisenberg Commutation Law $[x_j, \partial_j] = \mathrm{Id}$.

In the next problem, the Heisenberg inequality is extended to its natural domain. Define a Hilbert space H_j by

$$H_j \equiv \{u \in L^2(\mathbb{R}^d) : \int (1 + x_j^2)|u|^2\,dx + \int (1 + \xi_j^2)|\mathscr{F}u(\xi)|^2\,d\xi < \infty\}.$$

The square of the norm in H_j is the sum of the integrals in brackets.

3. (i) Prove that $\mathscr{S}$ is dense in H_j.
   (ii) Prove that the uncertainty relation is true for all $u \in H_j$ with $\int |u|^2\,dx = 1$.
   DISCUSSION. The space H_j is the natural space on which both $(x_j^2)^{\mathrm{av}}$ and $(\xi_j^2)^{\mathrm{av}}$ are defined. Equivalently, it is the natural space on which the dispersions in the Heisenberg relation are finite.

4. Suppose that $u \in C^2(\mathbb{R}^{1+d})$ satisfies $u_t = \nu\Delta u$ with $\nu \in \mathbb{C}\backslash 0$. Prove that if $t_1 < t_2$ and u vanishes on a neighborhood of $]t_1, t_2[\times \{x\}$, then u vanishes identically on $]t_1, t_2[\times \mathbb{R}^d$. *Hint*: Use Fritz John's Global Holmgren Theorem.

§3.3. Solutions of Schrödinger's Equation with Data in $\mathscr{S}(\mathbb{R}^d)$

The exponential solutions of the Schrödinger equation satisfy $\tau = -i|\xi|^2$. The exponential solutons are

$$e^{-it|\xi|^2 + ix\xi} = e^{i\xi(x - \xi t)},$$

which is a function of $x - \xi t$. The exponential solution of frequency ξ evolves by a translation at velocity ξ. For the equation $u_t = ic\Delta u$, the solutions are $e^{i\xi(x - c\xi t)}$ and the velocity is $c\xi$. The hypersurfaces of constant phase translate at speed $c|\xi|$ in the direction $\xi/|\xi|$. Consequently, $c\xi$ is called the *phase velocity* at frequency ξ. As we will see, the group velocity, equal to $2c\xi$ for the equation $u_t = ic\Delta u$, is a more important quantity. A hint at why this is so is provided by the thought experiment of trying to send a message at velocity ξ using the plane waves above.

One would like to say something of the sort, "When this maximum of the real part reaches you, turn on switch number three". As the maxima of the real parts translate at velocity ξ this sounds like a good strategy. However, the maxima are indistinguishable. The remedy is to consider a localized solution which has initial data equal to the product of a plane wave of frequency $\xi \gg 1$ and a cutoff function of width ≈ 1. The resulting solution does not travel with speed ξ but with speed $\approx 2c\xi$, as we will see in Example 2 below.

Returning to the case $c = 1$, and taking linear combinations of exponential solutions, yields the solutions

$$\int a(\xi) e^{-it|\xi|^2 + ix\xi}\, d\xi.$$

To solve the initial value problem

$$u_t = i\Delta u, \qquad u(0, \cdot) = f(\cdot), \tag{1}$$

set $t = 0$ and equate the resulting expression to f. This suggests $a = (2\pi)^{-d/2}\mathscr{F}f$, leading to the formula

$$u(t, x) = (2\pi)^{-d/2} \int e^{-it|\xi|^2} e^{ix\xi} \mathscr{F}f(\xi)\, d\xi. \tag{2}$$

An alternate derivation of the same formula starts by taking the Fourier transform of $u_t = i\Delta u$ with respect to x to obtain $\partial_t \hat{u} = -i|\xi|^2 \hat{u}$. This is an ordinary differential equation in time with ξ as parameter. Solving yields

$$\hat{u}(t, \xi) \equiv e^{-it|\xi|^2} \hat{f}(\xi).$$

Equivalently

$$u(t, x) \equiv \mathscr{F}^{-1}(e^{-it|\xi|^2} \mathscr{F} f), \tag{3}$$

which is the same as (2).

Next we show that formula (2), (3) yields solutions of (1). Rather than justify the steps in the derivations we give a direct verification. For $f \in \mathscr{S}$, the integral (2) can be differentiated arbitrarily often with respect to t and x. The differentiation is justified by the rapid decrease of $\mathscr{F}f$ since the t, x derivatives of the integrand are expressed as a finite linear combination of terms of the form

$$\text{polynomial}(t, \xi) e^{-it|\xi|^2} e^{ix\xi} \mathscr{F} f(\xi).$$

Thanks to the rapid decay of $\mathscr{F}f$, these are dominated by $c_{N,T}\langle\xi\rangle^{-N}$ on $[-T, T] \times \mathbb{R}^d_x \times \mathbb{R}^d_\xi$. We find that $u \in C^\infty(\mathbb{R} \times \mathbb{R}^d)$ and satisfies the initial value problem (1). In addition, the Fourier transform of $D^\alpha_{t,x} u(t, \cdot)$ has Fourier transform equal to a finite linear combination of terms of the form

$$\text{polynomial}(t, \xi) e^{-it|\xi|^2} \mathscr{F} f(\xi).$$

For t fixed, these lie in $\mathscr{S}(\mathbb{R}^d_\xi)$. Thus $t \mapsto u(t)$ is a map from $\mathbb{R}$ to $\mathscr{S}(\mathbb{R}^d)$. To show that this is continuous, it is sufficient to show that $\xi^\alpha \partial^\beta_\xi \mathscr{F} D^\gamma_{t,x}(u(t + h) - u(t))$ converges to zero in $L^\infty(\mathbb{R}^d_\xi)$ as h tends to zero. This difference is a finite sum of terms of the form

$$(p(t + h, \xi) e^{-i(t+h)|\xi|^2} - p(t, \xi) e^{-i(t)|\xi|^2}) \partial^\mu_\xi \mathscr{F} f,$$

where p is a polynomial. The Mean Value Theorem bounds the difference in parentheses by a multiple of $h(\langle\xi\rangle + \langle t\rangle)^N$. The rapid decay of $\partial^\mu_\xi \mathscr{F} f$ more than compensates for the polynomial growth.

We want to show that u is a differentiable function of time with values in $\mathscr{S}(\mathbb{R}^d)$. There are two reasonable definitions of this notion and they are equivalent. That is, the content of the next proposition whose proof is left to Problem 1.

Proposition 1. *Suppose that $u \in C(\mathbb{R} : \mathscr{S}(\mathbb{R}^d))$. Then the following are equivalent:*

(1) *$\mathscr{S}\text{-}\lim_{h\to 0}(u(t+h) - u(t))/h$ exists uniformly on bounded subsets of $\mathbb{R}_t$.*
(2) *The partial derivative $\partial u/\partial t$ exists at all t, x, $(\partial u/\partial t)(t, \cdot) \in \mathscr{S}$ for all t, and the map $t \mapsto (\partial u/\partial t)(t, \cdot)$ is a continuous map of $\mathbb{R}$ to $\mathscr{S}(\mathbb{R}^d_x)$.*

Taking the Fourier transform of

$$\mathscr{S}\text{-}\lim_{h\to 0} \frac{u(t+h) - u(t)}{h} = u_t$$

shows that $u \in C^1(\mathbb{R} : \mathscr{S}) \Leftrightarrow \mathscr{F}u \in C^1(\mathbb{R} : \mathscr{S})$. In that case, $\partial_t \mathscr{F} u = \mathscr{F}\partial_t u$.

Definition. For $k \geq 1$, $C^k(\mathbb{R} : \mathscr{S}(\mathbb{R}^d_x))$ is defined inductively as the set of functions $u \in C^{k-1}(\mathbb{R} : \mathscr{S})$ such that $\partial_t^{k-1} u \in C^1(\mathbb{R} : \mathscr{S})$, $C^\infty(\mathbb{R} : \mathscr{S}) \equiv \bigcap_k C^k(\mathbb{R} : \mathscr{S})$.

It is not hard to show that $C^k(\mathbb{R} : \mathscr{S})$ is exactly the set of $u \in C^k(\mathbb{R} \times \mathbb{R}^d)$ such that for $0 \leq j \leq k$, $t \mapsto \partial_t^j u(t) \in C(\mathbb{R} : \mathscr{S})$. In particular, the solution u constructed above belongs to $C^\infty(\mathbb{R} : \mathscr{S})$.

$C^\infty(\mathbb{R} : \mathscr{S})$ is a complete metric space, the metric derived from the sequence of norms

$$p_n(u) \equiv \sum_{|\alpha| + j + |\beta| \leq n} \| x^\alpha \partial_t^j \partial_x^\beta u \|_{L^\infty([-n,n] \times \mathbb{R}^d)}.$$

Finally, note that $u \in C^\infty(\mathbb{R} : \mathscr{S}) \Leftrightarrow \mathscr{F}u \in C^\infty(\mathbb{R} : \mathscr{S})$.

Theorem 2. *For any $f \in \mathscr{S}(\mathbb{R}^d)$, there is a unique $u \in C^\infty(\mathbb{R} : \mathscr{S}(\mathbb{R}^d))$ satisfying* (1). *The solution u is given by formula* (2).

PROOF. The existence is proved in the previous paragraphs. It remains to prove uniqueness. If $u \in C^1(\mathbb{R} : \mathscr{S}(\mathbb{R}^d))$ satisfies $u_t = i\Delta u$, then taking the Fourier transform of both sides yields $\partial_t \hat{u} = -i|\xi|^2 \hat{u}$, whence $\partial_t(e^{it|\xi|^2}\hat{u}) = 0$. It follows that $\hat{u} = e^{-it|\xi|^2}\mathscr{F}f(\xi)$. □

Note that $|\mathscr{F}u(t, \xi)|^2 = |\mathscr{F}f(\xi)|^2$ is independent of t. Thus the probability density for momentum is independent of t, a very strong form of conservation of momentum. Integrating over $\xi \in \mathbb{R}^d$ shows that the $L^2(\mathbb{R}^2)$ norm of $u(t)$ is independent of time. Thus the requirement (3.2.4) is satisfied as soon as it is satisfied at $t = 0$. The conservation of total momentum states that $\int \xi |\mathscr{F}u|^2 \, d\xi$ is independent of t. This is a weaker assertion.

EXAMPLE 1. Find the solution of the initial value problem (1) when $f(x) = e^{-a|x|^2/2}$, $a > 0$, a Gaussian with "width" $1/\sqrt{a}$ and height 1.

The formula $\mathscr{F}f(\xi) = a^{-d/2} e^{-|\xi|^2/2a}$ yields

$$\hat{u}(t) = e^{-it|\xi|^2}\mathscr{F}f(\xi) = a^{-d/2}e^{-(1/a+2it)|\xi|^2/2}, \tag{4}$$

again a Gaussian. As $\hat{u}$ is even, $u = \mathscr{F}^{-1}\hat{u} = \mathscr{F}\hat{u}$, and (2.4.7) yields

$$\begin{aligned} u &= a^{-d/2}(1/a + 2it)^{-d/2}e^{-|x|^2/(2(1/a+2it))} \\ &= (1 + 2ait)^{-d/2}e^{-a|x|^2/(2(1+4a^2t^2))}e^{ia^2t|x|^2/(1+4a^2t^2)}. \end{aligned} \tag{5}$$

This is one of the few explicit solutions of the Schrödinger equation. A good deal of intuitive content is hidden in the long formula. As $t \to \infty$, the width of u grows like $t\sqrt{a}$. The physical interpretation is that $(\delta x) \approx t\sqrt{a}$. Similarly, the momentum distribution is Gaussian with width $\delta p \approx \sqrt{a}$. The momentum of order $\sqrt{a}$ causes a spread like $t\sqrt{a}$ explaining the behavior of δx. This reasoning, based on the physical interpretations (3.2.2)–(3.2.3), should give you a little faith in these interpretations.

As $t \to \infty$, the amplitude of u decays like $(at)^{-d/2}$. The geometric explanation is that the wave spreads over $|x| \le t\sqrt{a}$. If a typical amplitude is M, then the square of the L^2 norm is like $M^2(t\sqrt{a})^d$ which must be independent of t, whence

$$M^2(t\sqrt{a})^d \approx \int e^{-a|x|^2/2}\,dx \sim a^{-d/2},$$

which yields $M^2 \sim (at)^{-d}$. These last phenomenological estimates are very useful in understanding some of the features of solutions of partial differential equations. They are not intended to be rigorous proofs.

EXAMPLE 2. A closely related example illustrates the propagation of oscillatory pulses. Take $f(x) = e^{ix\eta}e^{-|x|^2/2}$ with $\eta \in \mathbb{R}^d$. Then $\mathscr{F}f(\xi) = e^{-|\xi-\eta|^2/2}$ and

$$\begin{aligned} \hat{u}(t) &= e^{-it|\xi|^2}e^{-|\xi-\eta|^2/2} \\ &= e^{-it((\xi-\eta)^2+2(\xi-\eta)\eta+\eta^2)}e^{-(\xi-\eta))^2/2} \equiv \psi(t, \xi - \eta). \end{aligned} \tag{6}$$

Then

$$u = e^{ix\eta}v \qquad \text{where} \quad \hat{v} \equiv \psi(t, \xi) = e^{-it(\xi^2+2\xi\eta+\eta^2)}e^{-\xi^2/2}.$$

The factor $e^{-i2t\xi\eta}$ also corresponds to translation

$$v = \tau_{2\eta t}w \qquad \text{where} \quad \hat{w} = e^{-it\eta^2}e^{-(1+2it)\xi^2/2}.$$

Since the $\hat{w}$ is even, $w = \mathscr{F}^{-1}\hat{w} = \mathscr{F}\hat{w}$ is given by formula (2.4.7).

$$w = e^{-it\eta^2}(1 + 2it)^{-d/2}e^{-x^2/(2(1+2it))}.$$

Then

$$u = e^{i\eta(x-\eta t)}(1 + 2it)^{-d/2}e^{-(x-2\eta t)^2/(2(1+2it))}. \tag{7}$$

Let $g(t, x)$ be the solution with $\eta = 0$, so that g is an expanding Gaussian of Example 1. Then $u = e^{i\eta(x-\eta t)}g(x - 2\eta t)$ is the product of the plane wave solution with phase velocity η and an expanding pulse translating with the group velocity $2\eta t$. For $\eta \gg 1$, this justifies the description at the beginning of the section. Note that the phase velocity is slower than the group velocity so the pulse appears to overtake the plane wave. Note also that the speed of propagation depends on the frequency of the pulse. This gives another insight into the spreading of Gaussian pulses from Example 1. The "parts" of different frequencies move at different speeds which pulls the wave apart. Analogous phenomena explain the spliting, by a prism, of a beam of white light into colors. The pulling apart of localized waves and separation according to frequency are signatures of the phenomenon called *dispersion.*

The momentum interpretation (3.2.3) suggests that the momentum $\approx \eta$, since the Fourier transform is localized near η. If momentum is to equal the product of mass and velocity then the mass must equal $\frac{1}{2}$, since the velocity is equal to 2η. This again supports the physical interpretations of the introductory paragraphs. The equations for a mass m particle is $u_t = (i/2m)\Delta u$. The relations in the last two paragraphs will be analyzed in more detail in §4.4.

The operator $f \mapsto u(t, \cdot)$ is called the *propagator*, and is denoted $S(t)$ where S stands for solution. It is defined for all $t \in \mathbb{R}$. For each t, $S(t)$ is a continuous linear map of $\mathscr{S}$ to itself given by the formula $S(t) = \mathscr{F}^{-1} e^{-it|\xi|^2} \mathscr{F}$. It is a simple multiplication operator in the Fourier transformed unknowns. Physicists call the transformed unknowns the *momentum representation.*

From the definition of S, it follows that $S(t)\partial_x^\alpha f = \partial_x^\alpha(S(t)f)$. Using our Gaussians, it is then easy to compute, by induction on $|\alpha|$, formulas for $S(t)(x^\alpha e^{-a|x|^2/2})$ (Problem 3).

Problems

1. Prove Proposition 1.
2. Prove that the map $f \mapsto u$ from initial value to solution of the Schrödinger equation is a continuous map of $\mathscr{S}(\mathbb{R}^d)$ to $C^\infty(\mathbb{R}:\mathscr{S}(\mathbb{R}^d))$. *Hint.* If suffices to show that for every n, there is a c and N so that

$$p_n(u) \leq c \sum_{|\alpha| + |\beta| \leq N} \| x^\beta \partial_x^\alpha f \|_{L^\infty(\mathbb{R}^d)}.$$

Discussion. This second result asserts that the solution depends continuously on the initial data. For C^∞ regularity, one is forced out of the simple category of normed linear spaces into the category of countably normed spaces.

3. Following the hint in the last sentence of this section, compute $S(t)\,(x^\alpha e^{-a|x|^2/2})$ for $|\alpha| = 1$.

§3.4. Generalized Solutions of Schrödinger's Equation

In this section we construct solutions of the initial value problem

$$u_t = i\Delta u, \qquad u(0, \cdot) = f(\cdot) \in H^s(\mathbb{R}^d), \tag{1}$$

by approximating f by data in $\mathscr{S}$. To justify the passage to the limit requires H^s estimates for solutions. The special case of $f \in L^2(\mathbb{R}^d) = H^0(\mathbb{R}^d)$ is especially important from a physical point of veiw thanks to (3.2.4), while the case $f = \delta \in H^{-d/2-\varepsilon}$ is called the *fundamental solution* since it provides a formula in the general case (see §4.2).

For any $t > 0$, let $S(t)$: $\mathscr{S}(\mathbb{R}^d) \to \mathscr{S}(\mathbb{R}^d)$ be the map which sends f to $u(t)$. Then $\mathscr{F}(S(t)f) = e^{-it|\xi|^2}\mathscr{F}f$, so for any t_1, $t_2 \in \mathbb{R}$, the identity $S(t_1 + t_2) = S(t_1)S(t_2)$ holds. This identity captures a part of Huygen's ideas about secondary wavelets. To progress $t_1 + t_2$ units of time, one can first go t_2 units and then use the result as "source" for the next step of t_1 units.

The construction of generalized solutions amounts to extending the operators $S(t)$, by continuity, to larger spaces than $\mathscr{S}$.

Proposition 1. *For any* $s \in \mathbb{R}$, $t \in \mathbb{R}$, $f \in \mathscr{S}(\mathbb{R}^d)$,

$$\|S(t)f\|_{H^s} = \|f\|_{H^s}. \tag{2}$$

The case $s = 0$ is the conservation of probability in the physical interpretation.

PROOF. $\|S(t)f\|_{H^s} = \|e^{-it|\xi|^2}\langle\xi\rangle^s\mathscr{F}f\|_{L^2(\mathbb{R}^d)}$ is independent of t since $e^{-it|\xi|^2}$ is of modulus 1. □

Corollary 2. *For any* $s \in \mathbb{R}$, $t \in \mathbb{R}$, *the operator* $S(t)$ *extends uniquely to a unitary map of* $H^s(\mathbb{R}^d)$ *to itself. The extended operator satisfies* $\mathscr{F}(S(t)f) = e^{-it|\xi|^2}\mathscr{F}f$ *for any* $f \in H^s$.

PROOF. That S extends uniquely to an isometry is immediate. The identity $S(t_1 + t_2) = S(t_1)S(t_2)$ remains true since the two sides are continuous and are equal on the dense subset $\mathscr{S}$ of H^s, Thus $S(-t)$ is an inverse to $S(t)$ proving that $S(t)$ is unitary.

That $S(t) = \mathscr{F}^*e^{-it|\xi|^2}\mathscr{F}$ on H^s follows from the fact that both sides are bounded operators on H^s and they are equal on the dense subset $\mathscr{S}$. □

If $f \in H^{s_1} \cap H^{s_2}$, then $S(t)f$ is defined as an element of H^{s_1} if we view f as an element of H^{s_1}, and it is a well-defined element of H^{s_2} viewing f as an element of H^{s_2}. The Fourier transform formula shows that both ways yield the same answer. This leads to the following definition.

Definition. If $f \in \bigcup H^s$, then the function u: $\mathbb{R} \to \bigcup H^s(\mathbb{R}^d)$, defined by $\mathscr{F}u(t) = e^{-it|\xi|^2}\mathscr{F}f$, is called the *generalized solution* of the Schrödinger equation with initial value f.

If s is not sufficiently large the derivatives, $\partial_t u$, $\partial^2 u/\partial x_j^2$, which occur in the Schrödinger equation will not exist in the classical sense, whence the name generalized solution. In the next section, several equivalent characterizations of the generalized solution are given.

EXAMPLE. Find the generalized solution of the initial value problem (1) with $f = \delta \in \bigcap_{s<-d/2} H^s(\mathbb{R}^d)$.

By definition,

$$\mathscr{F}u(t) = e^{-it|\xi|^2}\mathscr{F}\delta = (2\pi)^{-d/2}e^{-it|\xi|^2}.$$

The right-hand side is even in ξ, so $\mathscr{F}$(rhs) $= \mathscr{F}^*$(rhs), so using formula (2.4.7) with $a = 2it$ yields for $t \neq 0$.

$$u(t) = (2\pi)^{-d/2}(2it)^{-d/2}e^{i|x|^2/4t} = (4\pi it)^{-d/2}e^{i|x|^2/4t}. \tag{3}$$

Note than the $H^s(\mathbb{R}^d)$ regularity of $u(t)$ is independent of t but $u(0) = \delta$ and $u(t) \in C^\infty$ for all $t \neq 0$. The fact that the initial disturbance is localized at $x = 0$ and propagates immediately to a solution of amplitude 1 uniform in space reflects one aspect of the uncertainty principle. Initially, $|\delta x| = 0$ so $|\delta p| = \infty$, and momenta of all sizes are present. This makes the instantaneous dispersion reasonable.

We have already mentioned that the initial value problem (1) has many solutions, a consequence of the fact that the hyperplane $t = 0$ is characteristic. The extension process described above singles out one of these. As $|x|$ tends to infinity the generalized solutions tend to zero and are square integrable in a weak sense, tested by measuring with devices moving to infinity.

Proposition 3. *If $f \in H^s(\mathbf{R})$ for some $s \in \mathbf{R}$ and $\phi \in C_0^\infty(\mathbf{R}^d)$, then $g(y) := \langle f, \tau_y\phi\rangle$ is a continuous function which tends to zero as $y \to \infty$. Moreover, g is square integrable.*

PROOF. The generalized Schwartz inequality implies that for any y

$$|g(y)| \leq \|f\|_{H^s}\|\tau_y\phi\|_{H^{-s}} = \|f\|_{H^s}\|\phi\|_{H^{-s}}.$$

Thus if $f_n \in C_0^\infty(\mathbf{R}^d)$ converges to f in the H^s norm, then the associated functions g_n converge uniformly to g on $\mathbf{R}^d$. Since $g_n(y) = (f_n * g)(-y)$ this represents g as the uniform limit of elements of $C_0^\infty(\mathbf{R})$, proving the first assertion.

Since $\hat{g}(-\xi) = c\hat{f}(\xi)\hat{\phi}(\xi)$ the last assertion follows from the Plancherel identity and the fact that $|\phi(\xi)| \leq c_s\langle\xi\rangle^s$. □

The basic estimate (2) shows that for $f \in H^s$, $u(t)$ is a bounded function with values in $H^s(\mathbb{R}^d)$. Thinking of the path $t \mapsto u(t)$ as the dynamics of our system it is important to have regularity in time. It is a general principle in linear analysis that boundedness estimates like (2) are often sufficient to prove continuity.

Proposition 4. *If $f \in H^s(\mathbb{R}^d)$, the generalized solution with initial value f is a continuous function of time with values in $H^s(\mathbb{R}^d)$.*

PROOF. Choose $f_n \in \mathscr{S}$, with $f_n \to f$ in H^s. Let u_n be the solution with initial value f_n. Estimate (2) applied to $u_n - u_m$ shows that $\|u_n(t) - u_m(t)\|_{H^s} =$

$\|f_n - f_m\|_{H^s}$. Thus, for any $T > 0$, u_n is a Cauchy sequence in the Banach space $C([-T, T] : H^s(\mathbb{R}^d))$. It follows that u_n converges in $C([-T, T] : H^s(\mathbb{R}^d))$ to a limit v. The generalized solution is defined by $u(t) = H^s\text{-}\lim u_n(t)$, and this limit is equal to $v(t)$. Thus $u = v \in C([-T, T] : H^s(\mathbb{R}^d))$. □

Proposition 4 describes exactly the H^s regularity of the generalized solution u with respect to x. Regularity in time is found by expressing time derivatives as spatial derivatives using the differential equations.

Proposition 5. *Suppose that $f \in H^s$ and u is the generalized solution of* (1). *Then, for any $j \geq 0$, $u \in C^j(\mathbb{R} : H^{s-2j}(\mathbb{R}^d))$, and for all $t \in \mathbb{R}$,*

$$\| \partial_t^j u(t) \|_{H^{s-2j}} \leq \|f\|_{H^s}. \tag{4}$$

PROOF. For $f \in \mathscr{S}$ we have

$$\| \partial_t^j u(t) \|_{H^{s-2j}} = \| (i\Delta)^j u \|_{H^{s-2j}} \leq \| u(t) \|_{H^s} = \|f\|_{H^s}. \tag{5}$$

Choose $\mathscr{S} \in f_n \to f$ in H^s and let u_n be the solution with initial data f_n. Then for any $T > 0$ and $N \in \mathbb{N}$, estimate (5) applied to $u_n - u_m$ shows that u_n is a Cauchy sequence in the Banach space $\bigcap_{j=1}^N C^j([-T, T] : H^{s-2j}(\mathbb{R}^d))$. Thus, u_n converges in this space to a limit v. As in Proposition 4, we must have $u = v \in \bigcap_{j=1}^N C^j([-T, T] : H^{s-2j}(\mathbb{R}^d))$. Since this holds for all T and N, the proof is complete. □

Corollary 6. *Suppose that u and f are as above, then*:

(i) $\partial_t^j \partial_x^\alpha u \in C(\mathbb{R} : H^{s-|\alpha|-2j}(\mathbb{R}^d))$.
(ii) *If $s - 2k > d/2$, then $u \in C^k(\mathbb{R}_t \times \mathbb{R}_x^d)$ and if $|\alpha| \leq k$ then as $|x| \to \infty$, $\lim \partial_x^\alpha u(t, x) = 0$ uniformly on compact subsets of $\mathbb{R}_t$.*

Remark. If $k \geq 2$, we find a classical solution. Naively, one might hope that the solution will be classical if $f \in C^2$ or if $f \in C^2 \cap L^2$. These conditions are *not* sufficient. Neither is the stronger condition, $\partial_x^\beta f \in C \cap L^2$ for $|\beta| \leq 2$. This insufficiency is demonstrated in Problem 3. A similar weakness of the spaces C^k for the Poisson equation is patched by working in the Hölder spaces C^β with $0 < \beta < 1$. This fix is effective for elliptic and parabolic equations but does not work for the Schrödinger equation. The H^s Sobolev space regularity results are the only sharp ones in the latter case.

PROBLEMS

For any $f \in H^s(\mathbb{R}^d)$, $S(t)f$ is a continuous function of $t \in \mathbb{R}$ with values in H^s. This property is called *strong continuity* of $S(t)$. Together with the property $S(t_1 + t_2) = S(t_1)S(t_2)$, this shows that S is a strongly continuous group of unitary operators on H^s. An even stronger notion of continuity is that the map $t \mapsto S(t)$ is continuous from $\mathbb{R}$ to $\text{Hom}(H^s, H^s)$, the bounded operators on H^s. If this were true, then as $h \to 0$, $\|S(t + h) - S(t)\|_{\text{Hom}(H^s, H^s)}$ would tend to 0.

1. Prove that $S(t)$ is not in $C(\mathbb{R}:\mathrm{Hom}(H^s, H^s)$ by showing that for any $t_1 \neq t_2$, $\|S(t_1) - S(t_2)\|_{\mathrm{Hom}(H^s,H^s)} = 2$.
DISCUSSION. The only groups which are norm continuous are solution operators of the equations $u_t = Au$ with A a bounded operator. The solution operator is e^{tA} in this case and A is called the *generator*. Strongly continuous groups may have unbounded generators (e.g. the Schrödinger equation where $i\Delta$ is unbounded on H^s). Unitary groups are generated by anti-self-adjoint operators. In our case $S(t) = e^{ti\Delta}$. The interested reader may consult texts on functional analysis. Semigroup methods have been particularly useful for parabolic equations. See, for example, the book of D. Henry [He].

In the next problem, you are asked to show that the Schrödinger propagator is continuous on L^p if and only if $p = 2$.

2. For $p \neq 2$ and $t \neq 0$ show that
$$\sup_{\varphi \in \mathscr{S}\setminus 0} \frac{\|S(t)\varphi\|_{L^p}}{\|\varphi\|_{L^p}} = \infty.$$
Hint. Consider $\varphi = e^{-(a+ib)|x|^2/2}$ with $a > 0$ and $b \in \mathbb{R}$. Let a tend to zero or infinity depending on the value of p.
DISCUSSION. There is a general principle. Suppose that S is an operator which respects the L^r norm in the sense that
$$c\|\varphi\|_{L^r} \leq \|S\varphi\|_{L^r} \leq C\|\varphi\|_{L^r}.$$
If there are functions such that φ_n and $S\varphi_n$ are spread regularly over their support and the $\mathrm{supp}(S\varphi_n)$ is incomparably smaller than $\mathrm{supp}(\varphi_n)$, then S cannot be bounded in L^p for any $p > r$. The reason is that if M_n (resp. m_n) is a typical magnitude for φ_n (resp. $S\varphi_n$), then respect for L^r yields M_n^r vol(supp φ_n) $\sim m_n^r$ vol (supp $S\varphi_n$). Thus $m_n/M_n \to \infty$, so S is not bounded on L^∞. Similarly, $\|S\varphi_n\|_{L^p}/\|\varphi_n\|_{L^p} \to \infty$ for any $p > r$.
If there are functions whose support is compressed, one finds that S is not bounded on L^p for any $p < r$. In our case $r = 2$, the L^2 norm is conserved and Gaussians of size $a^{-1/2}$ are spread by $S(t)$ over a region of size $a^{1/2}t$ which, letting a tend to infinity (resp. 0), is incomparably larger (resp. smaller) than the original spread. The conclusion is that S is unbounded on L^p for all $p \neq 2$.
The test functions of the hint are suggested by the solutions
$$S(t)\delta = (4\pi it)^{-d/2}e^{i|x|^2/4t}$$
and
$$S(t)((-4\pi it)^{-d/2}e^{-i|x|^2/4t}) = \delta,$$
which spread from a point to all of $\mathbb{R}^d$ and contract from all of $\mathbb{R}^d$ to a point. These are extreme examples of *dispersion*.
The operator $S(t)$ is the Fourier mutiplier $\mathscr{F}^* e^{-it|\xi|^2}\mathscr{F}$ and the multiplier $e^{-it|\xi|^2}$ is smooth and bounded but gives an operator which is unbounded on L^p for $p \neq 2$. This discontinuity is not obvious. Viewed from the point of view of the multiplier, the problem comes from the fact that $e^{-it|\xi|^2}$ oscillates faster and faster as $|\xi|$ tends to infinity.

3. (i) Use the case $p = \infty$ of Problem 2 to show that for any $t \neq 0$ there is an $f \in L^\infty$ such $S(t)f \equiv \mathscr{F}^* e^{-it|\xi|^2}\mathscr{F}f \notin L^\infty$. *Hint*. If $S(t)f \in L^\infty$ for all such f, use the Closed Graph Theorem to show that $S(t)$ is continuous from L^∞ to itself. Then use Problem 2.

(ii) Modify Problem 1 and part (i) to prove that for any $t \neq 0$ there is an $f \in L^\infty \cap L^2$ such that $S(t)f \notin L^\infty$.
(iii) Show that for any $t \neq 0$ there is an $f \in C^2(\mathbb{R}^d)$ with derivatives up to order 2 in $L^\infty \cap L^2$ and $S(t)f \notin C^2(\mathbb{R}^d)$. *Hint.* To the above ingredients add $[S, D^\alpha] = 0$.

4. Prove Corollary 6.

§3.5. Alternate Characterizations of the Generalized Solution

The propagator $S(t) = \mathscr{F}^{-1}e^{-it|\xi|^2}\mathscr{F} \in \text{Hom}(H^s)$ is characterized in terms of the Fourier transform. Such methods are not available for problems with variable coefficients, for example, the Schrödinger equation with potential $u_t = i(\Delta + V(x))u$. This section contains several ways of identifying the generalized solution u by looking at its action on test functions.

Regarding the test functions note that if $\psi \in \mathscr{S}(\mathbb{R}_t \times \mathbb{R}_x^d)$, then it is easy to verify that the map $t \mapsto \psi(t, \cdot)$ is $C^\infty(\mathbb{R}_t : \mathscr{S}(\mathbb{R}_x^d))$. In particular, for any $s \in \mathbb{R}$, $\psi \in C^\infty(\mathbb{R}_t : H^s(\mathbb{R}_x^d))$.

A related remark is that if $u \in C(\mathbb{R} : H^s(\mathbb{R}_x^d))$ for some $s \in \mathbb{R}$, then u defines a distribution on $\mathbb{R}_t \times \mathbb{R}_x^d$ by

$$u(\psi) = \int \langle u(t), \psi(t, \cdot)\rangle \, dt, \qquad \psi \in \mathscr{D}(\mathbb{R}^{1+d}).$$

Here $\psi \in C^\infty(\mathbb{R} : \mathscr{S})$, so $\langle u(t), \psi(t, \cdot)\rangle \in C_0(\mathbb{R})$ hence integrable.

A better estimate of the behavior of this $u \in \mathscr{D}'(\mathbb{R}_t \times \mathbb{R}_x^d)$ rests on the generalized Schwartz inequality which yields

$$|\langle u(t), \psi(t, \cdot)\rangle| \leq \|u(t)\|_{H^s(\mathbb{R}_x^d)} \|\psi(t, \cdot)\|_{H^{-s}(\mathbb{R}_x^d)}.$$

Integrating yields

$$|u(\psi)| \leq \|u\|_{L^\infty(I : H^s(\mathbb{R}_x^d))} \|\psi\|_{L^1(I : H^{-s}(\mathbb{R}_x^d))}, \qquad \text{for all} \quad \psi \in C_0^\infty(I \times \mathbb{R}^d). \quad (1)$$

Theorem 1. *Suppose that $s \in \mathbb{R}$, $f \in H^s(\mathbb{R}^d)$, and $u \in C(\mathbb{R} : H^s(\mathbb{R}^d))$ with $u(0) = f$. The following are equivalent:*

(i) *$u = S(t)f$ for all $t \in \mathbb{R}$.*
(ii) *$\hat{u}(t) = e^{-it|\xi|^2}\hat{f}$ for all $t \in \mathbb{R}$.*
(iii) *For any $\varphi \in C_0^\infty(\mathbb{R}^n)$, the function $t \mapsto \langle u(t), \varphi\rangle$ is continuously differentiable and $(d/dt)\langle u, \varphi\rangle = \langle u(t), (i\Delta)\varphi(x)\rangle$.*
(iv) *For any $\psi \in C_0^\infty(\mathbb{R}^{1+d})$, the function $t \mapsto \langle u(t), \psi(t, \cdot)\rangle$ is continuously differentiable with*

$$\frac{d}{dt}\langle u(t), \psi(t, \cdot)\rangle = \langle u(t), (\partial_t + i\Delta)\psi(t, \cdot)\rangle. \quad (2)$$

(v) *For any $\psi \in C_0^\infty(\mathbb{R} \times \mathbb{R}^d)$, $\langle u, (-\partial_t - i\Delta)\psi\rangle = 0$.*

(vi) *For any* $\psi \in C_0^\infty(\mathbb{R} \times \mathbb{R}^d)$ *and* $t_1 < t_2$

$$\int_{t_1}^{t_2} \langle u(t), (-\partial_t - i\Delta)\psi(t)\rangle \, dt = \langle u(t), \psi(t)\rangle \Big|_{t=t_1}^{t=t_2}.$$

Remarks. 1. Formal calculations from $u_t = i\Delta u$ yield (ii), (iii), (iv), (v), and (vi) immediately.

2. Part (v) is equivalent to the equation $(\partial_t - i\Delta)u = 0$ *in the sense of distributions.* Note that all derivatives have been passed to the test function ψ, and that the transpose of $\partial_t - i\Delta$ is equal to $-\partial_t - i\Delta$.

PROOF. (i) $\Leftrightarrow$ (ii) is the definition of generalized solutions.

(ii) $\Rightarrow$ (iii) For $\varphi \in C_0^\infty(\mathbb{R}^d) \subset \mathscr{S}(\mathbb{R}^d)$

$$\langle u(t), \varphi\rangle = \langle \hat{u}, \hat{\varphi}\rangle = \int e^{-it|\xi|^2} \hat{f}(\xi)\hat{\varphi}(\xi)\, d\xi.$$

Call the integrand $F(t, \xi)$. Then

$$|\partial_t F(t, \xi)| = (|\xi|^2 \langle\xi\rangle^{-s} |\hat{\varphi}(\xi)|)(\langle\xi\rangle^s |\hat{f}(\xi)|).$$

Since $\varphi \in \mathscr{S}$, the first term is square integrable and the second is in $L^2(\mathbb{R}^d)$ since $f \in H^s$. Thus $\partial_t F(t, \cdot)$ has an $L^1(\mathbb{R}^d)$ upper bound uniformly for $t \in \mathbb{R}$. It follows that differentiation under the integrand is justified and

$$\frac{d}{dt}\langle u(t), \varphi\rangle = \int -i|\xi|^2 e^{-it|\xi|^2} \hat{f}(\xi)\hat{\varphi}(\xi)\, d\xi$$

$$= \langle \mathscr{F}u, \mathscr{F}(i\Delta\varphi)\rangle = \langle u, i\Delta\varphi\rangle.$$

The right-hand side is continuous since $u \in C(\mathbb{R} : H^s)$ and the proof of (iii) is complete.

(iii) $\Rightarrow$ (iv) With the goal of differentating $\langle u(t), \psi(t, \cdot)\rangle$ at $t = \bar{t} \in \mathbb{R}$, write

$$\langle u(t), \psi(t, \cdot)\rangle = \langle u(t), \psi(\bar{t}, \cdot)\rangle + \langle u(t), \psi(t, \cdot) - \psi(\bar{t}, \cdot)\rangle.$$

(iii) implies that $(d/dt)\langle u(t), \psi(\bar{t})\rangle = \langle u(t), i\Delta\psi(\bar{t})\rangle$. The second term is equal to

$$\langle u(\bar{t}), \psi(t, \cdot) - \psi(\bar{t}, \cdot)\rangle + \langle u(t) - u(\bar{t}), \psi(t, \cdot) - \psi(\bar{t}, \cdot)\rangle.$$

Since $\psi \in C^1(\mathbb{R} : H^{-s}(\mathbb{R}_x^d))$ and $u(\bar{t}) \in H^s(\mathbb{R}_x^d)$, the first term is $C^1(\mathbb{R})$ with derivative equal to $\langle u(\bar{t}), \partial_t \psi(\bar{t})\rangle$. The second term is estimated, using the generalized Schwartz inequality,

$$|\langle u(t) - u(\bar{t}), \psi(t) - \psi(\bar{t})\rangle| \le \|u(t) - u(\bar{t})\|_{H^s} \|\psi(t) - \psi(\bar{t})\|_{H^{-s}}.$$

Since $u \in C(\mathbb{R} : H^s)$, the first factor is $o(1)$ as $t \to \bar{t}$. Since $\psi \in C^1(\mathbb{R} : H^{-s}(\mathbb{R}^d))$, the second factor is $O(|t - \bar{t}|)$. Thus the product is $o(|t - \bar{t}|)$ so its derivative at $t = \bar{t}$ exists and is equal to zero.

Adding the two contributions yields (iv). Since $(\partial_t - i\Delta)\psi \in \mathscr{D} \subset$

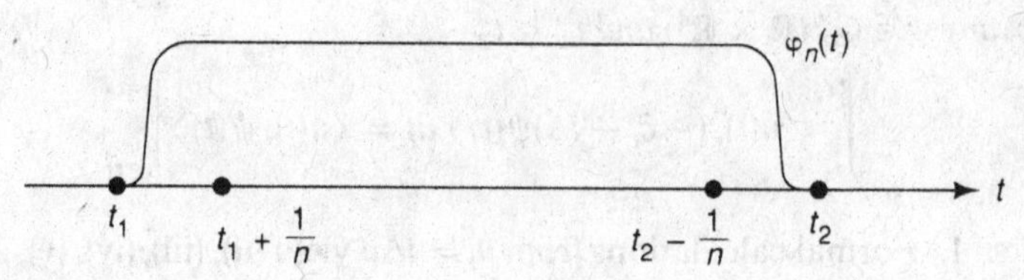

Figure 3.5.1

$C(\mathbb{R} : H^{-s}(\mathbb{R}^d))$, the right-hand side of (iv) is continuous which completes the proof that (iii) $\Rightarrow$ (iv).

(iv) $\Rightarrow$ (v) Since ψ has compact support integrating (iv) from $t = -\infty$ to $t = +\infty$ yields $\int \langle u(t), (\partial_t + i\Delta)\psi(t)\rangle \, dt = 0$, which is the desired identity.

(v) $\Rightarrow$ (iv) Fix ψ and $t_1 < t_2$. Choose $\varphi_n \in C_0^\infty(\mathbb{R}_t)$ approximating the characteristic function of $[t_1, t_2]$ (see Figure 3.5.1). Apply (v) to get

$$\int_{t_1}^{t_2} \langle u(t), (-\partial_t - i\Delta)(\varphi_n \psi)\rangle \, dt = 0.$$

Performing the differentations yields

$$0 = \int_{t_1}^{t_2} \langle u(t), \varphi_n(t)(-\partial_t - i\Delta)\psi\rangle \, dt + \int_{t_1}^{t_2} \langle u(t), -\varphi_n'(t)\psi\rangle \, dt \equiv I_1 + I_2.$$

The integrand in I_1 is dominated by

$$\|u(t)\|_{L^\infty((t_1,t_2):H^s)} \|(-\partial_t - i\Delta)\psi(t, \cdot)\|_{L^1((t_1,t_2):H^{-s})} < \infty$$

and converges pointwise to $\langle u(t), (-\partial_t - i\Delta)\psi\rangle$, so by Lebesgue's Dominated Convergence Theorem yields

$$I_1 \to \int_{t_1}^{t_2} \langle u(t), (-\partial_t - i\Delta)\psi(t)\rangle \, dt \qquad \text{as} \quad n \to \infty.$$

For I_2, note that $g(t) \equiv \langle u(t), \psi(t)\rangle$ is continuous since $u \in C(\mathbb{R} : H^{-s})$ and $\psi \in C(\mathbb{R} : H^{-s})$. The integrand is supported in $[t_1, t_1 + 1/n] \cup [t_2 - 1/n, t_2]$. The contribution of the first interval is

$$\int_{t_1}^{t_1+1/n} -\varphi_n'(t) g(t) \, dt = -g(t_1) + \int_{t_1}^{t_1+1/n} \varphi_n'(t)(g(t_1) - g(t)) \, dt.$$

The integral on the right is dominated by

$$\left(\max_{[t_1, t_1+1/n]} |g(t) - g(t_1)| \right) \int_{t_1}^{t_1+1/n} \varphi_n'(t) \, dt.$$

The maximum is $o(1)$ and the integral equals 1. Thus

$$-\int_{t_1}^{t_1+1/n} \varphi_n'(t) g(t) \, dt = -g(t_1) + o(1) \qquad \text{as} \quad n \to \infty.$$

Treating $[t_2 - 1/n, t_2]$ similarly yields the desired result.

(v) $\Rightarrow$ (i) Suppose that u satisfies (vi) and let $v \equiv S(t)f$ so v also satisfies (vi). Using (vi) with $t_1 = 0 < T$ and subtracting the identities for u and v yields

$$\int_0^T \langle (u - v)(t), (-\partial_t - i\Delta)\psi(t)\rangle \, dt = \langle (u - v)(T), \psi(T)\rangle \tag{3}$$

for all $\psi \in C_0^\infty(\mathbb{R} \times \mathbb{R}^d)$. More generally, if $\psi \in C^\infty(\mathbb{R} : \mathcal{S}(\mathbb{R}^d))$, choose $\chi \in C_0^\infty(\mathbb{R}_x^d)$ with $\chi(0) = 1$. Then $\chi(\varepsilon x)\psi$ has compact support in x, so applying (3) to $\chi\psi \in C_0^\infty(\mathbb{R}^{1+d})$ yields

$$\int_0^T \langle (u - v)(t), (-\partial_t - i\Delta)(\psi(t)\chi(\varepsilon x))\rangle \, dt = \langle (u - v)(T), \chi(\varepsilon x)\psi(T)\rangle.$$

As $\varepsilon \to 0$

$$(-\partial_t - i\Delta)\chi(\varepsilon x)\psi \to (\partial_t + i\Delta)\psi \quad \text{in } C(\mathbb{R} : H^s)$$

and

$$\chi(\varepsilon x)\psi(T) \to \psi(T) \quad \text{in } H^{-s}.$$

Thus, passing to the limit $\varepsilon \to 0$, using Lebesgue's theorem yields (3) for $\psi \in C^\infty(\mathbb{R} : \mathcal{S}(\mathbb{R}^d))$.

For any $h \in \mathcal{S}(\mathbb{R}^d)$, $\psi(t) \equiv \mathcal{F}^* e^{+i(t-T)|\xi|^2}\hat{h}$ satisfies $\psi \in C^\infty(\mathbb{R} : \mathcal{S})$ and

$$(\partial_t + i\Delta)\psi = 0, \qquad \psi(T) = h. \tag{4}$$

Plugging this into (3) yields $\langle (u - v)(T), h\rangle = 0$ for all $h \in \mathcal{S}$. Thus $u(T) = v(T)$. This is true for all $T > 0$, and a similar argument works for $T < 0$. Thus $u = v = S(t)f$ proving (i). □

Remark. The proof that (vi) $\Rightarrow$ (i) is like the proof of Holmgren's Theorem. The key is an existence theorem for the adjoint problem (4).

Corollary 2. *For $f \in H^s(\mathbb{R}^d)$, $\exists!\, u \in C(\mathbb{R} : H^s(\mathbb{R}^d))$ such that $u(0) = f$, and $(\partial_t - i\Delta)u = 0$ in the sense of distributions.*

Remark. The condition $u \in C(\mathbb{R} : H^s(R^d))$ has in it an L^2 growth condition as $|x| \to \infty$ which avoids the null solutions. Without such a condition there would be nonuniqueness.

Problem

The generalized Schwartz inequality plays a central role in the proofs of this section. In this problem we examine another aspect of that inequality. Any $\psi \in H^{-s}$ defines a linear functional $\langle \cdot, \psi\rangle : H^s \to \mathbb{C}$ of norm at most $\|\psi\|_{H^{-s}}$. In fact, every linear functional on H^s arises in this way.

1. Prove

Lax's Duality Theorem. *For any continuous linear functional $l: H^s(\mathbb{R}^d) \to \mathbb{C}$, there is a unique $\psi \in H^{-s}(\mathbb{R}^d)$ so that $l(\cdot) = \langle \cdot, \psi\rangle$. In addition, $\|l\| = \|\psi\|_{H^{-s}}$.*

Hint. First use the Riesz Representation Theorem to show that there is a unique $h \in H^s$ so that $l(\cdot) = (\cdot, h)_{H^s}$. Show that $\psi \doteq \mathscr{F}^{-1}\langle\xi\rangle^{2s}\mathscr{F}(\bar{h})$ does the trick. *Alternate Hint.* Use the isomorphism of $H^s \approx L^2(\mathbb{R}^d_\xi : \langle\xi\rangle^{2s}\, d\xi)$ and duality in that L^2 space. *Second Alternate Hint.* It is easy to show that the range of the map $\psi \mapsto l$ has closed range in $(H^s)'$. Show that the range is dense by a duality argument.
DISCUSSION. This theorem yields an algorithm for defining so-called negative norms when the Fourier transform is not available, for example, $H^{-s}(\Omega)$ when $\Omega \neq \mathbb{R}^d$.

The reader is warned that the Riesz Representation Theorem asserts that $(H^s)' \approx H^s$. The above result asserts that $(H^s)' \approx H^{-s}$. Clearly, the mappings from H^s and H^{-s} into $(H^s)'$ are different. They come from the bilinear forms $(\ ,\ )_{H^s}$ and $\langle\ ,\ \rangle$, respectively.

§3.6. Fourier Synthesis for the Heat Equation

The methods of §3.1 to §3.3 yield existence and uniqueness results for a variety of initial value problems. The strength of the technique is this generality, together with the fact that it distinguishes well-posed from ill-posed initial value problems. To appreciate the distinctions between the various well-posed problems requires some experience. This section is devoted to studying the heat equation

$$u_t = \nu\Delta u, \qquad \nu > 0, \tag{1}$$

for $u(t, x)$, $t, x \in \mathbb{R} \times \mathbb{R}^d$. Formal Fourier transformation yields

$$\hat{u}_t = -\nu|\xi|^2\hat{u}, \tag{2}$$

$$\hat{u} = e^{-\nu t|\xi|^2}\hat{f}(\xi), \tag{3}$$

where $u(0, \cdot) = f(\cdot)$. Note that for $f \in \mathscr{S}$ and $t > 0$ one has $\hat{u}(t) \in \mathscr{S}$ but *not* necessarily for $t < 0$. For example, if $\hat{f} \equiv e^{-\langle\xi\rangle}$, then $\hat{u}$ for $t < 0$ yields a function which is not even in $\mathscr{S}'(\mathbb{R}^d)$. Thus, for $f \in \mathscr{S}$, one has a reasonable recipe for $u(t)$ only for $t \geq 0$. With that change, the theory proceeds exactly as for the Schrödinger equation. In particular, the proof of the following theorem is just like the proof of Theorem 3.3.2 and so is omitted.

Theorem 1. *If $f \in \mathscr{S}(\mathbb{R}^n)$, then there exists one and only one $u \in C^\infty([0, \infty[: \mathscr{S}(\mathbb{R}^d))$ such that*

$$u_t = \nu\Delta u, \qquad u(0) = f. \tag{4}$$

For $t \geq 0$, u is given by formula (3).

EXAMPLE. Find the solution of (4) when $f = e^{-ax^2/2}$, $a > 0$.

Use formula (2.4.6) for the transform of f to find

$$\hat{u} = e^{-t\nu|\xi|^2}a^{-d/2}e^{-|\xi|^2/2a} = a^{-d/2}e^{-(2\nu t+1/a)|\xi|^2/2}.$$

This is an even function of ξ, so $u = \mathscr{F}^{-1}\hat{u} = \mathscr{F}\hat{u}$. Formula (2.4.7) yields

$$\begin{aligned} u &= a^{-d/2}\left(2vt + \frac{1}{a}\right)^{-d/2} e^{-|x|^2/2(2vt+1/a)} \\ &= (2avt+1)^{-d/2}e^{-|x|^2/2(2vt+1/a)}. \end{aligned} \tag{5}$$

Note that the width of the heat distribution grows like $\sqrt{vt}$. In contrast, the Schrödinger equation leads to growth of order $t\sqrt{v}$. Clearly, for heat propagation, $\hat{u}$ cannot be interpreted as a probability density for velocity or momentum, otherwise the spread would be linear in t.

The rate of decay as t tends to infinity is $t^{-d/2}$ as it was for the Schrödinger equation. Since the solution is spread over a much smaller region one finds that the L^2 norm tends to zero like $t^{-d/2}$. On the other hand, the integral of u is equal to $\mathscr{F}u(0) = \mathscr{F}f(0)$ so is independent of time. This is the law of *conservation of energy.*

To extend the solution operator to a more general class of data requires some estimates. To avoid confusion with the Schrödinger propagator we will denote the operator $u(0) \mapsto u(t)$ by $S_H(t)$, which we know is continuous from $\mathscr{S}$ to itself for $t \geq 0$. Fourier methods yield a host of Sobolev space estimates

$$\|u(t)\|^2_{H^s} = \|e^{-vt|\xi|^2}\langle\xi\rangle^s\mathscr{F}f\|_{L^2} \leq \|\langle\xi\rangle^s f\|_{L^2} = \|u(0)\|_{H^s}. \tag{6}$$

These H^s estimates imply that $S_H(t)$ extends to a continuous map of $H^s(\mathbb{R}^d)$ to itself, and $u = S_H(t)f$ for $f \in H^s$ defines a generalized solution u to the initial value problem. As for Schrödinger's equation, $u \in C([0, \infty[: H^s(\mathbb{R}^d))$, and more generally,

Theorem 2. *If $f \in H^s(\mathbb{R}^d)$, then the generalized solution satisfies $\partial_t^j u \in C([0, \infty[: H^{s-2j}(\mathbb{R}^n))$.*

EXAMPLE. Find the solution of (4) when $f = \delta$.

Formula (3) yields $\hat{u} = e^{-vt|\xi|^2}(2\pi)^{-d/2}$ which is again the transform of a Gaussian. Computing the inverse Fourier transform yields

$$u = (4\pi vt)^{-d/2}e^{-|x|^2/4vt}. \tag{7}$$

Note that an initially localized solution spreads over all of $\mathbb{R}^d$. The principle of Problem 3.4.2 shows that if S_H were defined for all t and conserved the L^2 norm, then the propagator S_H would not be bounded in any L^p for $p \neq 2$. The same argument would show that if the L^r norm were conserved for some r, then S_H would be unbounded in L^p for all $p \neq r$. For S_H no L^r norm is conserved, but positivity is preserved as is the L^1 norm of positive solutions. Nevertheless, S_H is bounded on all L^p. The failure of "reversability" explains the failure of the argument.

The solution with δ as initial data can be obtained from the solution, u_a, with data $(2\pi a)^{-d/2}\exp(-a|x|^2)$ from the first example upon letting a decrease to zero, noting that the initial data converges to δ in H^s for $s < -d/2$ and, therefore, $S_H(t)u_a \to S_H(t)\delta$ in the same H^s spaces.

As in the last section, the generalized solution has a variety of equivalent characterizations.

Theorem 3. *If $f \in H^s(\mathbb{R}^d)$, $u \in C([0, \infty[: H^s(\mathbb{R}^d))$, and $u(0) = f$, then the following are equivalent:*

(i) *u is the generalized solution with initial data f.*
(ii) $\mathcal{F}(u(t)) = e^{-t\nu|\xi|^2}\mathcal{F}f$, $t \geq 0$.
(iii) *For any $\varphi \in C_0^\infty(\mathbb{R}^d)$, $\langle u(t, x), \varphi(x)\rangle \in C^1([0, \infty[)$ and*

$$\frac{d}{dt}\langle u, \varphi\rangle = \langle u, \nu\Delta\varphi\rangle, \qquad \textit{for all} \quad t \geq 0.$$

(iv) *For any $\psi \in C_0^\infty(\mathbb{R}_t \times \mathbb{R}^d)$, $\langle u(t, \cdot), \psi(t, \cdot)\rangle$ belongs to $C^1([0, \infty[)$ and*

$$\frac{d}{dt}\langle u(t, \cdot), \psi(t, \cdot)\rangle = \langle u(t), \psi_t(t) + \nu\Delta\psi(t)\rangle.$$

(v) $(\partial_t - \nu\Delta)u = 0$ *in* $\mathcal{D}'(]0, \infty[\times \mathbb{R}^d$.
(vi) *For any $\psi \in C_0^\infty(\mathbb{R}_t \times \mathbb{R}^d)$ and $0 \leq t_1 < t_2$*

$$\int_{t_1}^{t_2} \langle u(t), (-\psi_t - \nu\Delta\psi)(t)\rangle\, dt = \langle u, \psi\rangle\Big|_{t_1}^{t_2}.$$

There are many estimates which are valid for the heat propagator which are false for the Schrödinger propagator. We derive some of these by what is called the *energy method*, though the inequalities obtained in this case are more closely associated with entropy increase than with energy balance.

The inequalities are proved by multiplying the equation by suitable, cleverly chosen, functions and then integrating by parts. For Schrödinger's equations, multiplication by $\bar{u}$ leads to conservation of the L^2 norm. Here multiplication by $\bar{u}$ leads to the estimate $\|u(t)\|_{L^2} \leq \|u(0)\|_{L^2}$. For Schrödinger's equation one cannot avoid complex numbers. In contrast, u solves the heat equation if and only if the real and imaginary parts of u satisfy the heat equation. Real initial data f yield real solutions (exercise). Physical temperatures are real. The next computations are a little easier for real solutions and we omit the modifications needed in the complex case.

As a first example, note that if $f \in \operatorname{Re}\mathcal{S}$ and u is the solution of Theorem 1, then multiplying the heat equation by u and integrating over $\mathbb{R}^d$ yields

$$\partial_t \int \frac{u^2(t, x)}{2}\, dx = \nu \int u\Delta u\, dx = -\nu \int |\nabla_x u(t, x)|^2\, dx \leq 0,$$

the last equality following upon integration by parts. In particular, the L^2

norm of $u(t)$ is a nonincreasing function of t on $[0, \infty[$. This gives an independent proof of the H^s estimate for $s = 0$. The $s = 1$ estimate can be derived by the energy method, upon multiplying the equation by Δu to find

$$0 = \int \Delta u(u_t - \nu \Delta u) = \int -\sum \frac{\partial u}{\partial x_j}\frac{\partial u_t}{\partial x_j} - \nu(\Delta u)^2 \, dx.$$

Thus

$$\partial_t \int |\nabla_x u(t, x)|^2 \, dx = -2\nu \int (\Delta u(t, x))^2 \, dx \le 0,$$

so $\int |\nabla u|^2 \, dx$ and $\int u^2 \, dx$ are nonincreasing. Adding proves that the H^1 norm is nonincreasing. For any $s \in \mathbb{N}_+$, multiplying by $\Delta^j u$, $j = 0, 1, 2, \ldots, s$ and adding the results yields the H^s estimate.

We obtain new estimates by multiplying by $\psi'(u)$ where $\psi \in C^2(\mathbb{R})$ is a convex function satisfying

$$\psi(0) = \psi'(0) = 0. \tag{8}$$

Then $\psi'(u)$ is rapidly decreasing as $x \to \infty$ and

$$\partial_t \int \psi(u) \, dx = \int \psi'(u) u_t \, dx = +\nu \int \psi'(u) \sum \partial_j \partial_j u \, dx,$$

using the differential equation for the last equality. The crucial step is to integrate by parts to find

$$= -\nu \int \psi''(u) \sum \left(\frac{\partial u}{\partial x_j}\right)^2 dx \le 0,$$

so $\int \psi(u) \, dx$ is nonincreasing.

Theorem 4. *If $u \in C^\infty([0, \infty[: \mathrm{Re}\, \mathscr{S}(\mathbb{R}^d))$ is a solution of the heat equation* (1), *then for any convex $\psi \in C^2(\mathbb{R})$ satisfying* (8), *$\int \psi(u(t, x)) \, dx$ is a nonincreasing function of t.*

EXAMPLES. 1. If $\psi(s) = s^2/2$, this yields the energy method proof of the decrease of the L^2 norm.

2. Taking $\psi(s) = |s|^p$ for $2 \le p < \infty$ we find that $\|u(t)\|_{L^p}$ is nonincreasing.

3. Passing to the limit $p \to \infty$ and noting that $u(t) \in \mathscr{S}$ yields

$$\|u(t)\|_{L^\infty} = \lim_{p \to \infty} \|u(t)\|_{L^p} \le \lim_{p \to \infty} \|u(0)\|_{L^p} = \|u(0)\|_{L^\infty},$$

so the sup norm, $\|u(t)\|_{L^\infty}$, is also nonincreasing.

4. For $p \in [1, 2[$, $\|u\|_{L^p}$ is also nonincreasing. The technical difficulty here is that $|s|^p$ is not twice differentiable so the integration by parts above is not trivial to justify. We regularize $|s|^p$ and pass to the limit. Let $\psi_\varepsilon(s) \equiv$

$(\varepsilon + |s|^2)^{p/2} - \varepsilon^{p/2}$, then

$$\int \psi_\varepsilon(u(t, x))\, dx \le \int \psi_\varepsilon(u(0, x))\, dx.$$

Passing to the limit $\varepsilon \to 0$ using Lebesgue's Dominated Convergence Theorem yields the desired result.

5. If $u \ge 0$ and $\psi(s) = s \ln(s) - s$, then again $\psi \notin C^2[0, \infty[$, but a simple regularization as above shows that $\int \psi(u)\, dx$ is nonincreasing. The integral $-\int \psi(u)\, dx$ is the entropy at time t.

6. If $\psi(s) = s$, the method reduces to simply integrating the equation over $\mathbb{R}^d_x$ and yields $\int u\, dx =$ constant, the conservation of energy.

The above multiplier or "energy" methods are very flexible, working for variable coefficient and nonlinear problems. The present case suggests multipliers to try in the more complicated cases.

Problems

1. Using the L^p estimates proved by the energy method, show that the heat propagator, $S_H(t)$ with $t \ge 0$, extends uniquely to a continuous linear operator on Re L^p for $1 \le p < \infty$. Show that for $f \in$ Re L^p, $t \mapsto S_H(t)u$ is continuous on $[0, \infty[$ with values in Re L^p.

   Discussion. Thus $S_H(t)$ is a strongly continuous semigroup on L^p. For $p = \infty$, this is no longer true. Using the L^∞ estimate and the fact that the closure of $\mathscr{S}$ in L^∞ is $\dot{C}$, one finds that $S(t)$ is a strongly continuous operator on $\dot{C}$. The next problem discusses L^∞.

2. (i) Show that for $t \ge 0$, there is a unique extension of $S(t)$ to a continuous linear operator from L^∞ to itself. *Hint*. Duality (L^∞ is the dual of L^1). Prove and use an identity $\langle Su, f\rangle = \langle u, Sf\rangle$ and the fact that $\mathscr{S}$ is sequentially dense in L^∞ with the weak-star topology.

   (ii) For $u(0) = \chi_{[0,1]} \in L^\infty(\mathbb{R}^1)$ show $u(t)$ is not continuous with values in L^∞ at $t = 0$. *Hint*. Show that for $t > 0$, $u(t) \in C(\mathbb{R})$. Conclude that $S(t)$ is not a strongly continuous semigroup on L^∞.

   (iii) Show that $S(t)$ is weakly continuous in the sense that for every $f \in L^\infty$, $S(t)f$ is continuous on $[0, \infty[$ with values in L^∞ endowed with the weak-star topology.

   Discussion. Such strong continuity in $\dot{C}$, together with weak-star continuity in L^∞, is quite common as $p = \infty$ replacements for L^p continuity.

3. Find and prove a complex analogue of Theorem 4.

4. Consider the initial value problem for Burgers' equation, $u_t + 2uu_x = 0$, $u(0, \cdot) = f(\cdot)$, which was analyzed using the method of characteristics in §1.9. Here we use the energy method and the "quasi-linear trick" to prove uniqueness. Precisely prove that if u and v are real $C^1([0, T] \times \mathbb{R})$ solutions which vanish for $|x| > R$, then $u = v$. *Hint*. Subtract the equations for u and v to prove an equation of the form $(u - v)_t + a(t, x)(u - v)_x + b(t, x)(u - v) = 0$ with a, $D_{t,x}a$, b in L^∞. Multiply by $(u - v)$, integrate by parts, and apply Gronwall's inequality to conclude that $\int (u - v)^2\, dx$ vanishes for all $0 \le t \le T$.

§3.7. Fourier Synthesis for the Wave Equation

This section follows the same path as for the Schrödinger and heat equations, namely:

(i) solve in $C^\infty(\mathbb{R} : \mathscr{S}(\mathbb{R}^d))$;
(ii) derive estimates; and
(iii) extend to generalized solutions.

The initial value problem is

$$u_{tt} - c^2 \Delta u = 0, \qquad c > 0, \tag{1}$$

$$u(0, \cdot) = f, \qquad u_t(0, \cdot) = g. \tag{2}$$

Note that this is a noncharacteristic initial value problem and that a good deal of information has already been found in §1.8 concerning finite speed of propagation and domains of influence and determinacy. In addition, the case $d = 1$ was solved explicitly.

The wave equation arises in many different areas of science. Each component of the electric and magnetic fields in free space is a solution with c equal to the speed of light. Small amplitude waves in a gas (acoustic waves) and small amplitude vibrations of an elastic medium (e.g. a membrane or jello) are also modeled by the wave equation. In the latter application $\int \rho u_t^2 \, dx$ represents the kinetic energy and $\int \kappa |\nabla u|^2 \, dx$ represents the potential energy where ρ is the density and κ is a physical constant. Then $c^2 = \kappa/\rho$ (see [Wh]). The principle of conservation of energy is that

$$\int u_t^2 + c^2 |\nabla u|^2 \, dx \quad \text{is independent of } t.$$

The corresponding conservation law for the electric field

$$\int |E_t|^2 + c^2 |\nabla_x E|^2 \, dx$$

does not have a straightforward physical interpretation. The energy in that case is $\int |E|^2 \, dx$.

Fourier transformation in x yields

$$\hat{u}_{tt} = -c^2 |\xi|^2 \hat{u}, \tag{3}$$

$$\hat{u}(t, \xi) = \hat{f}(\xi) \cos(c|\xi|t) + \hat{g}(\xi) \frac{\sin(c|\xi|t)}{c|\xi|}. \tag{4}$$

Note that $\cos(c|\xi|t)$ and $\sin(c|\xi|t)/c|\xi|$ are smooth bounded functions of ξ, as are all their ξ derivatives. Their time derivatives grow polynomially in ξ at most. Note the possible singularity at $\xi = 0$ does not occur since

$$\frac{\sin(ct|\xi|)}{c|\xi|} = \sum_n \frac{(-1)^n (c\xi t)^{2n}}{(2n+1)!}.$$

This yields

Theorem 1. *If $f, g \in \mathscr{S}(\mathbb{R}^d)$, then there is a unique $u \in C^\infty(\mathbb{R} : \mathscr{S}(\mathbb{R}^d))$ solving the initial value problem* (1), (2). *The solution is given by formula* (4).

Theorem 1.8.3 is a stronger uniqueness result. The evolution operator or propagator, $S_W(t)$, sends Cauchy data at time 0 to Cauchy data at time t, that is,

$$S_W(t)(f, g) \equiv (u(t), u_t(t)),$$

so $S_W(t): \mathscr{S} \times \mathscr{S} \to \mathscr{S} \times \mathscr{S}$. The first component of $S_W(t)\,(f, g)$ is the solution of the Cauchy problem.

The verification of the conservation of energy for these solutions is not difficult. For $f, g \in \mathrm{Re}(\mathscr{S})$ compute

$$\partial_t \int u_t^2 + c^2|\nabla u|^2 \, dx = 2 \int u_t u_{tt} + c^2 \nabla u \nabla u_t \, dx.$$

Integrate by parts in the second integral to find

$$= 2 \int u_t(u_{tt} - c^2 \Delta u) \, dx = 0.$$

Thus the multiplier u_t is appropriate for deriving this energy law. It can also be used to give an alternate proof of the theorems of finite speed and domain of influence (Problem 1). Such multiplier methods have the advantage of adaptability to variable coefficients and nonlinear problems.

Another derivation of the energy law is to note that

$$\widehat{u_t} = \cos(tc|\xi|)\hat{g} - \sin(tc|\xi|)c|\xi|\hat{f},$$

$$c|\xi|\hat{u} = \sin(tc|\xi|)\hat{g} + \cos(tc|\xi|)c|\xi|\hat{f}.$$

Take real parts, square, and add to find

$$(\mathrm{Re}\,\widehat{u_t})^2 + c^2|\xi|^2(\mathrm{Re}\,\hat{u})^2 = (\mathrm{Re}\,\hat{g})^2 + c^2|\xi|^2(\mathrm{Re}\,\hat{f}(\xi))^2.$$

A similar identity is valid for the imaginary parts. Adding yields

$$|\widehat{u_t}|^2 + c^2|\xi|^2|\hat{u}|^2 = |\hat{g}|^2 + c^2|\xi|^2|\hat{f}(\xi)|^2. \tag{5}$$

The left-hand side is the "energy at frequency ξ". It is the energy of the spring equation (3) satisfied by $\hat{u}(\cdot, \xi)$. Equation (5) shows that there is conservation at each ξ. Integrating $d\xi$ yields the global conservation of energy. More generally, one can multiply by $\langle \xi \rangle^{2(s-1)}$ to find that

$$\|u_t\|_{H^{s-1}}^2 + c^2 \|\nabla_x u\|_{H^{s-1}}^2 \quad \text{is independent of } t. \tag{6}$$

These estimates suggest the following extension of S_W.

Definition. D^s (D for Dirichlet) is the closure of $\mathscr{S}(\mathbb{R}^d)$ in the norm $\|u\|_{D^s} \equiv \|\nabla_x u\|_{H^{s-1}}$.

Then D^s is a decreasing scale of Hilbert spaces with $D^s \supset H^s$, and for any $\sigma \leq s$, $D^s \cap H^\sigma = H^s$.

Theorem 2. *For any $s \in \mathbb{R}$ and $t \in \mathbb{R}$, the map S_W extends uniquely to a unitary map of $D^s \times H^{s-1}$ to itself. The corresponding generalized solution $u \equiv (S_W(f, g))_{\text{first component}}$ satisfies $u \in C(\mathbb{R} : D^s)$ and for $|\alpha| \geq 1$, $\partial_{t,x}^\alpha u \in C(\mathbb{R} : H^{s-|\alpha|}(\mathbb{R}^d))$.*

The spaces D^s are naturally associated with the wave equation. The case $s = 1$ is the $\mathbb{R}^d$ analogue of the natural norm in the variational approach to the Dirichlet problem (see §5.2).

To estimate the H^s norm of the solution $u(t)$ in Theorem 2, note that, since $\langle\xi\rangle^s \leq c_s(|\xi|^s + \chi_{|\xi|<1})$,

$$\|u\|^2_{H^s(\mathbb{R}^d)} \leq c_s\left(\|u\|^2_{D^s(\mathbb{R}^d)} + \int_{|\xi|<1} |\hat{u}(t, \xi)|^2 \, d\xi \right).$$

Next use formula (4) for $\mathscr{F}u$ together with the estimates

$$|\cos(ct|\xi|)| \leq 1 \qquad \text{and} \qquad \frac{|\sin(ct|\xi|)|}{c|\xi|} \leq |t|$$

to show that for any $\sigma \leq s$ there is a constant $c(s, \sigma)$ so that

$$\int_{|\xi|<1} |\hat{u}(t, \xi)|^2 \, d\xi \leq c(s, \sigma)(|t| \, \|g\|_{H^\sigma} + \|f\|_{H^\sigma})$$

Thus

$$\|u(t)\|^2_{H^s} \leq c(s, \sigma)(\|g\|^2_{H^{s-1}} + \|f\|^2_{H^s} + |t| \, \|g\|^2_{H^\sigma}). \tag{7}$$

Thus u may grow linearly in time, but the growth depends only on very weak norms of the data.

Therefore, if $f, g \in H^s \times H^{s-1}$, then $u \in C(\mathbb{R} : H^s)$ and satisfies (7). As in previous sections, the generalized solutions have many equivalent characterizations.

Theorem 3. *If $f \in D^s$, $g \in H^{s-1}$, and $u \in C(\mathbb{R} : D^s) \cap C^1(\mathbb{R} : H^{s-1})$ with $u(0) = f$ and $u_t(0) = g$, then the following are equivalent:*

(1) *u is the generalized solution with Cauchy data f, g.*

(2) $\hat{u} = \hat{g} \dfrac{\sin(ct|\xi|)}{c|\xi|} + \hat{f} \cos(ct|\xi|)$.

(3) *For any $\varphi \in C_0^\infty(\mathbb{R}^d)$, $\langle u(t, \cdot), \varphi(\cdot)\rangle \in C^2(\mathbb{R}_t)$ and*

$$\frac{d^2}{dt^2}\langle u(t, x), \varphi(x)\rangle = \langle u(t, \cdot), c^2 \Delta\varphi(\cdot)\rangle.$$

(4) *For any $\psi \in C_0^\infty(\mathbb{R}^{1+d})$, $\langle u(t, \cdot), \psi(t, \cdot)\rangle \in C^2(\mathbb{R}_t)$ and*

$$\frac{d^2}{dt^2}\langle u(t, \cdot), \psi(t, \cdot)\rangle = \langle u(t, \cdot), (\psi_{tt} - c^2\Delta\psi)(t, \cdot)\rangle + \langle u_t(t, \cdot), \psi_t(t, \cdot)\rangle.$$

(5) *$(\partial_t^2 - c^2\Delta)u = 0$ in the sense of $\mathscr{D}'(\mathbb{R}^{1+d})$.*

(6) *For any* $\psi \in C_0^\infty(\mathbb{R} \times \mathbb{R}^d)$ *and* $t_1 < t_2$

$$\int_{t_1}^{t_2} \langle u(t), \psi_{tt}(t) - \Delta\psi(t)\rangle \, dt = (\langle u(t), \psi_t(t)\rangle - \langle \psi(t), u_t(t)\rangle)\Big|_{t_1}^{t_2}.$$

PROBLEMS

Let Γ be the dunce cap region $\{(t, x) \in \mathbb{R} \times \mathbb{R}^d : t \geq 0 \text{ and } |x| \leq R - ct\}$. Using the Global Holmgren Theorem, we have shown that any $u \in C^2(\Gamma)$ satisfying the wave equation, $u_{tt} - c^2\Delta u = 0$, and whose Cauchy data vanish on $\Gamma \cap \{t = 0\}$ must vanish in Γ. The same result can be proved by the energy method, which works for problems with nonanalytic coefficients and even for nonlinear problems. This type of calculation is very important and this problem is strongly recommended. Given such a u define

$$e(t) \equiv \int_{|x| \leq R - ct} (\partial_t u(t))^2 + c^2 |\text{grad}_x\, u(t)|^2 \, dx.$$

1. Prove that if $u \in C^2(\Gamma : \mathbb{R})$ satisfies $u_{tt} - c^2\Delta u = 0$, $c \geq 0$, then $e(t)$ is a nonincreasing function of t. If $u = u_t = 0$ at $t = 0$, conclude that u vanishes in Γ. *Hint.* For any $T > 0$, multiply the differential equation by u_t and integrate by parts in $\Gamma \cap \{0 < t < T\}$ motivated by the proof after Theorem 1.

In the next problem you are asked to analyze the wave equation with an elastic term bu, $b \geq 0$, and a viscous friction term au_t, $a \geq 0$. The initial value problem is

$$\partial_{tt}u - \Delta u + au_t + bu = 0 \quad \text{in } \mathbb{R}^{1+d},$$
$$u(0, \cdot) = f(\cdot), \qquad u_t(0, \cdot) = g(\cdot).$$

2. Prove that for $f \in H^s(\mathbb{R}^d)$, $g \in H^{s-1}(\mathbb{R}^d)$, there is a unique $u \in \bigcap_{j=0}^{j=\infty} C^j(\mathbb{R} : H^{s-j}(\mathbb{R}^d))$ solving the initial value problem.
3. Prove that if $a > 0$, then the solution satisfies

$$\lim_{t \to \infty} \|\partial_t^j u(t)\|_{H^{s-j}(\mathbb{R}^d)} = 0.$$

Hint. Express the desired quantity as an integral in ξ space and apply Lebesgue's Dominated Convergence Theorem.
DISCUSSION. This problem shows that the solution is driven to zero by the friction. The cases $j = 0, 1$ show that the classical energy tends to zero. The energy does not decay exponentially fast. The reason is that low frequencies are damped slowly. This is a general principle. Systems whose lowest frequency is strictly positive tend to be driven to zero exponentially fast by dissipative mechanisms (see Problem 5.7.5).

4. Prove that for $f \in C^\infty(\mathbb{R}^d)$, $g \in C^\infty(\mathbb{R}^d)$, there is a unique $u \in C^\infty(\mathbb{R} \times \mathbb{R}^d)$ solving the initial value problem $u_{tt} - c^2\Delta u = 0$, $u(0) = f$, $u_t(0) = g$. *Hint.* Uniqueness follows from Problem 1. For existence choose a locally finite partition of unity, $\{\varphi_i\}$, for $\mathbb{R}^d$ with $\varphi_i \in C_0^\infty(\mathbb{R}^d)$. Let u_i solve $\square u_i = 0$, $u_i(0) = f\varphi_i$, $\partial_t u_i(0) = \varphi_i g$. Show that the series $\sum u_i$ is locally a finite sum, and therefore is a solution of the initial value problem.
DISCUSSION. The same sort of patching argument can be used to prove solvability for $f, g \in \mathscr{D}'(\mathbb{R}^d)$.

5. If $u \in \bigcap C^j(\mathbb{R} : H^{s-j})$ satisfies $\square u = 0$ and $u(0, \cdot) = 0$ on $|x| < R$, then $u = 0$ in

$\{(t, x): |x| < R - c|t|\}$. *Hint*. Consider the solution u_ε to $\square u_\varepsilon = 0$, $u_\varepsilon(0) = j_\varepsilon * u(0)$, $\partial_t u_\varepsilon(0) = j_\varepsilon * u_t(0)$. Be careful to justify the passage to the limit.

6. Use Problem 5 to show that if $f \in H^s$, $g \in H^{s-1}$, and $u \in \bigcap C^j(\mathbb{R} : H^{s-j})$ is the solution to $\square u = 0$, $u(0) = f$, $u_t(0) = g$, then

$$\operatorname{supp} u \subset \{(t, x): (\exists y \in \Xi)(|x - y| \leq ct)\}, \qquad \Xi \equiv \operatorname{supp} f \cup \operatorname{supp} g.$$

§3.8. Fourier Synthesis for the Cauchy–Riemann Operator

In §1.1 we showed that the initial value problem

$$(\partial_t - i\partial_x)u = 0, \qquad u(0, \cdot) = f(\cdot), \tag{1}$$

is badly set, in the sense that there exists a solution only if f is real analytic, in which case the solution is the holomorphic extension of f. The Fourier transform gives us complementary information and gives precise assertions concerning general ill-posed initial value problems. The first goal is to show that the set $f \in \mathscr{S}(\mathbb{R}^d)$ for which a solution exists is "thin". After that, following Hadamard, we prove a strong version of discontinuous dependence on initial data.

Suppose that for some s, perhaps very negative, $u \in C(\mathbb{R} : H^s(\mathbb{R}^d))$ satisfies (1), then in the sense of $\mathscr{D}'(\mathbb{R}^{1+d})$, $u_t = iu_x$, and the right-hand side is continuous with values in $H^{s-1}(\mathbb{R}^d)$. It follows (Problem 1) that $u \in C^1(\mathbb{R} : H^{s-1}(\mathbb{R}^d))$. Taking the Fourier transform yields $\partial_t \hat{u} = -\xi\hat{u}$. View this as an identity in $\mathscr{D}'(\mathbb{R}_t \times \mathbb{R}^d)$. It follows that in $\mathscr{D}'(\mathbb{R}_t \times \mathbb{R}^d_\xi)$, $\partial_t(e^{t\xi}\hat{u}) = 0$, so $\hat{u}(t) = e^{-t\xi}\hat{f}$. The key observation is that $\hat{f}$ is multiplied by a factor which is exponentially large as $\xi \to (\operatorname{sgn} t)\infty$. Since $f \in \mathscr{S}$, polynomial growth is tolerable, being compensated by the rapid decrease of $\hat{f}$. If $\hat{f}$ is not exponentially small, for example, $\hat{f} = e^{-\langle\xi\rangle^{1/2}}$, then $e^{-t\xi}\mathscr{F}f$ is not the Fourier transform of an element of $\mathscr{S}'(\mathbb{R}^d)$.

In order that $e^{-T\xi}\mathscr{F}f$ lie in H^s it is necessary and sufficient that

$$\int \langle\xi\rangle^{2s}(e^{-2T\xi} + 1)|\hat{f}(\xi)|^2 \, d\xi < \infty. \tag{2}$$

Proposition 1. *If $f \in \mathscr{S}(\mathbb{R}^d)$ and there is an $s \in \mathbb{R}$ and a solution $u \in C([0, T]:$ $H^s(\mathbb{R}^d_x))$ of* (1), *then* (2) *holds. Conversely, if f satisfies* (2), *then $\hat{u}(t) = e^{-t\xi}\mathscr{F}f$ defines a $C([0, T] : H^s)$ solution.*

Proof. The first statement is proved above and the converse is straightforward. □

If one views u as a holomorphic function of $x + it$, the above result identifies the set of H^s functions of x which are boundary values of holomorphic functions in the strip $\text{Im } z < T$, and such that the map $t \mapsto u(\cdot + it)$ is continuous on $[0, T]$ with values in H^s. This is a result of Paley–Weiner type.

EXAMPLES. 1. If $\hat{f}$ has compact support K, then $u \in C^\infty(\mathbb{R}_t : \mathscr{S}(\mathbb{R}_x))$ and $\mathscr{F}u(y)$ is a smooth function with support in K for all y. The set of such data is dense in $\mathscr{S}(\mathbb{R}_x)$. Nevertheless, the set is thin as shown in Theorem 2.

2. If $\hat{f} = e^{-\langle\xi\rangle^{1/2}}$, then (2) holds for no $s \in \mathbb{R}$. The point is that (2) is rarely satisfied for $f \in \mathscr{S}$. In fact, using the Baire Category Theorem one can show that

$$\{f \in \mathscr{S}(\mathbb{R}^d) \colon (\exists s \in \mathbb{R} \text{ and } T > 0) \text{ such that (2) holds}\}$$

is of first category in the complete metric space $\mathscr{S}$ (Problem 2).

Theorem 2. *The set of* $f \in \mathscr{S}(\mathbb{R}^d)$, *such that there is a* $T > 0$ *and an* $s \in \mathbb{R}$ *so that* (1) *has a solution* $u \in C([0, T] : H^s)$ *or* $C([-T, 0] : H^s)$, *is a set of first category in* $\mathscr{S}$.

Next turn to the construction of examples showing strongly discontinuous dependence on initial data. The Fourier transform construction yields solutions as linear combinations of the exponential solutions $e^{-t\xi}e^{i\xi x}$, $\xi \in \mathbb{R}$. The idea is that as $\xi \to -\infty$ the initial values of $e^{-t\xi}e^{ix\xi}$ are bounded but the solutions explode exponentially for any $t > 0$. For $k \in \mathbb{N}$ fixed, the value of the solutions, $\langle\xi\rangle^{-k}e^{-t\xi}e^{ix\xi} \equiv u_\xi$ at $(t, 0)$, $t > 0$, grow exponentially as ξ tends to $-\infty$, but their initial data converge to zero in the sense that the derivatives of u_ξ of order less than or equal to $k - 1$ converge uniformly to zero in $\mathbb{R}$. This is Hadamard's construction.

A skeptic might think that the source of this problem is that the data do not converge to zero as $|x| \to \infty$. By taking a superposition over a range of ξ, one vitiates this criticism and proves a much stronger discontinuity theorem.

Theorem 3. *There is a family* $u_\varepsilon \in C^\infty(\mathbb{R} : \mathscr{S}(\mathbb{R}^d))$ *of solutions to* $(\partial_t - i\partial_x)u_\varepsilon = 0$ *and a* $\varphi \in \mathscr{S}(\mathbb{R}^d)$, *so that as* $\varepsilon \to 0$,

$$\mathscr{S}\text{-lim } u_\varepsilon(0) = 0, \tag{3}$$

and

$$\text{for all } t \neq 0, \qquad \lim \langle u_\varepsilon, \varphi\rangle = \infty. \tag{4}$$

The explosive behavior (4) implies that for any $s \in \mathbb{R}$, no matter how negative, $\|u_\varepsilon(t)\|_{H^s} \to \infty$ as $\varepsilon \to 0$.

PROOF. Define $v \in C^\infty(\mathbb{R} : \mathscr{S}(\mathbb{R}^d))$ by $\mathscr{F}v_\varepsilon(t) \equiv e^{-(\varepsilon\xi)^2}e^{-t\xi}e^{-\langle\xi\rangle^{1/2}}$. Then $(\partial_t - i\partial_x)v_\varepsilon = 0$, and as $\varepsilon \to 0$, v_ε converges formally to the misbehaved solution with data $\mathscr{F}^* \exp(-\langle\xi\rangle^{1/2})$. Define φ by $\mathscr{F}\varphi \equiv \exp(-\langle\xi\rangle^{1/4})$. Then

$$\langle v_\varepsilon(t), \varphi\rangle = \int e^{-t\xi}e^{-(\varepsilon\xi)^2}e^{-\langle\xi\rangle^{3/4}}\, d\xi.$$

The integrand is strictly positive and integrable. For $t > 0$, the integral is larger than the integral over the interval of length one with center at $-1/\varepsilon$. In this interval the integrand is bounded below by $ce^{t/\varepsilon}$, $c > 0$, as ε tends to zero. Thus

$$\langle v_\varepsilon(t), \varphi \rangle \geq ce^{t/\varepsilon}.$$

For $t < 0$, an interval about $1/\varepsilon$ yields the same lower bound. Let $u_\varepsilon \equiv \varepsilon v_\varepsilon$ to complete the proof. □

PROBLEMS

1. Suppose that $u \in \mathscr{D}'(\mathbb{R}_t \times \mathbb{R}_x^d) \cap C(\mathbb{R} : H^s(\mathbb{R}^d))$ and the distribution derivative $\partial u/\partial t$ lies in $C(\mathbb{R} : H^s(\mathbb{R}^d))$. Prove that $u \in C^1(\mathbb{R} : H^s(\mathbb{R}^d))$.

2. Prove that for any $s \in \mathbb{R}$ and $t \in \mathbb{R}\backslash 0$

$$\{f \in \mathscr{S}(\mathbb{R}^d) : \int \langle \xi \rangle^{-2s}(e^{-2t\xi} + 1)|\hat{f}(\xi)|^2 \, d\xi < \infty\}$$

is of first category in the sense of Baire. *Hint*. Use the uniform boundedness principle as sketched in the next paragraph.

Suppose that E and F are complete countably normed vector spaces ($\equiv$ Fréchet spaces). A subset is called bounded if and only if it is bounded with respect to each of the countable number of defining seminorms. This is equivalent to requiring that the set be bounded in the natural metrics defining the topology. A sequence of continuous linear maps $A_n: E \to F$ is called uniformly bounded if and only if for any bounded subset $B \subset E$, the set $\bigcup A_n(B)$ is bounded in F. The Uniform Boundedness Theorem asserts that if A_n is not uniformly bounded, then the set of $x \in E$ such that $\{A_n(x)\}_{n=1}^{\infty}$ is bounded in F is a set of first category. Apply this result to the family of maps $f \mapsto \langle \xi \rangle^{-s}(e^{-t\xi} + 1)\hat{f}(\xi)\chi_{[-n,n]}(\xi)$ from $\mathscr{S}(\mathbb{R}^d)$ to $L^2(\mathbb{R}^d)$.

DISCUSSION. Call the set in Problem 2, $B_{s,t}$. Then B is increasing in s and decreases as t moves away from 0. Thus

$$\bigcup_{s \in \mathbb{R}, t \neq 0} B_{s,t} = \bigcup_{1 \leq n < \infty} (B_{-n,1/n} \cup B_{-n,-1/n})$$

is a countable union of sets of first category, hence of first category. This proves the conclusion of Theorem 2.

§3.9. The Sideways Heat Equation and Null Solutions

Sometimes initial value problems arise with a time parameter which is not the physical time. Consider, for example, the following inverse problem for the heat equation. An observer at the origin $x = 0$ in $\mathbb{R}_x^1$ observes the temperature $u(t, 0) = f(t)$ and the heat flux $u_x(t, 0) = g(t)$ of a solution to the heat equation $u_t = u_{xx}$ in $x \geq 0$. The problem is to recover the full temperature field $u(t, x)$. The boundary value problem is then

$$u_{xx} = u_t, \tag{1}$$

$$u(t, 0) = f, \qquad u_x(t, 0) = g. \tag{2}$$

This is a noncharacteristic Cauchy problem with "time" variable x and "space" variable t. Since the initial line $x = 0$ is noncharacteristic the observations determine all the derivatives of u at $x = 0$, so that the observer has all the information which could possibly be obtained at $x = 0$.

The exponential solutions are given by the roots of $-\xi^2 = i\tau$, that is, $\xi = \pm(\tau/i)^{1/2}$. The solutions $\exp(i\tau t \pm (i\tau)^{1/2}x)$ for τ real are bounded on $x = 0$ and grow exponentially like $\exp(\pm|\tau|^{1/2}x)$ suggesting that the initial value problem (1), (2) is not well-posed since the rate of growth increases without bound as $\tau \to \infty$.

To analyze further, take the Fourier transform in the "space variable" t to obtain

$$\hat{u}_{xx} = i\tau\hat{u}(\tau, x), \tag{3}$$

where τ denotes the transform variable associated to t. Equation (3) has the solution

$$\hat{u}(\tau, x) = \frac{\sin\sqrt{i\tau}x}{\sqrt{i\tau}}\hat{g}(\tau) + (\cos\sqrt{i\tau}x)\hat{f}(\tau). \tag{4}$$

Here z^α for $\alpha \in]0, 1[$ is defined for $z \in \mathbb{C}\backslash]-\infty, 0]$ as the branch with $1^\alpha = 1$.

Note that the *amplification factors*, $\sin\sqrt{i\tau}x/\sqrt{i\tau}$ and $\cos\sqrt{i\tau}x$, grow like $\exp(|\tau|^{1/2}x)$ as $\tau \to \pm\infty$. Thus, for typical $f, g \in \mathscr{S}$, the products in (4) will not belong to $\mathscr{S}'(\mathbb{R}^d)$ which yields nonexistence and discontinuous dependence results as in the last section.

Theorem 1

(a) *The set of $f, g \in \mathscr{S} \times \mathscr{S}$, with the property that for some $\underline{x} > 0$ and $s \in \mathbb{R}$ there is a $u \in C([0, \underline{x}] : H^s(\mathbb{R}_t^d))$ satisfying (1) for $0 < x < \underline{x}$ and (2), is a set of first category in $\mathscr{S} \times \mathscr{S}$.*

(b) *For any $\underline{x} > 0$, there exists a sequence of solutions $u_n \in C^\infty(\mathbb{R}_x^1 : \mathscr{S}(\mathbb{R}_t))$ to (1) and $\varphi \in \mathscr{S}(\mathbb{R}_t)$ such that $u_n(t, 0)$, $\partial_x u_n(t, 0)$ converge to zero in $\mathscr{S}(\mathbb{R}^d)$ and*

$$\lim \int u_n(t, \underline{x})\varphi(t)\, dt = \infty.$$

Part (b) shows that in spite of the unique determination of u (Holmgren) the determination is so discontinuous as to be practically useless.

For the Cauchy–Riemann operator, the amplification factors grow exponentially in the Fourier variables, so only Cauchy data whose transform decay exponentially have a chance for existence. These data are real analytic and therefore possess unique continuation properties. In the present situation data whose transform decays more rapidly than $e^{-\sqrt{|\tau|}}$ suffice. For example,

$$\hat{u}(\tau, x) = \sqrt{2\pi}e^{x\sqrt{i\tau}}e^{-(i\tau)^\alpha}, \qquad \alpha \in]\tfrac{1}{2}, 1[$$

yields the solution

$$u(t, x) = \int_{-\infty}^{\infty} e^{-(i\tau)^\alpha}e^{x\sqrt{i\tau}}e^{i\tau t}\, d\tau. \tag{5}$$

As $|\tau| \to \infty$ the first factor in the integrand decays like $\exp(-|\tau|^\alpha)$ and the second grows at most like $\exp(|x|\,|\tau|^{1/2})$, so the integral is absolutely convergent for all t, x since $\alpha > 1/2$. As $\mathscr{F}u(t, 0) = (2\pi)^{1/2}\exp(-(i\tau)^\alpha)$, u is not identically equal to zero.

Let Γ_a denote the contour $\mathrm{Re}(z) = a \geq 0$ oriented in the direction of the increasing imaginary part. Then

$$u(t, x) = \int_{\Gamma_0} e^{x(z)^{1/2}} e^{-z^\alpha} e^{zt}\, dz. \tag{6}$$

The integrand is holomorphic in $\mathrm{Re}(z) > 0$ and continuous in $\mathrm{Re}(z) \geq 0$. Starting from (6) we prove that u vanishes in $t < 0$. To do this, shift the contour toward the right.

Denote by $F(t, x, z)$ the integrand in (6). Since $e^{-z^\alpha} \leq ce^{-|z|^\alpha}$ for $\mathrm{Re}\, z \geq 0$ and

$$M_R \equiv \sup_{|x| \leq R,\, 0 \leq \mathrm{Re}(z) < \infty} |e^{x(z)^{1/2}} e^{-z^\alpha/2}| < \infty.$$

it follows that $F = 0(\exp(-|\mathrm{Im}\, z|^\alpha/2)$ uniformly for $0 \leq \mathrm{Re}(z) < \infty$. This suffices to justify a contour shift using Cauchy's Theorem to show that

$$u(t, x) = \int_{\Gamma_a} e^{x(z)^{1/2}} e^{-z^\alpha} e^{zt}\, dz. \tag{7}$$

For any $a > 0$, $R > 0$, the derivatives of the integrand in (7) satisfy

$$|\partial_t^\alpha \partial_x^\beta F| \leq c(R, a, \alpha, \beta)(1 + |z|^{N(\alpha,\beta)})e^{-|z|^\alpha/2}, \qquad \text{for all} \quad |t, x| \leq R.$$

Thus differentiation with respect to t, x under the integral sign in (7) is justified and proves $u \in C^\infty(\mathbb{R} \times \mathbb{R})$.

For $|x| \leq R$ estimate

$$|u(t, x)| \leq e^{at} M_R \int_{\Gamma_a} |e^{-z^\alpha/2}|\,|dz|.$$

The integral is largest for $a = 0$, so $|u| \leq c(R)e^{at}$. If $t < 0$, pass to the limit $a \to \infty$ to show that $u \equiv 0$ for $t < 0$. This proves the following theorem.

Theorem 2. *There is a nontrivial* $u \in C^\infty(\mathbb{R}_t \times \mathbb{R}_x)$ *satisfying the heat equation,* $u_t = u_{xx}$, *and vanishing identically for* $t < 0$.

Such solutions are called *null solutions*. This result is typical of characteristic initial value problems. In Problem 3 you are asked to construct similar null solutions for Schrödinger's equation. The reason that such solutions do not violate the uniqueness theorems of §3.6 is that the solutions are not continuous functions of time with values in $H^s(\mathbb{R}^d)$. They grow exponentially fast as $|x| \to \infty$ and are not tempered distributions in x. The interested reader is encouraged to study the asymptotics of the solution (5) when x grows with t fixed.

Problems

In the first problem you are asked to prove a result about the time $T \geq 0$, when a null solution of the heat equation, as in Theorem 2, ignites. Suppose that $u \in C^\infty(\mathbb{R}^1_t \times \mathbb{R}^1_x)$ satisfies

$$u_t = u_{xx}, \qquad u = 0 \qquad \text{for} \quad t < 0.$$

Define the ignition time T by

$$T \equiv \sup\{s: u = 0 \text{ for } t < s\}.$$

1. Prove that all points (T, x), $x \in \mathbb{R}$, lie in the support of u. *Hint*. Use the Global Holmgren Theorem.
   Discussion. This shows that the solutions ignite at all points simultaneously.
   It is a nontrivial theorem of Widder [W] that there are no nonnegative solutions vanishing for $t < 0$.
2. For any $\alpha \in]\frac{1}{2}, 1[$, show that $T = 0$ for the solution (5).
3. Construct null solutions for Schrödinger's equation. *Hint*. Consider the sideways problem.

§3.10. The Hadamard–Petrowsky Dichotomy

The previous sections present examples of a dichotomy between initial value problems with a satisfactory existence theory and those without. The idea is simple. Consider a constant coefficient operator,

$$P(D_t, D_x) = \sum_{j=1}^{m} A_j(D_x) D_t^j, \tag{1}$$

of degree m with respect to t. The initial value problem we consider is

$$Pu = 0 \qquad \text{in} \quad t \geq 0, \tag{2}$$

$$D_t^j u|_{t=0} = f_j \in \mathcal{S}(\mathbb{R}^d), \qquad 0 \leq j \leq m - 1. \tag{3}$$

In all the examples we have discussed so far, A_m was constant, but there are examples in mathematical physics with $A_m(D)$ nonconstant. For example, the linearized BBM (Benjamin–Bona–Mahoney) operator

$$P \equiv (1 - \partial_x^2)\partial_t + \partial_x^3 \tag{4}$$

arises in a model of long waves.

We assume throughout this section that

$$A_m(\xi) \neq 0 \qquad \text{for all} \quad \xi \in \mathbb{R}^d. \tag{5}$$

The Seidenberg–Tarski Theorem (see Hormander II, Appendix 2, Ex. A.2.7) shows that there is a $c > 0$ and $\beta \in \mathbb{Q}$ such that

$$\min\{|A_m(\xi)|: \xi \in \mathbb{R}^d \text{ and } |\xi| = \rho\} = c\rho^\beta(1 + o(1)) \qquad \text{as} \quad \rho \to \infty. \tag{6}$$

In particular, there is a $c > 0$ and $N \in \mathbb{N}$ such that

$$|A_m(\xi)| \geq c\langle\xi\rangle^{-N} \qquad \text{for all} \quad \xi \in \mathbb{R}^d. \tag{7}$$

Thus $f \mapsto A_m(D)f$ is a 1–1 onto map of $\mathcal{S}$ to itself (Problem 2.1.3). In examples, (7) is usually obvious. For example, the BBM operator has $A_m(\xi) = 1 + |\xi|^2$.

In an attempt to construct a solution, take the Fourier transform in x to find

$$P(D_t, \xi)\hat{u}(t, \xi) = 0, \tag{8}$$

$$D_t^j \hat{u}(0, \xi) = \hat{f}_j(\xi), \qquad j \leq m - 1. \tag{9}$$

For each ξ this is a Cauchy problem for an mth order ordinary differential equation in time, thanks to (5). Solving determines $\hat{u}(t, \xi) \in C^\infty(\mathbb{R}_t \times \mathbb{R}_\xi^d)$.

To perform an inverse Fourier transform we need to know that $\hat{u}$ does not grow too fast as $|\xi| \to \infty$. As the $\mathcal{F}f_j$ are rapidly decreasing as $|\xi| \to \infty$, we must look at how large solutions of $P(D_t, \xi)v = 0$ can be compared to the size of their Cauchy data. The key is that the general solution is a linear combination of exponential solutions $e^{it\tau}$ where $P(\tau, \xi) = 0$ (if there are multiple roots, polynomials in t appear as factors). Now, for such an exponential solution and $t > 0$, the solution is of magnitude $e^{-(\operatorname{Im}\tau)t}$ times the magnitude at time zero. As the $\mathcal{F}f$ decay faster than $\langle\xi\rangle^{-n}$ for all n we can tolerate polynomial growth in ξ for t fixed. This leads us to the following condition.

Definition 1. The differential operator satisfies the *Hadamard–Petrowsky condition for forward evolution* if and only if there exists c, n such that

$$\xi \in \mathbb{R}^d, \quad \tau \in \mathbb{C}, \qquad P(\tau, \xi) = 0 \quad \Rightarrow \quad e^{-(\operatorname{Im}\tau)} \leq c\langle\xi\rangle^n. \tag{10}$$

We call this the H–P *condition* for brevity. Taking logarithms shows that (10) is equivalent to

$$\xi \in \mathbb{R}^d, \quad \tau \in \mathbb{C}, \qquad P(\tau, \xi) = 0 \quad \Rightarrow \quad -\operatorname{Im}(\tau) \leq c' \ln(2 + |\xi|). \tag{10$'$}$$

That is, $-\operatorname{Im}\tau$ can grow at most logarithmically in $|\xi|$. As the τ are roots of algebraic equations only power law growth is possible. Precisely, the Seidenberg–Tarski Theorem shows that

$$\max\{-\operatorname{Im}(\tau) : P(\tau, \xi) = 0 \text{ for some } \xi \in \mathbb{R}^d \text{ with } |\xi| = \rho\} = c\rho^\alpha(1 + o(1)) \tag{11}$$

as $\rho \to \infty$. Thus the only way that (10)$'$ can be satisfied is if $\alpha \leq 0$, in which case $-\operatorname{Im}(\tau)$ is bounded above.

The sufficiency of the logarithmic bound for solvability was proved by Petrowsky. That at most logarithmic growth implies boundedness was proved by Garding [Gard] in the case that $t = 0$ is noncharacteristic. In that case operators satisfying the Hadamard–Petrowsky condition are called *hyperbolic*. The utility of the Seidenberg–Tarski Theorem in this and other contexts in the theory of partial differential equations was discovered by Hormander.

Theorem 2. *Suppose that $A_m(\xi) \neq 0$ for all $\xi \in \mathbb{R}^d$ and that the* H–P *condition for forward evolution is satisfied. Then, for any $f_j \in \mathscr{S}(\mathbb{R}^d)$, $0 \leq j \leq m-1$, there is a unique solution $u \in C^\infty([0, \infty[: \mathscr{S}(\mathbb{R}^d))$ of* (2), (3). *The map $(f_0, f_1, \ldots, f_{m-1}) \to u$ is continuous from $\mathscr{S}(\mathbb{R}^d)^m$ to $C^\infty([0, \infty[: \mathscr{S}(\mathbb{R}^d))$.*

The natural backward and forward–backward versions are true. For evolution into the past, $t < 0$, the H–P condition is $\operatorname{Im}(\tau) \leq c'$. For evolution forward and backward in time, the condition is $|\operatorname{Im} \tau| \leq c''$.

Proof of Theorem 2. If $u \in C^\infty([0, \infty[: \mathscr{S}(\mathbb{R}^d))$ is a solution, we have shown that $\hat{u}(t, \xi) \in C^\infty([0, \infty[\times \mathbb{R}^d)$ must satisfy (8), (9). Since $\mathscr{F}u$ is determined, this proves uniqueness.

For $0 \leq j \leq m-1$, define $M_j(t, \xi) \in C^\infty(\mathbb{R} \times \mathbb{R}^d)$ to be the solution of

$$P(D_t, \xi) M_j = 0, \tag{12}$$

$$D_t^k M_j(0, \xi) = \begin{cases} 0 & \text{if} \quad k \neq j, \\ 1 & \text{if} \quad k = j. \end{cases} \tag{13}$$

Then we have

$$\hat{u} = \sum M_j(t, \xi) \mathscr{F} f_j(\xi). \tag{14}$$

If we can show that the right-hand side of (14) belongs to $C^\infty([0, \infty[: \mathscr{S})$, then the inverse Fourier transform yields a solution $u \in C^\infty([0, \infty[: \mathscr{S})$. Thus, it suffices to show that for any $\alpha \in \mathbb{N}^{1+d}$ and $T > 0$ there is a $c = c(\alpha, T)$ and $N = N(\alpha, T)$ such that

$$|\partial_{t,\xi}^\alpha M_j| \leq c \langle \xi \rangle^N \tag{14$_j$}$$

on $[0, T] \times \mathbb{R}_\xi^d$.

The first step is to observe that all the M_j can be expressed in terms of M_{m-1}. Toward that end, note that for $j < m-1$ and $t = 0$,

$$\partial_t^k \partial_t M_{j+1} = \partial_t^{k+1} M_{j+1} = \begin{cases} 1 & \text{if } k = j, \\ 0 & \text{if } k \neq j \text{ and } k < m-1, \\ -A_m^{-1}(\xi) A_{j+1}(\xi) & \text{if } k = m-1. \end{cases}$$

The last identity follows from (12) upon solving for the D_t^m term and using (13). Thus

$$M_j = \partial_t M_{j+1} + A_m(\xi)^{-1} A_{j+1}(\xi) M_{m-1}.$$

$m - 1 - j$ applications of this yields an expression for M_j in terms of M_{m-1} which shows that it suffices to prove $(14)_j$ for $j = m-1$.

For M_{m-1} we use an explicit solution formula.

Lemma 3. *The solution of the constant coefficient initial value problem*

$$0 = P(D_t)u \equiv (a_m D_t^m + a_{m-1} D_t^{m-1} + \cdots + a_0)u,$$

$$D_t^j u(0) = \begin{cases} 0 & \textit{for} \quad 0 \leq j \leq m-2, \\ 1 & \textit{for} \quad j = m-1, \end{cases}$$

is given by

$$u = \frac{a_m}{2\pi i} \oint_\Gamma \frac{e^{i\tau t}}{P(\tau)}\, d\tau,$$

where Γ *is any rectifiable arc in* $\mathbb{C}$ *which winds once about each root of* $P(\tau) = 0$.

PROOF. Differentiating under the integral defining u yields

$$P(D_t)u = \frac{a_m}{2\pi i} \oint_\Gamma \frac{P(\tau)e^{i\tau t}}{P(\tau)}\, d\tau = 0,$$

by Cauchy's Theorem.

The same theorem shows that if all the roots of $P(\tau)$ lie in the disc $|\tau| < R$, then

$$u \equiv \frac{a_m}{2\pi i} \oint_{|\tau|=R} \frac{e^{i\tau t}}{P(\tau)}\, d\tau.$$

For $j \le m - 1$,

$$D_t^j u(0) = \frac{a_m}{2\pi i} \oint_{|\tau|=R} \tau^j/P(\tau)\, d\tau.$$

For $j \le m - 2$, the integrand is $O(R^{-2})$, so the integral is $O(R^{-1})$. Letting $R \to \infty$ yields $D_t^j u(0) = 0$. On the other hand, as $R \to \infty$,

$$D_t^{m-1} u(0) = \frac{a_m}{2\pi i} \oint_{|\tau|=R} \frac{\tau^{m-1}}{a_m \tau^m (1 + O(1/R))}\, d\tau.$$

The integrand is equal to $(a_m\tau)^{-1} + O(R^{-2})$. Thus, as $R \to \infty$, the right-hand side converges to 1. □

The lemma yields the following explicit formula

$$M_{m-1}(t, \xi) = \frac{A_m(\xi)}{2\pi i} \oint_\Gamma \frac{e^{i\tau t}}{P(\tau, \xi)}\, d\tau,$$

where $\Gamma = \Gamma(\xi)$ is a contour in $\mathbb{C}$ enclosing all the roots, τ, of $P(\tau, \xi) = 0$. A good choice of Γ is needed. We take Γ to be equal to the boundary of Ω_ξ where Ω_ξ is the union of the discs of radius 1 with centers at the roots $\tau_j = \tau_j(\xi)$ of $P(\tau, \xi) = 0$.

The length of Γ is at most $2\pi m$ since there are at most m circles. The H–P condition shows that $-\operatorname{Im} \tau \le c$ on Γ, which implies that for $\tau \in \Gamma(\xi)$, $|e^{\tau t}| \le e^{ct}$. Finally, on Γ,

$$|P(\tau, \xi)| = |A_m(\xi)| \prod |\tau - \tau_j| \ge |A_m(\xi)| \ge c\langle\xi\rangle^{-N},$$

which bounds the denominator of the integrand away from zero. It follows that

$$|M_{m-1}(t, \xi)| \le c' e^{ct} \langle\xi\rangle^N,$$

which is the $\alpha = 0$ case of $(14)_{m-1}$.

Differentiating under the integral sign yields

$$D^{\alpha}_{t,\xi} M = \oint_{\Gamma} \frac{Q_{\alpha}(\tau, \xi) e^{i\tau t}}{P(\tau, \xi)^{|\alpha|+1}} \, d\tau$$

with a polynomial Q. Since the coefficients of P are polynomial in ξ and $|A_m| \geq c\langle\xi\rangle^{-N}$, the roots of P are $O(\langle\xi\rangle^{N'})$ for some N'. Thus, for $\tau \in \Gamma$, $|\tau| \leq c\langle\xi\rangle^{N'}$. This together with our previous estimates yields $(14)_{m-1}$. □

If the H–P condition is violated, then (11) shows that suitable exponential solutions $e^{i\tau t}e^{ix\xi}$, $\xi \in \mathbb{R}^d$, $|\xi| \to \infty$, explode like $e^{t|\xi|^a}$ for some $a > 0$. This yields discontinuous dependence and general nonexistence theorems in the style of Theorems 3.8.2, 3.8.3, and 3.9.1.

Theorem 4. *Suppose that* $A_m(\xi) \neq 0$ *for all* $\xi \in \mathbb{R}^d$ *and that the* H–P *condition for forward evolution is* not *satisfied. Then:*

(a) Generic Nonexistence. *The set of* $(f_0, \ldots, f_{m-1}) \in \mathcal{S}(\mathbb{R}^d)^m$, *such that there is a* $T > 0$, $s \in \mathbb{R}$, *and* $u \in C^m([0, T] : H^s(\mathbb{R}^d))$ *satisfying* (2), (3), *is a set of first category in* $\mathcal{S}(\mathbb{R}^d)^m$.

(b) Discontinuous Dependence. *There is a sequence of solutions* $u_n \in C^{\infty}(\mathbb{R} : \mathcal{S}(\mathbb{R}^d))$ *to* $Pu_n = 0$ *and a* $\varphi(t) \in C^{\infty}([0, \infty[: \mathcal{S})$ *such that the Cauchy data,* $\partial_t^j u_n(0, \cdot)$, $0 \leq j \leq m-1$, *converge to zero in* $\mathcal{S}$, *and as* $n \to \infty$, $\lim \langle u_n(t), \varphi(t)\rangle = \infty$ *uniformly on compact time intervals in* $]0, \infty[$.

Proof. (a) If $u \in C^m([0, T] : H^s(\mathbb{R}^d))$ is a solution with data f, then $\mathcal{F}u$ is given by (14). Thanks to (11) we have

$$\max_{|\xi|=\rho} |M_j(T, \xi)| \geq c' e^{c\rho^a T} \qquad \text{with} \quad a > 0. \tag{15}$$

Then $u(T) \in H^s$ yields

$$\int \left| \sum M_j(T, \xi) \mathcal{F} f_j(\xi) \right|^2 \langle\xi\rangle^{2s} \, d\xi < \infty.$$

As in Problem 3.8.2, (15) implies that the set of such f_j is of first category in $\mathcal{S}$.

(b) The Seidenberg–Tarski Lemma implies that $-\mathrm{Im}\, \tau \leq c\langle\xi\rangle^b$ for some $b \in \mathbb{Q}$. Define u_n by

$$\hat{u}_n = \varepsilon e^{-\langle\varepsilon\xi\rangle^c} e^{-\langle\xi\rangle^{a/3}} M_{m-1}(t, \xi), \qquad \varepsilon \equiv 1/n,$$

with $c > b$ and a as in (15). Define φ by

$$\hat{\varphi} = e^{-\langle\xi\rangle^{a/3}} \overline{M_{m-1}}(t, \xi)/(1 + |M_{m-1}(t, \xi)|^2)^{1/2}.$$

To show that $\langle u(t), \varphi(t)\rangle$ explodes compute in ξ variables, noting that the integrand in $\langle \hat{u}(t), \hat{\varphi}(t)\rangle$ is nonnegative. Equation (11) shows that the integrand is bounded below by $\varepsilon c' \exp(ct|\xi|^a) \exp(-|\xi|^{-2a/3})$ at points $\xi(\varepsilon)$ on the spheres $|\xi| = 1/\varepsilon$. Using the bound (11) and the formula for $\partial_\xi M_{m-1}$ one estimates,

$$|\partial_\xi M(t, \xi)| \le c''\langle\xi\rangle^k \exp(ct|\xi|^a).$$

It follows that the lower bound for the integrand holds, with new constants, on a ball $B_r(\xi(\varepsilon))$ with radius r decreasing as a negative power of $|\xi|$. This bounds the integral below by a multiple of $\varepsilon^{k'} \exp(ct/\varepsilon^{a/3})$. □

The criterion of H–P is easily and broadly applicable. It is one of the main recipes of the theory of partial differential equations.

Summary. To check if the initial value problem for $P(D_t, D_x)$ is well-posed in $t \ge 0$, find the roots of $P(\tau, \xi) = 0$ for $\xi \in \mathbb{R}^d$. The good problems are those for which $-\operatorname{Im}\tau$ is bounded from above.

EXAMPLES. 1. $P = \partial_t + \sum a_j \partial_j + b$. Then $P(\tau, \xi) = i\tau + i\sum a_j \xi_j + b$. The root τ is given by $\tau = -\sum a_j \xi_j - b/i$. If at least one of the a, say a_j, is not real, then by choosing $\xi_k = 0$ for $k \ne j$, the imaginary part of τ can be forced to be arbitrarily large, positive, and negative. Thus the H–P condition is violated for both forward and backward evolution. On the other hand, if the a_j are real then $|\operatorname{Im}\tau| \le |b|$, and the condition is satisfied in both directions. In this case, the initial value problem is explicitly solvable by integrating along the integral curves of the vector field $\partial_t + \sum a_j \partial_j$ as in §1.1. The same method works when the coefficients a_j are smooth real-valued functions of (t, x), and is called the *method of characteristics*.

2, 3, 4. The heat equation, with $i\tau = -|\xi|^2$, satisfies H–P for forward evolution and not backward evolution, while the wave and Schrödinger equations satisfy H–P for forward and backward evolution. For the wave equation, $\tau = \pm|\xi|$, and for the Schrödinger equation, $\tau = -|\xi|^2$. In both cases $\operatorname{Im}\tau = 0$. This is typical of conservative equations.

5, 6. The Laplace equation $u_{tt} + u_{xx} = 0$ yields $\tau = \pm i|\xi|$ with imaginary parts large in both directions, so violates H–P both forward and backward. In fact, if $d \ge 2$, no elliptic operator can satisfy the H–P condition (Problem 4). On the other hand, for $d = 1$, $P(D_t)$ is elliptic and satisfies the H–P condition (there are no $\xi \ne 0$ so the condition is automatic).

7. For any $m \ge 1$, ∂_t^m satisfies H–P forward and backward since roots are $\tau = 0$. For $m \ge 2$, this example is unstable, being destroyed by lower-order perturbations. For example, $\partial_t^2 - \partial_x$ is the sideways heat equation of §3.9. The roots are $\tau = \pm(i\xi)^{1/2}$ and the H–P condition is violated in both directions.

The example ∂_t^m also illustrates a phenomenon of *weak well-posedness* or *loss of derivatives*. To estimate derivatives of a solution up to order k requires more than k derivatives at $t = 0$. To see this note that the solution of the initial value problem is given by

$$u = \sum_0^{m-1} (t^j/j!) f_j(x).$$

Thus

$$\|u(t)\|_{H^s} \le c(t) \sum \|f_j\|_{H^s} = c(t) \sum \|\partial_t^j u(0)\|_{H^s}.$$

To show that $u(t) \in H^s$, one needs $\partial_t^j u(0) \in H^s$ for $j \le m-1$. One needs $s + m - 1$ derivatives at $t = 0$ to guarantee s derivatives at time t. The estimate reflects a loss of $m - 1$ derivatives. In analogy with $\Box u = 0$, set

$$e_s(t) \equiv \sum_{j \le m-1} \|\partial_t^j u\|_{H^{s-j}}^2.$$

Then $e_s(t) \le c(t) e_{s+m-1}(0)$, and no better. Here the loss of $m - 1$ derivatives is clear. The initial value problem is said to be weakly well-posed. Such weak well-posedness is often destroyed by small perturbations. For example, if $m \ge 2$, $\partial_t^m + \varepsilon\partial_x$ does not satisfy the H–P condition (Problem 1). Similarly, variable coefficient problems whose "frozen" problems are weakly well-posed often do not inherit the well-posedness. Ditto for nonlinear problems whose linearization have frozen problems only weakly well-posed. For this reason, it is important to identify, among the H–P good problems, those which are more stably well-posed. The heat, Schrödinger, and wave equations are examples.

A variety of ill-posed initial value problems appears in the descriptions of *instabilities* in physical systems. The ill-posedness is then desirable and is used to study the modes of explosion. Sometimes nonlinear problems with well set initial value problems have ill-posed linearizations. The linear theory then predicts the manner in which solutions grow until the linearization hypothesis is no longer appropriate. The simplest example is the ordinary differential equation of Van der Pol which models self-excited periodic oscillations.

Problems

1. For each of the following operators determine whether the H–P condition for forward/backward evolution is satisfied:
   (i), (ii) $\partial_{tt} \pm (\Delta)^2$.
   (iii), (iv) $\partial_t \pm (\Delta)^2$.
   (v) $\partial_t + (\partial_x)^m$, $m \in \mathbb{N}$.
   (vi) The linearized BBM equation (4).
   (vii) $\partial_t^m + \varepsilon\partial_x$, $m \ge 2$, $\varepsilon \in \mathbb{R}\backslash 0$.

2. Find necessary and sufficient conditions on the real coefficients, a and b, so that $\partial_{tt} + a\partial_{tx} + b\partial_{xx}$ satisfies the H–P condition for forward evolution.
   Discussion. It is interesting to take this as a first step in finding the most general homogeneous operator of order 2 on $\mathbb{R}^{1+d}$ satisfying the H–P condition.

3. For the wave operator $\partial_{tt} - \Delta_x$ with $x \in \mathbb{R}^d$, consider the sideways Cauchy problem: $(\partial_{tt} - \Delta)u = 0$, $u|_{x_1=0} = f$, $u_{x_1}|_{x_1=0} = g$, where f and g belong to $\mathscr{S}(\mathbb{R}_t \times \mathbb{R}^{d-1}_{x_2,\ldots,x_d})$. This corresponds to reconstructing the solution from observations at $x_1 = 0$. Prove that for $d = 1$, the sideways Cauchy problem satisfies the H–P condition for both forward and backward evolution. Prove that for $d > 1$, the condition is violated in both directions.

DISCUSSION. Analogous inverse problems are very common in geophysics where one observes, for example, seismic waves, at the surface of the Earth ($x_1 = 0$) and tries to find out what is going on below the surface. There are many very good algorithms for $d = 1$. Unfortunately, $d = 2$ is more reasonable and the ill-posedness from Problem 3 renders this case extremely difficult. Current methodology is not very good.

4. Prove that if $\dim(x) \equiv d \geq 1$ and $P(D_t, D_x)$ is elliptic, then it can satisfy the H–P condition for neither forward nor backward evolution.

§3.11. Inhomogenous Equations, Duhamel's Principle

If $P(D_t, D_x)$ satisfies the Hadamard–Petrowsky (H–P) condition for forward evolution, then we can also solve the inhomogeneous initial value problem,

$$Pu = F \in C^\infty([0, \infty[: \mathscr{S}(\mathbb{R}_x^d)), \tag{1}$$

$$\partial_t^j u(0, \cdot) = 0, \qquad 0 \leq j \leq m - 1. \tag{2}$$

In many practical problems, the F in $Pu = F(t, x)$ represents external sources or stimuli while the homogeneous, $F = 0$, problem is free motion.

EXAMPLES. 1. $u_t - \nu\Delta u = F(t, x)$ represents heat flow with external source of heat $F(t, x)$ calories per unit volume per unit time.

2. The wave equation $u_{tt} - c^2\Delta u = F(t, x)$ with $x \in \mathbb{R}^2$ models the small vibrations of a membrane with external force $F(t, x)$ per unit mass per unit time. For example, a membrane in the presence of a constant gravitation field yields $u_{tt} - c^2\Delta u = g/\rho$ ($\rho \equiv$ density).

If $u \in C^\infty([0, \infty[: \mathscr{S}(\mathbb{R}^d))$ is a solution of (1), (2), then Fourier transformation in x yields an inhomogeneous ordinary differential equation for $\hat{u}(t, \xi)$,

$$P(D_t, \xi)\hat{u}(t, \xi) = \mathscr{F}F(t, \xi), \tag{3}$$

$$\partial_t^j \hat{u}(0, \xi) = 0, \qquad 0 \leq j \leq m - 1. \tag{4}$$

Lemma 1 (ODE Duhamel). *If $P(D_t) \equiv a_m D_t^m + a_{m-1} D_t^{m-1} + \cdots + a_0$ is a constant coefficient ordinary differential operator, $a_m \neq 0$, and $F(t) \in C^\infty([0, \infty[)$, then the solution $w \in C^\infty([0, \infty[)$ to*

$$Pw = F, \qquad \partial_t^j w(0) = 0 \leq j \leq m - 1,$$

is given by

$$w(t) = \int_0^t G(t - s)F(s)\, ds,$$

where G is the solution to

$$PG = 0, \qquad D_t^j G(0) = 0 \quad \text{for} \quad j \leq m - 2, \qquad D_t^{m-1}G(0) = i/a_m.$$

PROOF. Define H to be G for $t \geq 0$ and zero in $t < 0$. Then $H \in C^{m-2}(\mathbb{R})$ and $D_t^{m-1}H(0)$ has a jump discontinuity with jump $[D_t^{m-1}H] = i/a_m$ at $t = 0$. Then in $\mathscr{D}'(\mathbb{R})$, $\partial_t^j H = (\partial_t^j G)\chi_{[0,\infty[}$ for $j \leq m-1$, and $\partial_t^m H = (\partial_t^m G)\chi_{[0,\infty[} + [\partial_t^{m-1}H]\delta$. Thus

$$P(D_t)H = \frac{a_m}{i}[D_t^{m-1}H]\delta = \delta.$$

Choose $\psi \in C^\infty(\mathbb{R})$ with $\psi(t) = 0$ for $t < 1$ and $\psi = 1$ for $t > 2$. Let $F_n \equiv \psi(nt)F$.

Then $P(D_t)(H * F_n) = \delta * F_n = F_n$. In addition, $H * F_n$ converges to $H * (F\chi_{[0,\infty[})$ in $C^{m-1}(\mathbb{R})$ (exercise). Passing to the limit $n \to \infty$ yields the desired result. □

Apply the lemma to $\hat{u}(t, \xi)$, recalling the definition (3.10.11), (3.10.12) of M_{m-1} to find

$$\hat{u}(t, \xi) = \int_0^t M_{m-1}(t-s, \xi)\mathscr{F}F(s, \xi)\, d\xi. \tag{5}$$

One finds that $\hat{u} \in C^\infty([0, \infty[: \mathscr{S}(\mathbb{R}^d))$ and that $\hat{u}$ is a solution of the initial value problem (3), (4). For this, the estimates (3.10.14) for M_{m-1} suffice to justify differentiation under the integral sign (exercise).

Theorem 2. *If P satisfies the Hadamard–Petrowsky condition for forward evolution, then for any $F \in C^\infty([0, \infty[: \mathscr{S}(\mathbb{R}^d))$ there is a unique solution $u \in C^\infty([0, \infty[: \mathscr{S}(\mathbb{R}^d))$ to (1), (2). The solution is given by formula (5).*

Formula (5) has several alternate descriptions. For example, in the last section we saw that $\mathscr{F}^* M_{m-1}(t, \xi)\mathscr{F}f \equiv S_P(t)f$ is the value at time t of the solution of the initial value problem

$$Pv = 0, \qquad \partial_t^j v(x) = \begin{cases} 0 & \text{if } j \leq m-2, \\ f & \text{if } j = m-1. \end{cases}$$

S_P is a propagator for the evolution equation $Pv = 0$. Then (5) becomes *Duhamel's formula*

$$u(t) = \int_0^t S_P(t-s)F(s)\, ds. \tag{6}$$

As a typical example of how Duhamel's formula can be used to solve inhomogeneous equations with less regular F, consider $P = \partial_t - \nu\Delta$, $\mathrm{Re}(\nu) \geq 0$, thereby treating the heat and Schrödinger equations simultaneously.

For $F \in C^\infty([0, \infty[: \mathscr{S})$ the solution to

$$(\partial_t - \nu\Delta)u = F, \qquad u|_{t=0} = 0, \tag{7}$$

is given by (6). Let S_ν denote the propagator of $\partial_t - \nu\Delta$. The H^s norm of u is estimated by

$$\|u(t)\|_{H^s} \leq \int_0^t \|S_\nu(t-\sigma)F(\sigma)\|_{H^s}\, d\sigma \leq \int_0^t \|F(\sigma)\|_{H^s}\, d\sigma. \tag{8}$$

This suggests that for $F \in L^1_{\text{loc}}([0, \infty[: H^s)$ the initial value problem (7) has a solution in $C([0, \infty[: H^s)$. Before proving this we show that several ways of defining the notion of solution to (7) are equivalent.

Proposition 3. *For $F \in L^1_{\text{loc}}([0, \infty[: H^s)$ and $u \in C([0, \infty[: H^s)$ with $u|_{t=0} = 0$, the following are equivalent:*

(i) *For all $t \geq 0$, Duhamel's formula (6) holds. Here the integrand is in $L^1([0, t] : H^s)$.*

(ii) *For any function $\varphi \in C_0^\infty(\mathbb{R}^d)$ the function $t \mapsto \langle u(t), \varphi\rangle$ is absolutely continuous on $[0, \infty[$ and*

$$\langle u(t), \varphi\rangle' = \nu\langle u(t), \Delta\varphi\rangle + \langle F(t), \varphi\rangle. \tag{9}$$

(iii) *For any $\psi \in C_0^\infty(\mathbb{R} \times \mathbb{R}^d)$ and $t_1 < t_2$,*

$$\int_{t_1}^{t_2} \langle u(t), (-\partial_t - \nu\Delta)\psi\rangle - \langle F(t), \psi(t)\rangle\, dt = -\langle u(t), \psi(t)\rangle\Big|_{t_1}^{t_2}.$$

(iv) *$(\partial_t - \nu\Delta)u = F$ in the sense of distributions on $]0, \infty[\times \mathbb{R}^d$.*

This result is like Theorems 3.5.1, 3.6.4, and 3.7.3. The proof is omitted. When the equivalent conditions are satisfied we say that u is a solution of (7).

Theorem 4. *If $\operatorname{Re}(\nu) \geq 0$, $s \in \mathbb{R}$, and $F \in L^1_{\text{loc}}([0, \infty[: H^s)$, then there is a unique solution $u \in C([0, \infty[: H^s)$ to the initial value problem (7). The solution satisfies the estimate*

$$\|u(t)\|_{H^s} \leq \int_0^t \|F(\sigma)\|_{H^s}\, d\sigma.$$

PROOF. Choose $F_n \in C^\infty([0, \infty[: \mathscr{S})$ such that $F_n \to F$ in $L^1_{\text{loc}}([0, \infty[: H^s)$. The estimate (8) applied to $u_n - u_m$ shows that u_n is a Cauchy sequence in $C([0, \infty[: H^s)$. Let $u \equiv \lim u_n$.

Then for $\varphi \in C_0^\infty(\mathbb{R}^d)$, $\langle u_n, \varphi\rangle \to \langle u, \varphi\rangle$ in $C([0, \infty[)$, and

$$\frac{d}{dt}\langle u_n, \varphi\rangle = \langle u_n, \nu\Delta\varphi\rangle + \langle F_n(t), \varphi\rangle \to \langle u, \nu\Delta\varphi\rangle + \langle F(t), \varphi\rangle$$

in $L^1_{\text{loc}}([0, \infty[)$. It follows that $\langle u, \varphi\rangle$ is absolutely continuous and that (9) holds. This completes the proof of existence.

Uniqueness is a consequence of uniqueness for the homogeneous equation since the difference of two solutions is a solution of the homogeneous equation with zero initial value. □

Remark. If $\nu \in i\mathbb{R}$ then $[0, \infty[$ can be replaced by $\mathbb{R}$, that is, the solution exists in both the forward and backward directions of time.

Theorem 3 gives an adequate response to the question "What is the regularity of a solution with respect to x?". For regularity in t, one uses the differential equation to express the time derivatives of u in terms of x derivatives of u and derivatives of F. The equalities are as distributions on $]0, \infty[\times \mathbb{R}^d$. The first such identity is the equation itself

$$u_t = \nu\Delta u + F. \tag{10}$$

For the next, differentiate with respect to t to find

$$u_{tt} = \nu\Delta u_t + F_t = \nu\Delta(\nu\Delta u + F) + F_t = (\nu\Delta)^2 u + \nu\Delta F + F_t. \tag{11}$$

To use (10) note that with F as in Theorem 4, we know that $u \in C([0, \infty[: H^s)$, so $\Delta u \in C([0, \infty[: H^{s-2})$. If F lies in the same space we conclude that $\partial_t u \in C([0, \infty[: H^{s-2})$.

Similarly, in (11) the first two terms on the right belong to $C([0, \infty[: H^{s-4})$. Thus, if F_t belongs to this space we find that

$$\text{and} \quad \left.\begin{array}{l} F_t \in C([0, \infty[: H^{s-4}) \\ F \in C([0, \infty[: H^{s-2}) \end{array}\right\}, \quad \text{imply} \quad u_{tt} \in C([0, \infty[: H^{s-4}).$$

Continuing in this manner one derives the regularity of all the time derivatives of u in terms of the regularity of F.

Theorem 5. *If for $0 \le j \le N$, $\partial_t^j F \in C([0, \infty[: H^{s-2j})$, then $\partial_t^j u \in C([0, \infty[: H^{s-2j})$ for $0 \le j \le N$. The derivatives satisfy the estimate* (12) *below.*

Proof. The key is an estimate for the derivatives $\partial_t^j u$ of solutions $u \in C^\infty([0, \infty[: \mathscr{S})$

$$\sup_{0 \le t \le T} \|\partial_t^j u(t)\|_{H^{s-2j}} \le C_{j,T} \sum_{k \le j} \int_0^t \|\partial_t^k F(\sigma)\|_{H^{s-2j}}\, d\sigma, \qquad j \le N. \tag{12}$$

To prove the estimate with $j = 1$ or 2 use (10), (11) and estimate the right-hand sides crudely. The general case is similar.

Given (12) one retraces the proof of Theorem 4 using (12) to control the convergence. □

Problems

1. Find an explicit formula for the solution of the initial value problem $v'' + 4v = f$, $v(0) = v'(0) = 0$.
2. Show that Lemma 1 in the special case $P = D_t^n$ implies Taylor's Theorem with remainder.
3. Give a detailed proof of Theorem 5.
4. Formulate and prove analogues of Theorems 3, 4, and 5 for the inhomogeneous wave equation $\Box u = F$.

CHAPTER 4

Propagators and x-Space Methods

§4.1. Introduction

This chapter continues the use of the Fourier transform but the point of view is different. The main estimates in the last chapter were in ξ space. This section concentrates on behavior in x.

One method which works explicitly in the x variables is the energy method. This technique was applied several times in the last chapter, for example, to derive L^p estimates for the heat equation. As membership in L^p for $p \neq 2$ is not easily read from the Fourier transform, such results are often clearer in the x variables. For example, the fact that the Schrödinger propagator $\mathcal{F}^{-1}e^{-it|\xi|^2}\mathcal{F}$ is not bounded in L^p is not obvious in ξ space but becomes so upon studying the Gaussians in x space (Problem 3.4.2).

The methods of this chapter are analogous to the latter success. The Fourier transform is used to derive formulas in x space which are then analyzed.

§4.2. Solution Formulas in x Space

Consider the propagator for the heat equation

$$\mathcal{F}(S_{\mathrm{H}}(t)f) = e^{-\nu t|\xi|^2}\mathcal{F}f, \qquad f \in \mathcal{S}(\mathbb{R}^d). \tag{1}$$

Let $K_{\mathrm{H}}(t) \in \mathcal{S}(\mathbb{R}^d)$ be the function defined by

$$\mathcal{F}(K_{\mathrm{H}}(t)) = (2\pi)^{-d/2}e^{-\nu t|\xi|^2} \in \mathcal{S}(\mathbb{R}^d), \tag{2}$$

then (1) is equivalent to

$$S_{\mathrm{H}}(t)f = K_{\mathrm{H}}(t) * f = \int K_{\mathrm{H}}(t, x - y)f(y)\,dy. \tag{3}$$

Happily, K_H is a Gaussian already computed in §2.2,

$$K_H(t, x) = (4\pi\nu t)^{-d/2} e^{-|x|^2/4\nu t}. \tag{4}$$

Thus, for $t > 0$, $S_H(t)f$ is given by the explicit formula

$$u(t, x) = (4\pi\nu t)^{-d/2} \int e^{-|x-y|^2/4\nu t} f(y)\, dy. \tag{5}$$

A similar argument for the Schrödinger equation yields

$$S_S(t)f = K_S(t) * f, \tag{6}$$

$$\mathscr{F}(K_S(t)) = (2\pi)^{-d/2} e^{-it|\xi|^2}. \tag{7}$$

In this case, $K_S \in \mathscr{S}'$, and the convolution is that between an element of $\mathscr{S}'$ and an element of $\mathscr{S}$. K_S is given by formula (2.4.7)

$$K_S(t, x) = (4\pi i t)^{-d/2} e^{-|x|^2/4it}. \tag{8}$$

It is the same formula as the heat propagator with $\nu = i$,

$$u(t, x) = (4\pi i t)^{-d/2} \int e^{i|x-y|^2/4t} f(y)\, dy. \tag{9}$$

For the wave equation one has

$$\mathscr{F}(S_W(t)(f, g)) = \hat{f}\cos(ct|\xi|) + \hat{g}\,\frac{\sin(ct|\xi|)}{c|\xi|}. \tag{10}$$

Define $K_W(t)$ by

$$\mathscr{F}(K_W(t)) \equiv (2\pi)^{-d/2}\,\frac{\sin(ct|\xi|)}{c|\xi|}, \tag{11}$$

then $K_W(t) \in \mathscr{S}'$ and

$$u = \partial_t(K_W(t) * f) + K_W(t) * g. \tag{12}$$

The computation of K_W for $d = 1, 2, 3$ is postponed to §4.5–§4.8. For $d \geq 3$, K_W is not locally integrable. As d increases, K_W is a distribution of increasing order.

Formulas (3), (6), and (12) extend to generalized solutions with data f, $g \in \mathscr{E}'(\mathbb{R}^d)$. To prove this, choose approximate data f_n, g_n with support in a fixed compact set and converging to f, g in $H^s(\mathbb{R}^d)$ for $s \ll 0$. Then consider the formulas with f, g replaced by f_n, g_n. For fixed t, the left-hand sides converge in H^s for s very negative (and therefore in $\mathscr{S}'(\mathbb{R}^d)$) to the generalized solution at time t. Since $\mathscr{S}'(\mathbb{R}^d) * \mathscr{E}'(\mathbb{R}^d) \subset \mathscr{S}'(\mathbb{R}^d)$, the right-hand sides converge in $\mathscr{S}'(\mathbb{R}^d)$. Equating the $\mathscr{S}'$ limits yields the desired identity.

Taking $f = \delta$ in (3) and (6), or $f = 0$, $g = \delta$, in (12), identifies K_H, K_S, and K_W as the generalized solutions of

$$(\partial_t - \nu\Delta)K_H = 0, \qquad K_H(0) = \delta, \tag{13}$$

$$(\partial_t - i\Delta)K_S = 0, \qquad K_S(0) = \delta, \tag{14}$$

$$(\partial_t^2 - c^2\Delta)K_W = 0, \qquad K_W(0) = 0, \qquad \partial_t K_W(0) = \delta. \tag{15}$$

In particular,

$$K_W \in C^1(\mathbb{R} : H^{-d/2-\varepsilon}(\mathbb{R}^d)) \tag{16}$$

for any $\varepsilon > 0$, which serves to justify the differentiation in (12).

The distributions K are sometimes called *fundamental solutions*, *propagators*, or *Green's functions*. The representations, (3), (6), (12), sometimes reveal information which is not obvious from the formulas involving the Fourier transform.

Each of the three equations has a natural scaling property. For example, $u(t, x)$ satisfies the heat equation if and only if $u_\lambda(t, x) \equiv u(\lambda^2 t, \lambda x)$ is a solution for $\lambda > 0$. The same scaling works for the Shrödinger equation. For the wave equation $u_\lambda \equiv u(\lambda t, \lambda x)$ is the corresponding transformation. A solution is called *self-similar* if for every $\lambda > 0$, u_λ is a multiple of u. Each of the three fundamental solutions is self-similar. For example $K_H(\lambda^2 t, \lambda x)$ is a solution of the heat equation which belongs to $C([0, \infty[: H^{-d/2-\varepsilon}(\mathbb{R}^d))$ and whose initial value is $\delta(\lambda x) = \lambda^{-d}\delta$. Thus by uniqueness of solutions $K_H(\lambda^2 t, \lambda x) = \lambda^{-d} K_H$. Another expression of self similarity is the scaling laws

$$K_H(t, x) = t^{-d/2} K_H\left(1, \frac{x}{\sqrt{t}}\right), \tag{17}$$

$$K_S(t, x) = t^{-d/2} K_S\left(1, \frac{x}{\sqrt{t}}\right), \tag{18}$$

$$K_W(t, x) = t^{-d/2} K_W\left(1, \frac{x}{t}\right). \tag{19}$$

The precise versions of the right-hand side use the dilation operator from §2.2 but are harder to read, for example, $K_H(t) = t^{-d/2}\sigma_{t^{-1/2}} K_H(1)$. These formulas show that the values of the functions K at $t = 1$ are sufficient to determine K everywhere.

Problems

The next problems introduce you to the propagator for the Airy equation

$$u_t + u_{xxx} = 0.$$

The root of $P(\tau, \xi) = 0$ is $\tau = \xi^3$, so the Hadamard–Petrowsky condition is satisfied for forward and backward evolution. The evolution operator is given by the Fourier multiplier

$$S_A(t) = \mathscr{F}^* e^{i\xi^3 t} \mathscr{F}.$$

1. Use the energy method and the Fourier transform to give two proofs of each of the following two conservation laws for solutions $u \in C^\infty(\mathbb{R} : \mathscr{S})$ to Airy's equation

$$\int u(t, x)\, dx \text{ is independent of } t,$$

$$\int |u(t, x)|^2\, dx \text{ is independent of } t.$$

Discussion. More generally, $S_A(t)$ is unitary on H^s for all s, t.

The fundamental solution, $K_A(t) = S_A(t)\delta$ has Fourier transform equal to

$(2\pi)^{-1/2}\exp(i\xi^3 t)$. For $t = 1$, K_A is a multiple of the special function called the *Airy function*, $\text{Ai}(x)$.

2. (i) Find a complex number c such that

$$\left(\frac{d}{dx}\right)^2 \text{Ai} + cx\,\text{Ai} = 0$$

in the sense of $\mathscr{S}'(\mathbb{R})$.

(ii) Prove that $K_A(t) = t^{-1/3}\sigma_{t^{-1/3}} K_A(1)$.

(iii) Find the scaling law $u \mapsto u_\lambda$ so that the identity in (ii) expresses the fact that K_A is self-similar.

DISCUSSION. The ordinary differential equation for Ai suggests that Ai is a smooth function. This is true and all the derivatives of Ai are bounded on $\mathbb{R}$. The general principle in Problem 3.4.2 suggests that for $t \neq 0$, S_A is unbounded on L^p for $p \neq 2$. This is correct and as for the Schrödinger equation gives a Fourier multiplier which is unbounded on L^p and is not obviously so.

Self-similar solutions provide many important examples for nonlinear equations, for example, the rarefaction waves at the end of §1.9.

3. (i) Find a power α such that if u satisfies the inviscid Burgers equation $u_t + (u^2)_x = 0$, then $u_\lambda = \lambda^\alpha u(\lambda t, \lambda x)$ is also a solution.

(ii) Find the analogous scaling law $u_\lambda = \lambda^b u(\lambda^2 t, \lambda x)$ for the viscous Burgers equation $u_t + (u^2)_x = u_{xx}$.

DISCUSSION. For linear problems such multiplicative prefactors are inessential. In the nonlinear case they are essential.

§4.3. Applications of the Heat Propagator

Our first applications rest on the fact that for $t > 0$

$$K_H(t) \in \mathscr{S}(\mathbb{R}^d), \qquad K_H(t, x) > 0, \qquad \text{for all} \quad t > 0,\ x \in \mathbb{R}^d,$$

and

$$\int K_H(t, x)\,dx = 2\pi^{d/2}\mathscr{F}K_H(t, 0) = 1. \tag{1}$$

Young's inequality for convolutions implies that for $f \in \mathscr{S}(\mathbb{R}^d)$ and $p \in [1, \infty]$

$$\|S_H(t)f\|_{L^p} \le \|K_H(t)\|_{L^1}\|f\|_{L^p} = \|f\|_{L^p}, \tag{2}$$

which is an inequality derived by the energy method in §3.6 and applied in Problem 3.6.1 to generalized solutions with data in L^p.

In the same vein, the solution formula (4.2.5) extends to general $f \in L^p$ as follows.

Theorem 1. *If $p \in [1, \infty]$ and $f \in L^p(\mathbb{R}^d)$, then the solution $u = S_H(t)f$ belongs to $C^\infty(]0, \infty[\times \mathbb{R}^d)$. The derivatives are given by the absolutely convergent integrals*

$$D^\alpha_{t,x} u(t, x) = \int (D^\alpha_{t,x} K_H)(t, x - y) f(y)\,dy, \qquad t > 0. \tag{3}$$

PROOF. If $1 \le p < \infty$, choose $f_n \in \mathscr{S}(\mathbb{R}^d)$ with $f_n \to f$ in L^p. Then $u_n \equiv S_H f_n$ converges to u in $C([0, \infty[: L^p(\mathbb{R}^d))$. In particular, u_n converges to u in $\mathscr{D}'(]0, \infty[\times \mathbb{R}^d)$. For $p = \infty$, the convergence of f_n is taken in the weak-star topology. Then u_n converges to u uniformly on compact sets in $]0, \infty[\times \mathbb{R}^d$ and therefore in $\mathscr{D}'(]0, \infty[\times \mathbb{R}^d)$.

The solutions u_n are smooth with

$$D^\alpha_{t,x} u_n(t, x) = \int (D^\alpha_{t,x} K_H)(t, x - y) f_n(y)\, dy.$$

For any $\varepsilon > 0$, the functions $D^\alpha_{t,x} K_H(t, \cdot)$ are uniformly bounded in $L^q(\mathbb{R}^d)$ for $t \ge \varepsilon$ where $1/q + 1/p = 1$.

Hölder's inequality shows that the right-hand side converges uniformly to the right-hand side of (3), which is therefore bounded and continuous on $t \ge \varepsilon$.

The left-hand side converges in $\mathscr{D}'(t > 0)$ to $D^\alpha_{t,x} u$, which proves the desired result. □

In §3.6, the operator $S_H(t)$ was extended so that $S_H(t)f$ is well defined if $f \in H^s$ for some $s \in \mathbb{R}$, or if $f \in L^p$ for some $p \in [1, \infty]$. For $f \in \mathscr{S}'$, we say that $f \ge 0$ if for all $\varphi \in \mathscr{S}$ with $\varphi \ge 0$, $\langle f, \varphi \rangle \ge 0$. For $f \in L^p$, this is equivalent to $f \ge 0$ a.e. For $f \in \mathscr{S}'$ it implies that f is a nonnegative Radon measure.

Theorem 2. *Suppose that $f \in H^s(\mathbb{R}^d)$ for some s or that $f \in L^p(\mathbb{R}^d)$ for some $p \in [1, \infty]$. If $f \ge 0$, then for all $t \ge 0$, $S_H(t)f \ge 0$.*

PROOF. Let $j_\varepsilon \ge 0$ and χ_ε be the usual nonnegative mollifiers and plateau functions. Then

$$f_\varepsilon \equiv j_\varepsilon * (\chi_\varepsilon f) \ge 0,$$

and for $\varphi \in \mathscr{S}$, $\varphi \ge 0$,

$$\langle S_H(t)f, \varphi \rangle = \lim_{\varepsilon \to 0} \langle S_H(t) f_\varepsilon, \varphi \rangle.$$

This follows since $S_H(t) f_\varepsilon \to S_H(t) f$ in H^s or in L^p.

Now $S_H(t) f_\varepsilon = K_H(t) * f_\varepsilon \ge 0$ since $K_H \in \mathscr{S}$ and $f_\varepsilon \in C_0^\infty$ are nonnegative, so the convolution is equal to an integral with nonnegative integrand.

It follows that the limit in nonnegative. □

Corollary 3 (Comparison Theorem). *If $X = H^s(\mathbb{R}^d)$, $s \in \mathbb{R}$ or $X = L^p(\mathbb{R}^d)$, $p \in [1, \infty]$, and $f, g \in X$ with $f \ge g$, then $S_H(t)f \ge S_H(t)g$.*

PROOF. $S_H(t)(f - g) \ge 0$ by Theorem 2. □

For $f \in L^\infty$ we may take g to be the constant function ess inf(f) for which $S_H(t)g$ is independent of t, x. Similarly, ess sup(f) is an upper bound for u.

Corollary 4. *If $f \in L^\infty(\mathbb{R}^d : \mathbb{R})$ and $u = S_H(t)f$, then for all $t \geq 0$*

$$\text{ess inf}(u(0)) \leq u(t) \leq \text{ess sup}(u(0)). \tag{4}$$

Thus the temperature is always between its initial extremes, consistent with the intuition that heat flows from hot to cold.

An alternate derivation of the corollary goes as follows. Let $\Psi(s) \equiv [\max(s - M, 0)]^4$. Then Ψ is C^2 and convex, so by the energy method (Theorem 3.6.4), $\int \Psi(u(t, x))\, dx$ is a nonincreasing function of t. If $f \in \mathcal{S}$ and $f \leq M$, this proves that $u \leq M$ for all $t \geq 0$. Approximating a bounded f by elements $f_n \in \mathcal{S}$ with $f_n \leq \text{ess sup} f$ yields an independent proof.

Next we examine the large time behavior of $u = S_H(t)f$ when $f \in L^1$. Recall that the L^1 norm of nonnegative solutions is the physical energy. We have

$$\begin{aligned} |u(t, x)| &= \left| \int K_H(t, x - y) f(y)\, dy \right| \\ &\leq \|K_H(t)\|_{L^\infty} \|f\|_{L^1} \\ &= (4\pi\nu t)^{-d/2} \|f\|_{L^1}. \end{aligned}$$

Thus finite energy solutions decay uniformly as $t^{-d/2}$, the same rate of decay as Gaussian solutions. More generally, we have a sharp rate of decay for the L^p norm for any $1 \leq p \leq \infty$.

Theorem 5. *If $f \in L^1(\mathbb{R}^d)$, $p \in [1, \infty]$, $1/q + 1/p = 1$, and $u(t) = S_H(t)f$, then*

$$\|u(t)\|_{L^p} \leq (4\pi\nu t)^{-d/2q} \|f\|_{L^1(\mathbb{R}^d)}. \tag{5}$$

PROOF. For $p = 1$, (5) expresses the decrease of the L^1 norm proved in (2) and by the energy method in §3.6. The estimate for $p = \infty$ was proved immediately before the statement of the theorem. The Riesz–Thorin Theorem completes the proof. □

The behavior as $t \to \infty$ can be described even more precisely if f decays sufficiently rapidly as $|x| \to \infty$. For this, the argument is simpler using the Fourier transform, where $\hat{u}(t, \xi) = e^{-\nu t|\xi|^2}\hat{f}(\xi)$ decays exponentially fast as $t \to \infty$, except at the origin, $\xi = 0$. This suggests replacing $\hat{f}$ by a Taylor expansion at $\xi = 0$

$$\hat{u} = e^{-\nu t|\xi|^2} \sum_{|\alpha| \leq N} \frac{\hat{f}^\alpha(0)\xi^\alpha}{\alpha!} + R_N(t, \xi).$$

The case $N = 0$ is the most interesting,

$$\hat{u} = \hat{f}(0)e^{-\nu t|\xi|^2} + e^{-\nu t|\xi|^2}(\hat{f}(\xi) - \hat{f}(0)).$$

Taking the inverse Fourier transform yields

$$u = (2\pi)^{d/2}\hat{f}(0)K_H(t) + \mathcal{F}^{-1}(e^{-\nu t|\xi|^2}(\hat{f} - \hat{f}(0))). \tag{6}$$

To estimate the last term note that

$$|\hat{f}(\xi) - \hat{f}(0)| \le |\xi| \, \|\nabla_\xi \hat{f}\|_{L^\infty(\mathbb{R}^d_\xi)} \le (2\pi)^{-d/2} |\xi| \, \|xf\|_{L^1(\mathbb{R}^d_x)},$$

the last estimate following from $\nabla_\xi \hat{f} = \mathscr{F}(ixf)$. The $L^\infty(\mathbb{R}^d_x)$ norm of the second term in (6) is therefore dominated by

$$\|e^{-\nu t|\xi|^2}(\hat{f} - \hat{f}(0))\|_{L^1(\mathbb{R}^d_\xi)} \le (2\pi)^{-d/2} \|xf\|_{L^1(\mathbb{R}^d_x)} \|\xi e^{-\nu t|\xi|^2}\|_{L^1(\mathbb{R}^d_\xi)}.$$

Compute the L^1 norm on the right using polar coordinates

$$\int |\xi| e^{-\nu t|\xi|^2} \, d\xi = \omega_d \int_0^\infty r e^{-\nu t r^2} r^{d-1} \, dr.$$

The change of variable $\rho \equiv (\nu t)^{1/2} r$ yields

$$= \omega_d (\nu t)^{-(d+1)/2} \int_0^\infty \rho^d e^{-\rho^2} \, d\rho \equiv c(\nu t)^{-(d+1)/2}. \tag{7}$$

This proves the following theorem.

Theorem 6. *If* $\langle x \rangle f \in L^1(\mathbb{R}^d_x)$ *and* $u = S_{\mathrm{H}}(t)f$, *then there is a* $c = c(d)$ *such that for* $t > 0$

$$\left\| u(t) - \left(\int f(x) \, dx \right) K_{\mathrm{H}}(t) \right\|_{L^\infty(\mathbb{R}^d_x)} \le c(\nu t)^{-(d+1)/2} \|xf\|_{L^1(\mathbb{R}^d_x)}.$$

Note that the error on the right decays faster, by a factor $t^{-1/2}$, than the solution itself. We known that the energy, $\int u(t, x) \, dx$, is independent of time. The above theorem shows that for t large u is close to the Gaussian with the same energy. For large time, solutions with the same initial energy are essentially indistinguishable. Thus for t large, there is only one observable, the energy.

This is a rather striking degradation of information. For example, one could code the *Encyclopaedia Britannica* as a sequence of 0's and 1's and encode that as a step function

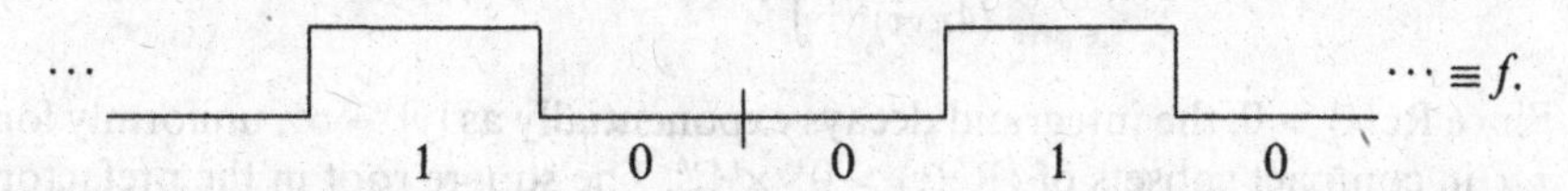

Then asymptotically, all one could measure is the number of bits of information rather than the information itself. This is a strongly irreversible process. Time's arrow is clearly visible.

Our next observations concern the *smoothing property* of the solution operator S_{H}. Theorem 1 shows that in the C^k categories the solutions are immediately smoothed. Looking in Fourier we sill show that the same is true for H^s regularity. Write

$$\hat{u}(t) = e^{-\nu t|\xi|^2} \hat{f}.$$

If $f \in H^\sigma(\mathbb{R}^d)$ for some $\sigma \in \mathbb{R}$ and s is any real number, perhaps much larger than σ, then

$$\langle\xi\rangle^s \hat{u}(t) = (\langle\xi\rangle^s \langle\xi\rangle^{-\sigma} e^{-\nu t|\xi|^2})(\langle\xi\rangle^\sigma \hat{f}).$$

The last factor belongs to $L^2(\mathbb{R}^d_\xi)$ and the first factor is uniformly bounded on $[\eta, \infty[\times \mathbb{R}^d_\xi$ for any $\eta > 0$. Thus $u(t) \in H^s$ and

$$\|u(t)\|_{H^s} \leq c(s - \sigma, t)\|f\|_{H^\sigma},$$

$$c(r, t) = \max_{\xi \in \mathbb{R}^d} \langle\xi\rangle^r e^{-\nu t|\xi|^2}.$$

This is used together with the estimate

$$\|\partial_t^j \partial_x^\alpha u(t)\|_{L^\infty} = \|\partial_x^\alpha (\nu\Delta)^j u(t)\|_{L^\infty} \leq c\|u(t)\|_{H^{2j+\alpha+d/2+\varepsilon}},$$

to prove the following theorem.

Theorem 7. *If $f \in H^\sigma(\mathbb{R}^d)$ for some σ and $u = S_H f$, then for any s, $u \in C^\infty(]0, \infty[: H^s(\mathbb{R}^d))$ and for any j, α, and $\eta > 0$, there is a $c > 0$ so that for all $t \geq \eta$*

$$\|\partial_t^j u(t)\|_{H^s(\mathbb{R}^d)} \leq c(s - \sigma, \eta)\|f\|_{H^\sigma(\mathbb{R}^d)}.$$

This regularizing illustrates again the degradation of information and non-reversability in heat propagation.

Actually, much more is true than that $u \in C^\infty$. Using the propagator K_H we show that u is real analytic.

Theorem 8. *If $f \in L^p(\mathbb{R}^d)$ for some $1 \leq p \leq \infty$ and $u = S_H(t)f$, then $u(t, x) \in C^\omega(]0, \infty[\times \mathbb{R}^d)$. In fact, u is the restriction to $]0, \infty[\times \mathbb{R}^d_x$ of the holomorphic function $u(\tau, \zeta)$ on $\{\mathrm{Re}(\tau) > 0\} \times \mathbb{C}^n$ from (8) below.*

PROOF. Let

$$u(\tau, \zeta) \equiv \frac{1}{(4\pi\nu\tau)^{1/2}} \int e^{-\sum(\zeta_j - y_j)^2/4\nu\tau} f(y)\, dy. \tag{8}$$

Since $\mathrm{Re}(\tau) > 0$, the integrand decays exponentially as $|y| \to \infty$, uniformly for τ, ζ in compact subsets of $\{\mathrm{Re}(\tau) > 0\} \times \mathbb{C}^n$. The square root in the prefactor is the branch, in $\mathrm{Re}(\tau) > 0$, which is real for τ positive and real.

That (8) defines a holomorphic function is verified by differentiation under the integral sign (justify!). This shows that u satisfies the Cauchy–Riemann equations in τ, ζ since the integrand does.

Theorem 1 shows that (8) is an extension of $u(t, x)$. □

The same conclusion is valid if $f \in H^s(\mathbb{R}^d)$ for $s < 0$, even if f is not locally integrable so the formula using integration is not valid.

Instead, observe that the function $\exp(-\sum(\zeta_j - y_j)^2/4\nu\tau) \in \mathcal{S}(\mathbb{R}^d)$, so if

$\langle\ ,\ \rangle$ denotes the pairing of $\mathscr{S}$ and $\mathscr{S}'$ we can define $u(\tau, \zeta)$ by

$$u(\tau, \zeta) \equiv \frac{1}{(4\pi\nu\tau)^{1/2}} \left\langle f, \exp\left(-\frac{\sum(\zeta_j - y_j)^2}{4\nu\tau} \right) \right\rangle. \tag{9}$$

In Problem 4, you are asked to prove that this is holomorphic. This suffices to extend Theorem 8 to H^s data. These arguments also show that S_H extends naturally to $\mathscr{S}'$ and that the conclusion is valid for any data from $\mathscr{S}'$.

Corollary 9. *If $u = S_H(t)f$ and there is a $t > 0$ and an open set $\omega \subset \mathbb{R}^d_x$ so that $u(t, x) = 0$ for $x \in \omega$, then $u \equiv 0$.*

Proof. u is smooth in $\{t > 0\} \times \mathbb{R}^d$ and $\partial_t^j \partial_x^\alpha u = \partial_x^\alpha (\nu\Delta)^j u$ vanishes on $\{t\} \times \omega$. The unique continuation property for holomorphic functions implies that $u(\tau, \zeta)$ is identically zero on $\{\mathrm{Re}(\tau) > 0\} \times \mathbb{C}^d$. □

In general, u will not be real analytic at $t = 0$. In order that u be real analytic at $(0, \underline{x})$, a necessary condition is that f be real analytic at $\underline{x}$. In §1.3 we observed that this condition is not sufficient, since the time derivatives of u of order k grow like the space derivatives of order $2k$. The result is that the regularity in t, for real analytic f, is described by the Gevrey class G^2 defined in [H2, p. 281].

As a final application, consider the decay of the derivatives of $S_H(t)f$ as $t \to \infty$. The formula $\hat{u}(t) = \exp(-\nu t|\xi|^2)\hat{f}$ shows that the high frequencies decay faster than the low frequencies. This suggests that derivatives of u may decay faster than u itself.

The rate of decay of the L^2 norm of derivatives of order k is estimated as follows:

$$\|D_x^\alpha u(t)\|_{L^2} = \|\xi^\alpha e^{-\nu t|\xi|^2} \hat{f}\|_{L^2} \le \left(\sup_\xi |\xi|^k e^{-\nu t|\xi|^2} \right) \|\hat{f}\|_{L^2}.$$

The change of variable, $\eta \equiv (\nu t)^{1/2}\xi$, shows that

$$\sup_{\xi \in \mathbb{R}^d} |\xi|^k e^{-\nu t|\xi|^2} = \sup_{\eta \in \mathbb{R}^d} \left| \frac{\eta}{(\nu t)^{1/2}} \right|^k e^{-\eta^2} = \frac{c_k}{(\nu t)^{k/2}}.$$

Thus, if $|\alpha| = k$

$$\|D_x^\alpha u\|_{L^2} \le c_k (\nu t)^{-k/2} \|f\|_{L^2}.$$

Note that the higher the order of the derivative is, the faster the decay is.

For estimates based on the energy, $\|f\|_{L^1}$, and for sharp rates of decay in L^p for p other than 2 (e.g., L^∞) use the fundamental solution K_H. For $f \in L^1$ and $t > 0$, the derivatives of $u = S_H(t)f$ are given by

$$D_x^\alpha u = (D_x^\alpha K_H(t)) * f.$$

For $t > 0$, $D_x^\alpha K_H(t) \in \mathscr{S}$, and

$$\|D_x^\alpha u(t)\|_{L^p} \le \|D_x^\alpha K\|_{L^p} \|f\|_{L^1}.$$

Lemma 10. $\|D_x^\alpha K_{\mathrm{H}}(t)\|_{L^p(\mathbb{R}^d)} = c(\alpha, p)/t^{(|\alpha|+d/q)/2}$ *where* $1/p + 1/q = 1$.

The lemma implies the basic estimate for the decay of derivatives.

Theorem 11. *For* $\alpha \in \mathbb{N}^d$ *and* $p \in [1, \infty]$, *there is a constant c such that for all* $t > 0$ *and* $f \in L^1(\mathbb{R}^d)$

$$\|D^\alpha S_{\mathrm{H}}(t)f\|_{L^p} \le \frac{c}{t^{(|\alpha|+d/q)/2}} \|f\|_{L^1(\mathbb{R}^d)}.$$

PROOF OF LEMMA 10. Differentiating $K = (4\pi\nu t)^{-d/2}e^{-|x|^2/4\nu t}$ yields $\partial_j K = (x_j/2\nu t)K$. Continuing, we see that $\partial_x^\alpha K$ is a linear combination of terms of the form $(x^\beta/t^l)K$.

More precisely, notice that when one differentiates $(x^\gamma/t^k)K$, if the derivative falls on K, then $|\gamma|$ increases by 1 and $|k|$ also increases by 1. If the derivative falls on x^γ, then $|\gamma|$ decreases by 1 and k remains the same. In both cases, $2k - |\gamma|$ increases by 1. Thus $\partial_x^\alpha K$ is a linear combination of terms $(x^\beta/t^l)K$ with $2l - |\beta| = |\alpha|$.

Write such terms in the form

$$\left(\frac{x}{\sqrt{t}}\right)^\beta \left(\frac{1}{\sqrt{t}}\right)^{|\alpha|} e^{-(x/(4\nu t)^{1/2})^2}.$$

The change of variables $y = x/(4\nu t)^{1/2}$ in the integral for $\|(x^\beta/t^l)K\|_{L^p}^p$ yields the desired result. □

PROBLEMS

1. Prove that if $f \in L^\infty(\mathbb{R}^d)$ is uniformly continuous on $\mathbb{R}^d$, then $u = S_{\mathrm{H}}(t)f$ is uniformly continuous on $[0, \infty[\times \mathbb{R}_x^d$.
   DISCUSSION. This is a good addition to the information in Problem 3.6.2.

2. For $f \in L^r(\mathbb{R}^d)$ and $p > r$, find the rate of decay as t tends to infinity of $\|S_{\mathrm{H}}(t)f\|_{L^p(\mathbb{R}^d)}/\|f\|_{L^r(\mathbb{R}^d)}$. *Hints.* Start with the inequality $\|S_{\mathrm{H}}(t)f\|_{L^p(\mathbb{R}^d)} \le \|f\|_{L^p}$. Then derive $\|S_{\mathrm{H}}(t)f\|_{L^\infty(\mathbb{R}^d)} \le \|K_{\mathrm{H}}(t)\|_{L^q}\|f\|_{L^p}$. Finally, compute $\|K_{\mathrm{H}}(t)\|_{L^q}$ and use interpolation.

3. If $\langle x\rangle^2 f \in L^1$, use the degree one Taylor polynomial of $\mathscr{F}f$ at 0 to compute the asymptotic behavior of $S_{\mathrm{H}}(t)f$ as t tends to infinity up to errors
   $$O((\nu t)^{(-d-2)/2}\|\langle x\rangle^2 f\|_{L^1}).$$
   Find a statement analogous to that of Theorem 6 in the sense that the principal terms and error estimate are given by simple expressions involving $f(x)$.

4. Show that for any $f \in \mathscr{S}'(\mathbb{R}^d)$, the function $u(\tau, \zeta)$ defined on $\{\mathrm{Re}(\tau) > 0\} \times \mathbb{C}^d$ by
   $$u(\tau, \zeta) \equiv (4\pi\nu\tau)^{-1/2}\langle f, e^{-\sum(\zeta_j - y_j)^2/4\nu\tau}\rangle$$
   is holomorphic.

5. In analogy with Problem 2, estimate the rate of decay as $t \to \infty$ of $\|\partial_x^\alpha S_{\mathrm{H}}(t)f\|_{L^p}$ for $f \in L^r$ and $p > r \ge 1$.

6. Prove that if $f \in L^p(\mathbb{R}^d)$ with $p > 1$, then $\lim_{t\to\infty} \|S_{\mathrm{H}}(t)f\|_{L^p} \to 0$.

§4.4. Applications of the Schrödinger Propagator

The solution formula for the Schrödinger equation is

$$u(t, x) = (4\pi it)^{-d/2} \int e^{-|x-y|^2/4it} f(y)\, dy. \tag{1}$$

Note that conservation of the L^2 norm is anything but obvious from this expression. Even boundedness in L^2 is not clear.

Our first estimate relies on the fact that $\|K_S(t)\|_{L^\infty} = |4\pi t|^{-d/2}$, the same as the L^∞ norm of the heat propagator.

Theorem 1. *If $f \in L^1(\mathbb{R}^d)$ and $u(t) = S_S(t)f$, then for all $t \in \mathbb{R}$*

$$\|u(t)\|_{L^\infty} \le |4\pi t|^{-d/2} \|u(0)\|_{L^1}. \tag{2}$$

Though this is the same rate of pointwise decay as the heat equation, other L^p norms behave differently. For example, the L^2 norm is conserved for the Schrödinger equation and decays for the heat equation as in Theorem 4.3.5 and Problem 4.3.6.

Corollary 2. *If $f \in L^q(\mathbb{R}^d)$, $1 \le q \le 2$, $1/p + 1/q = 1$, and $u = S_S f$, then*

$$\|u(t)\|_p \le |4\pi t|^{-d(1/2-1/p)/2} \|u(0)\|_{L^q}. \tag{3}$$

PROOF. Interpolate between the cases $q = 1$ and 2. □

Corollary 3. *If $f \in L^2(\mathbb{R}^d)$, $u(t) = S_S(t)f$, and $\Omega \subset \mathbb{R}^d$ has finite Lebesgue measure, then*

$$\lim_{t\to\infty} \operatorname{Prob}(x \in \Omega) \to 0.$$

PROOF. Given $\varepsilon > 0$ choose $\varphi \in \mathscr{S}$, $\|\varphi - f\|_{L^2} < \varepsilon$. Then

$$\begin{aligned}\left(\int_\Omega |S_S f|^2\, dx\right)^{1/2} &\le \left(\int_\Omega |S_S f|^2\, dx\right)^{1/2} + \left(\int_\Omega |S_S(\varphi - f)|^2\, dx\right)^{1/2} \\ &\le O(|t|^{-d/2}) + \varepsilon.\end{aligned}$$

□

A better idea of the behavior as $t \to \infty$ comes from expanding the exponent in

$$e^{-|x-y|^2/4it} = e^{-|x|^2/4it} e^{+ixy/2t} e^{-|y|^2/4it}.$$

For f of compact support, $e^{-|y|^2/4it}$ is close to 1 on $\operatorname{supp}(f)$ and

$$\int e^{-ixy/2t} f(y)\, dy = (2\pi)^{d/2} \hat{f}(x/2t).$$

This suggests that

$$\mathscr{S}(t)f \equiv \frac{e^{-|x|^2/4it}}{(2it)^{d/2}}\hat{f}(x/2t) \tag{4}$$

is an approximation to $u = S_S(t)f$.

Theorem 4. *For $f \in L^2(\mathbb{R}^d)$ and $u(t) = S_S(t)f$ we have*

$$\lim_{|t|\to\infty} \left\| u(t) - \frac{e^{-|x|^2/4it}}{(2it)^{d/2}}\hat{f}\left(\frac{x}{2t}\right)\right\|_{L^2(\mathbb{R}^d_x)} = 0. \tag{5}$$

Note that $e^{-|x|^2/4it}$ is of modulus 1, so the relative probability of being at t, x is approximately proportional to $\hat{f}(x/2t)$, which is consistent with the interpretation of $|\hat{f}|^2$ as a momentum density if we recognize that $m = \frac{1}{2}$ for our equation.

The asymptotics (5) give a physical intuition into the decay rates for the L^p norms. If the momentum density decays rapidly at infinity, then the solution u is concentrated over a regions which dilates linearly with time, so has volume growing like $|t|^d$. The amplitude decays like $|t|^{-d/2}$, which is to be expected given conservation of the L^2 norm and is verified in formulas (2) and (5). Amplitude $t^{-d/2}$ over a region of size t^d yields the rates of decay (3). Exactly such spread is present in the explicit Gaussian solutions computed in §3.3.

Proof of Theorem 4. Since $e^{-|x|^2/4it}$ is of modulus 1, the change of variable $\zeta \equiv x/2t$ shows that $\|\mathscr{S}(t)\|_{L^2\to L^2} = 1$ for all $t \neq 0$.

For $f \in L^2(\mathbb{R}^d)$ and any $\varepsilon > 0$, choose $g \in C_0^\infty(\mathbb{R}^d)$ with $\|g - f\|_{L^2} < \varepsilon$. Then

$$\begin{aligned}\mathscr{S}(t)g &= S(t)(e^{+|y|^2/4it}g(y)) \\ &= S(t)g + S(t)((e^{+|y|^2/4it} - 1)g).\end{aligned}$$

Now $\|(e^{-|y|^2/4it} - 1)g\|_{L^2} = O(1/|t|)$. Thus

$$\begin{aligned}\|S(t)f - \mathscr{S}(t)f\|_{L^2} &\le \|S(t)g - \mathscr{S}(t)g\|_{L^2} + \|S(t)(f-g)\|_{L^2} + \|\mathscr{S}(t)(f-g)\|_{L^2} \\ &= O\left(\frac{1}{|t|}\right) + 2\varepsilon.\end{aligned}$$

Thus

$$\limsup_{|t|\to\infty} \|(S(t) - \mathscr{S}(t))f\| \le 2\varepsilon,$$

and the proof is complete. □

Corollary 5 (Dollard). *If $f \in L^2$, $\Gamma \subset \mathbb{R}^d$ is a measurable cone, and $u(t) = S_S(t)f$, then*

$$\lim_{t\to\infty} \int_\Gamma |u(t,x)|^2\,dx = \int_\Gamma |\hat{f}(\xi)|^2\,d\xi.$$

PROOF. For any $t > 0$

$$\int_\Gamma |\mathscr{S}(t)f|^2\,dx = \int_\Gamma |\hat{f}(\xi)|^2\,d\xi.$$

Then since

$$\int_\Gamma |u(t,x) - \mathscr{S}(t)f|^2\,dx \to 0$$

as $t \to \infty$, the proof is complete. □

This corollary shows that for large positive time the probability that a particle lies in Γ converges to the probability that its momentum lies in Γ. Similarly, the probability that a particle lies in $-\Gamma$ for large negative time converges to the probability that the momentum lies in Γ. The fact that these two probabilities are equal shows that there is no change in direction of motion in the scattering of particles by the Schrödinger equation.

Dispersive phenomena like those for Schrödinger's equation can also be studied directly from the Fourier representation. From that point of view, one is lead to oscillatory integrals which are estimated using integration by parts in what is called the *method of (non)stationary phase*. Consider the solution with data $f \in C_0^\infty(\mathbb{R}^d)$

$$u(t,x) = (2\pi)^{-d/2} \int e^{ix\xi} e^{-it|\xi|^2} \hat{f}(\xi)\,d\xi = (2\pi)^{-d/2} \int e^{i\psi} \hat{f}(\xi)\,d\xi,$$

with phase $\psi \equiv x\xi - t|\xi|^2$. Where $\nabla_\xi \psi \neq 0$, the integral is oscillating and one expects cancellation, so a special role is played by the points where $\nabla_\xi \psi$ vanishes. These are the points of stationary phase and are given by $x = 2\xi t$. Thus the values of $\hat{f}$ near $\bar{\xi}$ are expected to play an important role where $x/t = 2\bar{\xi}$, that is, the points observed by an observer moving with velocity $2\bar{\xi}$. Note that this is the group velocity at frequency ξ encountered in Example 2 of §3.4. Theorem 4 is a precise result capturing part of this idea. Using the method of nonstationary phase, we will show that if $\hat{f}$ vanishes on a neighborhood of $\bar{\xi}$, then an observer moving with speed $2\bar{\xi}$ will make observations decaying faster than any power of $1/t$ as $t \to \infty$.

Theorem 6. *Suppose that $f \in \mathscr{S}(\mathbb{R}^d)$ and $\Lambda \subset \mathbb{R}^d$ satisfies $\delta \equiv \mathrm{dist}(\Lambda, 2\,\mathrm{supp}\,\hat{f}) > 0$. Then for any $n \in \mathbb{N}$, $u(t,x) = O((t + |x|)^{-n})$ as $t \to \infty$ in the set $\{(t,x): x/t \in \Lambda\}$.*

PROOF. For such x, t and $\xi \in \mathrm{supp}\,\hat{f}$, $|\nabla_\xi \psi| = |2\xi - x/t|t \geq t\delta$. Let L be the differential operator

$$L \equiv -i \sum \frac{\partial\psi/\partial\xi_j}{|\nabla_\xi \psi|^2} \frac{\partial}{\partial\xi_j},$$

which satisfies the crucial identity $Le^{i\psi} = e^{i\psi}$. Then integration by parts yields

$$u = (2\pi)^{-d/2} \int L^n(e^{i\psi})\hat{f}\, d\xi = (2\pi)^{-d/2} \int e^{i\psi}(L')^n \hat{f}\, d\xi,$$

where L' is the transpose of L (L^t might lead to confusion with the time variable t). The proof is completed by showing that for $x/t \in \Lambda$ and $\xi \in \operatorname{supp} \hat{f}$

$$(L')^n = \sum a_\alpha(t, x, \xi)\partial_\xi^\alpha \quad \text{and} \quad |a_\alpha| \le c(t + |x|)^{-n} \tag{6}$$

as $t \to \infty$ in $\{(t, x) : x/t \in \Lambda\}$ (Problem 1). □

PROBLEMS

1. Complete the proof of Theorem 6. *Hint.* Show that the coefficients of $(L')^n$ are homogeneous of degree $-n$ in x, t.
2. Prove Corollary 5 without using the explicit formula (1) but starting from Theorem 6 instead.
3. Prove that if $f \in H^s(\mathbb{R}^d)$ has compact support, then $u = S_S(t)f$ belongs to $C^\infty(\{t \ne 0\})$.
DISCUSSION. This is an example of *dispersive smoothing.* The fact that different frequencies correspond to distinct velocities "tears apart" singularities of compactly supported data.

Consider the solution of Airy's equation $u_t + u_{xxx} = 0$ (see the problems in §4.2)

$$u(t, x) = (2\pi)^{-1/2} \int e^{ix\xi + it\xi^3}\hat{f}(\xi)\, d\xi = (2\pi)^{-1/2} \int e^{i\varphi}\hat{f}(\xi)\, d\xi \tag{7}$$

with phase $\varphi = x\xi + t\xi^3$. The points of stationary phase satisfy $x/t = -3\xi^2 \le 0$. This suggests that the frequency ξ is associated with the group velocity equal to $-3\xi^2$. Since all these velocities are nonpositive there is a phenomenon of one-way propagation. Prove the following precise version.

4. **Theorem.** *Suppose that* $f \in \mathcal{S}(\mathbb{R}^d)$ *and* u *is the solution* (7) *of Airy's equation with* $u(0, \cdot) = f(\cdot)$. *Then for any* $\varepsilon > 0$ *and* $n > 0$, $u(t, x) = O((t + |x|)^{-n})$ *as* $t \to \infty$ *in the region* $\{(t, x) : x/t \ge \varepsilon\}$.

DISCUSSION. The Schrödinger equation has exponential solutions $e^{i(x\xi - \omega(\xi)t)}$ with $\omega(\xi) = |\xi|^2$. The Airy equation has $\omega(\xi) = -\xi^3$. The initial value problem is solved by

$$u(t, x) = (2\pi)^{-d/2} \int e^{i\varphi}\hat{f}(\xi)\, d\xi, \qquad \varphi \equiv x\xi - \omega(\xi)t.$$

Points of stationary phase satisfy $x/t = \nabla_\xi \omega$. Such real-valued ω are called *dispersion relations.* The velocity $\nabla_\xi \omega$ giving the points of stationary phase is called the *group velocity at frequency* ξ. There are formulas analogous to (5) asserting that for t large $u(t, x)$ is approximately equal to $m(t, \xi)\hat{f}(\xi)$, where the group velocity at ξ is equal to x/t and the factor m is determined by the method of stationary phase (see [Whitham]).

§4.5. The Wave Equation Propagator for $d = 1$

Recall that the solution of

$$\Box u = 0, \qquad u(0, \cdot) = f, \qquad u_t(0, \cdot) = g, \tag{1}$$

is given by

$$u(t) = K_W(t) * g + (\partial_t K_W(t)) * f, \tag{2}$$

where

$$\mathscr{F}(K_W(t)) = (2\pi)^{-d/2} \frac{\sin(ct|\xi|)}{c|\xi|},$$

$$\partial_t^j K_W \in C(\mathbb{R} : H^{s+1-j}), \qquad \text{for all} \quad s < -d/2.$$

When $d = 1$, the fact that $(\sin x)/x = \sin|x|/|x|$ yields

$$c\xi\mathscr{F}(K_W(t)) = (2\pi)^{-1/2} \sin ct\xi = (2\pi)^{-1/2} \frac{e^{ict\xi} - e^{-ict\xi}}{2i}.$$

Thus

$$\partial_x K_W(t) = \frac{\delta_{-ct} - \delta_{ct}}{2c}.$$

Happily, there is a simple distribution whose derivative is equal to the right-hand side. Note that $\partial_x h(x - a) = \delta_a$, take $a = \pm ct$ and subtract to find $\partial_x \chi_{[-ct,ct]} = \delta_{-ct} - \delta_{ct}$. For t fixed, this shows that $K - \chi/2c$ has distribution derivative equal to zero, so must be independent of x,

$$K_W(t) - \frac{1}{2c}\chi_{[-ct,ct]} = \text{constant}(t).$$

However, Plancherel's Theorem shows that the left-hand side is in $L^2(\mathbb{R}_x)$ so the constant must vanish and we have

$$K_W(t) = \frac{1}{2c}\chi_{[-ct,ct]} \qquad \text{when} \quad d = 1. \tag{3}$$

Then, for $\varphi \in \mathscr{S}(\mathbb{R}_x)$,

$$\frac{d}{dt}\langle K_W(t), \varphi\rangle = \frac{d}{dt}\frac{1}{2c}\int_{-ct}^{+ct} \varphi\, dx = \frac{\varphi(ct) + \varphi(-ct)}{2} = \frac{\langle \delta_{ct} + \delta_{ct}, \varphi\rangle}{2}.$$

Thus formula (2) becomes

$$u(t, x) = \frac{f(x - ct) + f(x + ct)}{2} + \frac{1}{2c}\int_{x-ct}^{x+ct} g\, dx, \tag{4}$$

which is *D'Alembert's formula*. An independent proof was given in §1.8 where a variety of consequences were analyzed. The reader is invited to give the generalizations made possible by using the language of Distribution Theory.

In particular, the results on domains of determinacy and domains of dependence are true when the data f, g belong to $H^s(\mathbb{R}^d) \times H^{s-1}(\mathbb{R}^d)$ for some s.

PROBLEM

1. Find the propagator K for the equation $\partial_t u + c\partial_x u = 0, c \in \mathbb{R}$, by Fourier transform techniques.

The next problems use D'Alembert's formula to study the *reflection* of waves at a boundary. Examples modeled by differential equations in $x > 0$, supplemented by boundary conditions at $x = 0$, are a semi-infinite string fixed at one end (the boundary condition is $u(t, 0) = 0$) or free at one end ($u_x(t, 0) = 0$), and acoustic waves in a semi-infinite narrow pipe closed at one end ($u_x(t, 0) = 0$, where u is the pressure).

Suppose that in the half-space $x > 0$, waves satisfy

$$u_{tt} - c^2 u_{xx} = 0 \qquad \text{for} \quad t \in \mathbb{R} \quad \text{and} \quad x > 0. \tag{5}$$

Suppose that for $t < 0$, u represents a wave approaching the boundary, that is,

$$\text{for } t < 0, \qquad u = \varphi(x + ct) \qquad \text{with} \quad \varphi \in C_0^\infty(]0, \infty[). \tag{6}$$

2. (i) Find a function $u(t, x) \in C^\infty(\mathbb{R}_t \times [0, \infty[)$ satisfying (5), (6) and the boundary condition $u(t, 0) = 0$ for $t \in \mathbb{R}$. *Hint*. Look for u as in formula (1.8.3).
   (ii) Solve the same problem with boundary condition $u_x(t, 0) = 0$.
   (iii) Sketch solutions with incoming wave consisting of a single bump.

   DISCUSSION. The solution u is uniquely determined by these conditions. This is proved, as in Problem 3.7.1, with integration by parts over $\Gamma \cap \{x \geq 0\}$. Because of the boundary conditions, the $\{x = 0\}$ boundary terms from integration by parts vanish. See also Problem 5.7.5.

   The term $\psi(x - ct)$ is called the *reflected wave*. For reasons which should be clear from the answers, the *reflection coefficient* is equal to one for the boundary condition $u_x = 0$ and equal to minus one for the condition $u = 0$.

§4.6. Rotation-Invariant Smooth Solutions of $\square_{1+3} u = 0$

The strategy for computing $K_W(t)$ when $d = 3$ is to approximate by the solutions to

$$\square u_\varepsilon = 0, \qquad u_\varepsilon(0, \cdot) = 0, \qquad \partial_t u_\varepsilon(0, \cdot) = j_\varepsilon(\cdot), \tag{1}$$

where j_ε is a rotation-invariant smooth approximant to δ. Such a u_ε is smooth and rotation-invariant. Since $j_\varepsilon \to \delta$ in $H^{-d/2-\varepsilon}(\mathbb{R}^d)$ for any $\varepsilon > 0$, it follows that $u_\varepsilon \to K_W$ in $C(\mathbb{R} : H^{1-d/2-\varepsilon})$.

When $d = 3$, explicit formulas for spherically symmetric solutions allow a painless passage to the limit $\varepsilon \to 0$.

Definition. If $u \in C^\infty(\mathbb{R}^d)$ and $A \in GL(\mathbb{R}^d)$, then $u_A \in C^\infty(\mathbb{R}^d)$ is defined by $u_A(x) \equiv u(A^{-1}x)$. The function u is called A *invariant* if and only if $u_A = u$.

The definition of u_A asserts that the value of u_A at Ax is equal to the value of u at x.

Definition. A mapping $P: C_0^\infty(\mathbb{R}^d) \to C^\infty(\mathbb{R}^d)$ is called *A invariant* if and only if for all $u \in C_0^\infty$, $P(u_A) = (Pu)_A$.

EXAMPLE. Problem 2.2.3 asked you to show that $\mathscr{F}: \mathscr{S}(\mathbb{R}^d) \to \mathscr{S}(\mathbb{R}^d)$ is A invariant if and only if A is orthogonal.

Proposition 1. *If $P = P(D)$ is a constant coefficient partial differential operator, the following are equivalent:*

(i) *$P(D)$ is A invariant.*
(ii) *$P(\xi)$ is A^t invariant.*
(iii) *$P(\xi) = \sum P_j(\xi)$, P_j homogeneous of degree j, and each $P_j(\xi)$ is A^t invariant.*

PROOF. The equivalence (ii) $\Leftrightarrow$ (iii) is immediate.

To prove (i) $\Leftrightarrow$ (ii), take $u = e^{ix\xi}$ and compute

$$P(u_A) = P(D)e^{i\langle x,(A^{-1})^t\xi\rangle} = P((A^{-1})^t\xi)e^{i\langle x,(A^{-1})^t\xi\rangle},$$
$$(Pu)_A = (P(\xi)e^{i\langle x,\xi\rangle})_A = P(\xi)e^{i\langle x,(A^{-1})^t\xi\rangle}.$$

That (i) $\Rightarrow$ (ii) follows immediately.

On the other hand, if (ii) holds, then $P(u_A) = (Pu)_A$ holds for $u = e^{ix\xi}$. For $u \in \mathscr{S}(\mathbb{R}^d)$ the identity then follows by superposition upon writing $u = \int e^{ix\xi}\hat{u}(\xi)\, d\xi/(2\pi)^{d/2}$. □

EXAMPLES. 1. $A \in GL(\mathbb{R}^d)$ is orthogonal if and only if Δ is A invariant.

2. $L \in GL(\mathbb{R}^{1+d})$ is a Lorentz transformation (e.g. preserves the bilinear form $c^2t - |x|^2$) if and only if $\square$ is L invariant.

3. No nonzero first-order scalar operator, $\sum a_j\partial_j$, is orthogonal invariant.

Next we extend the map $\mathscr{S} \ni u \overset{\Lambda}{\mapsto} u_A \in \mathscr{S}$. To show that Λ extends uniquely to $\mathscr{S}'$ we use Proposition 2.4.4. For $T \in \mathscr{S}'$, $\langle \Lambda' T, \varphi\rangle \equiv \langle T, \Lambda\varphi\rangle$ for all $\varphi \in \mathscr{S}$. Thus, if $T \in \mathscr{S}$,

$$\langle \Lambda' T, \varphi\rangle = \int T(x)\varphi(A^{-1}x)\, dx.$$

Make the change of variable, $y = A^{-1}x$, $dx = |\det(A)|\, dy$, to find

$$\langle \Lambda' T, \varphi\rangle = \int T(Ay)|\det(A)|\varphi(y)\, dy,$$

whence $\Lambda' T = |\det(A)|\, T_{A^{-1}}$ belongs to $\mathscr{S}$. Thus Λ extends, and for any $U \in \mathscr{S}'$, $\varphi \in \mathscr{S}$,

$$\langle \Lambda U, \varphi\rangle = \langle U, \Lambda'\varphi\rangle = \langle U, |\det(A)|\varphi_{A^{-1}}\rangle.$$

EXAMPLES. 1. $\delta_A = |\det(A)|\delta$, so δ is A invariant if and only if $|\det(A)| = 1$. For proof, compute

$$\langle \delta_A, \varphi \rangle \equiv \langle \delta, |\det(A)| \varphi_{A^{-1}} \rangle = |\det(A)| \varphi_{A^{-1}}(0) = |\det(A)| \varphi(0).$$

2. If A is orthogonal on $\mathbb{R}^d_x$ and if u satisfies (4.5.1), then u_A satisfies

$$\Box u_A = 0, \qquad u(0) = f_A, \qquad u_t(0) = g_A.$$

In particular, if $f_A = f$ and $g_A = g$, then $u_A = u$. Thus if $f, g \in \bigcup H^s$ are rotation-invariant, then the solution to (4.5.1) is rotation-invariant. Thus

$$(K_{\mathrm{W}}(t))_A = K_{\mathrm{W}}(t) \quad \text{for all orthogonal } A.$$

Similarly, the approximations u^ε defined in (1) are rotation-invariant whenever j_ε is. Thus u^ε is a smooth rotation-invariant solution to $\Box u = 0$.

If $w(x)$ is a smooth rotation-invariant function, then $w(x)$ depends only on $r = |x| \geq 0$. Thus $w(x) = W(|x|) = W(r)$, $W \in C^\infty(]0, \infty[)$. Since $W(r) = w(r, 0, 0, \ldots, 0)$, it follows that W extends uniquely to a smooth even function of $r \in \mathbb{R}$. Thus, the map $w(r, 0, \ldots, 0) = W(r)$ establishes a one-to-one correspondence between the smooth rotation-invariant functions on $\mathbb{R}^d$ and the smooth even functions on $\mathbb{R}$. This identification is usually taken for granted and we write $w = w(r)$ and abuse notation by writing $\partial_r w$ or w_r.

For $r > 0$ and w rotationally symmetric

$$\Delta w = \frac{1}{r^{d-1}} \partial_r (r^{d-1} \partial_r w) = w_{rr} + \frac{d-1}{r} w_r. \tag{2}$$

When $d = 3$, multiplication by r yields

$$r\Delta w = r w_{rr} + 2 w_r = (rw)_{rr} \qquad \text{when} \quad d = 3.$$

Thus, if u is a smooth rotation-invariant solution of $\Box_{1+3} u = 0$, then $v \equiv ru(t, r)$ satisfies $v_{tt} - c^2 v_{rr} = 0$ for $r \geq 0$. Since the left-hand side is a smooth odd function of r the equation holds on all of $\mathbb{R}_t \times \mathbb{R}_r$. Thus v is a smooth odd solution of the one-dimensional wave equation.

Conversely, if v is a smooth odd solution of $\Box_{1+1} v = 0$, then

$$u(t, x) \equiv \begin{cases} v(t, |x|)/|x| & \text{if } \ x \neq 0, \\ v_r(t, 0) & \text{if } \ x = 0, \end{cases} \tag{3}$$

is a smooth spherically symmetric solution of $\Box_{1+3} u = 0$. That u solves the wave equation in $x \neq 0$ is a simple calculation. On the other hand, u is smooth on $\mathbb{R}^{1+3}$ so $\Box u$ is a smooth function vanishing on $x \neq 0$ and therefore identically zero. This proves the following proposition.

Proposition 2. *The correspondence $v(t, r) = ru(t, r)$ defines a one-to-one correspondence between smooth rotationally symmetric solutions u to $\Box_{1+3} u = 0$ and smooth odd solutions to $\Box_{1+1} v = 0$.*

The most general such v is easily described. First, it must be a solution of $\Box_{1+1}v = 0$ whence

$$v = \varphi(ct + x) + \psi(ct - x), \qquad \varphi, \psi \in C^\infty(\mathbb{R}).$$

Since v is odd, v must be equal to the odd part of the right-hand side

$$2v = [\varphi(ct + x) - \varphi(ct - x)] + [\psi(ct - x) - \psi(ct + x)].$$

Let $F \equiv (\varphi - \psi)/2$, then

$$v = F(ct + x) - F(ct - x). \tag{4}$$

Conversely, for every $F \in C^\infty(\mathbb{R})$, (4) defines a smooth odd solution. Note that adding a constant to F does not change the value of the right-hand side. The function v is determined by F in the equivalence class of F in $C^\infty(\mathbb{R})/\mathbb{R}$.

Proposition 3. *The map $F \mapsto v$ given by* (4) *defines a one-to-one correspondence between $C^\infty(\mathbb{R})/\mathbb{R}$ and the smooth odd solutions of $\Box_{1+1}v = 0$.*

PROOF. It remains to show that the map is injective. Equivalently, it suffices to show that if v defined by (4) vanishes, then F must be constant. If v vanishes, differentiate (4) with respect to x and set $x = 0$ to find that $F' = 0$, whence F is constant. □

Combining the last two results yields the main result of this section.

Theorem 4. *The map $F \mapsto u$*

$$u(t, x) \equiv \begin{cases} \dfrac{F(ct + |x|) - F(ct - |x|)}{|x|}, & x \neq 0, \\ 2F'(ct), & x = 0, \end{cases} \tag{5}$$

defines a one-to-one correspondence between $C^\infty(\mathbb{R})/\mathbb{R}$ and the rotation-invariant smooth solutions to $\Box_{1+3}u = 0$.

The rotationally symmetric solutions of the theorem are key examples in forming an intuition into the behavior of multi-dimensional wave equations.

As a first example, consider the behavior of smooth radial solutions with Cauchy data supported in the ball $|x| < \rho$. Let $f(r) \equiv ru(0, r)$ and $g(r) \equiv ru_t(0, r)$, both smooth odd functions of r. The Cauchy data for u defined by (5) yield the equations

$$F(r) - F(-r) = f(r) \qquad \text{and} \qquad F'(r) - F'(-r) = \frac{g(r)}{c}.$$

Differentiating the first with respect to r and adding yields $F' = f' + g/c$. As F is only determined up to a constant, we may take

$$F(r) = \int_{-\infty}^{r} f'(s) + \frac{g(s)}{c}\, ds.$$

Since g is odd and f has compact support the integral vanishes for $r > \rho$. Thus F is supported in $[-\rho, \rho]$.

Consider formula (5) with $r \geq 0$. For t large, $ct + r > \rho$, so $F(ct + r) = 0$. Then $u(t, r) = -F(ct - r)/r$ is an outgoing spherical wave with profile $-F/r$. In particular, it decays like t^{-1}.

In the distant past, $ct - r < -\rho$ so $F(ct - r) = 0$. Then $u(t, r) = F(ct + r)/r$ is an incoming spherical wave with cross section like F/r. The big picture is an incoming spherical wave which emerges as an outgoing spherical wave with profile changed only by a factor of -1.

Finally, notice that u vanishes unless at least one of $ct \pm |x|$ belongs to $[-\rho, \rho]$. The set of such t, $|x|$ is sketched in Figure 4.6.1. For the support of u, this yields

$$\operatorname{supp} u \subset \{(t, x) \colon -\rho \leq |x| - c|t| \leq \rho\}. \tag{6}$$

For $t > 0$ this is the region between the two light cones $|x| - ct = \pm\rho$ sketched in Figure 4.6.2.

The derivative F' in formula (5) causes an interesting loss of regularity in the classical C^k spaces. If $F \in C_0^k(\mathbb{R})$ with $k \geq 3$ and F vanishes on a neighborhood of the origin, then the corresponding solution is $C^{k-1}(\mathbb{R}^{1+3})$ because of the derivative F' in (5). On the other hand, the Cauchy data satisfy $u(0) \in C^k$,

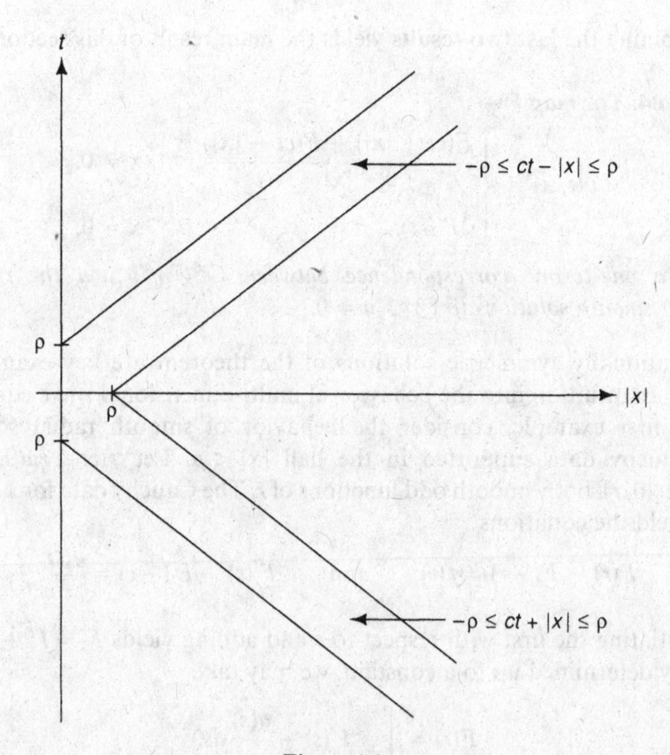

Figure 4.6.1

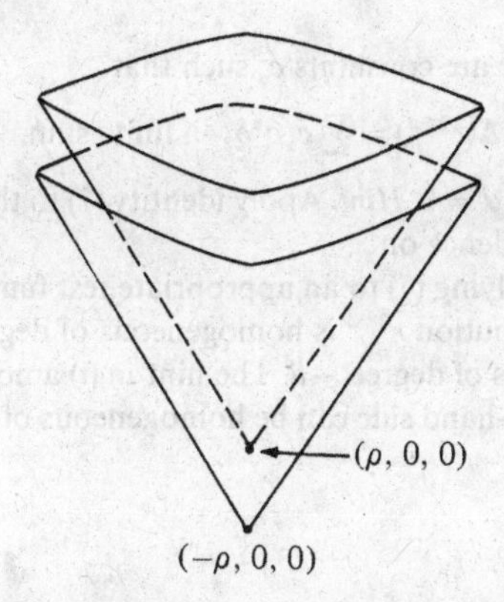

Figure 4.6.2. The support of u is between the two light cones

$u_t(0) \in C^{k-1}$, that is, the Cauchy data have the natural regularity for a solution of class C^k. In the C^k category there is a loss of one derivative. In the spaces H^s there was no such loss. The loss of regularity in the C^k category is due to *focusing.*

A bound on the number of continuous derivatives which may be lost in dimension d can be found as follows. If the data lie in $C_0^k \times C_0^{k-1}$, then they lie in $H^k \times H^{k-1}$. Then Theorem 3.7.2 implies that $\partial_t^j u \in C(\mathbb{R} : H^{k-j}(\mathbb{R}^d))$. The Sobolev Lemma implies that $u \in C^m(\mathbb{R}^{1+d})$ provided $m < k - d/2$. Thus there is a loss of no more than $d/2 + \varepsilon$ derivatives. This indicates that we might expect more loss as d increases and that is the case. The rule of thumb is one-half of a derivative per space dimension above 1.

For $F \in C^\infty$, the presence of the derivative in (5) causes no loss of regularity but is felt in the phenomenon of amplification due to focusing. A wave initially supported away from the origin will have amplitudes of size F' at later times and F' may be much larger than $\sup|F|$. This can happen even if the initial data for $\partial_t u$ vanish. Thus, measured in sup norm the map from data at time zero to data at time t is not continuous. This discontinuous dependence and the loss of derivatives are two aspects of the same physical phenomenon. The theme is discussed again in Problem 4.7.1.

Problems

The next problems consider rotationally symmetric solutions of the Laplace equation.

1. Find all rotationally symmetric smooth harmonic functions on $\mathbb{R}^d \backslash \{0\}$.
2. In Problem 2.5.5 you showed that when $d = 3$, $(a^2 - \Delta)(e^{-ar}/r) = c(a)\delta$. Compute $\Delta(1/r)$ by passing to the limit $a \to 0$.
   Discussion. Problems 3 and 5.10.3 give independent proofs.

From Problem 1 you know that for $d \geq 3$, r^{2-d} is harmonic on $\mathbb{R}^d \backslash 0$. Since r^{2-d} is locally integrable it follows that $\Delta(r^{2-d})$ is a well-defined distribution with support at the origin. Therefore there are constants c_α such that

$$\Delta(r^{2-d}) = \sum c_\alpha \partial^\alpha \delta \quad \text{a finite sum.} \tag{7}$$

3. (i) Prove that $c_\alpha = 0$ if $\alpha \neq 0$. *Hint*. Apply identity (7) to the test function $\psi(\lambda x)$ and consider the dependence on λ.
   (ii) Evaluate c_0 by applying (7) to an appropriate test function in $\mathscr{S}(\mathbb{R}^d)$.
   DISCUSSION. The distribution r^{2-d} is homogeneous of degree $2 - d$. It follows that $\Delta(r^{2-d})$ is homogeneous of degree $-d$. The hint in (i) amounts to showing that the only way that the right-hand side can be homogeneous of degree $-d$ is if all $c_\alpha = 0$ for $\alpha \neq 0$.

§4.7. The Wave Equation Propagator for $d = 3$

To compute the solution K_W to

$$\square_{1+3} K = 0, \qquad K(0, \cdot) = 0, \qquad \partial_t K(0, \cdot) = \delta, \tag{1}$$

we analyze the limit $\varepsilon \to 0$ in (4.6.1) where

$$j \in C_0^\infty(|x| < 1), \qquad j \geq 0, \qquad \int j\, dx = 1, \qquad j_\varepsilon \equiv \varepsilon^{-d} j\left(\frac{x}{\varepsilon}\right), \tag{2}$$

and j is rotationally symmetric. Since $j_\varepsilon \to \delta$ in $H^s(\mathbb{R}^d)$ for all $s < -d/2$, we have for any t

$$H^s\text{-lim}\ u_\varepsilon(t) = K_W(t). \tag{3}$$

In particular, $u_\varepsilon(t) \to K_W(t)$ in $\mathscr{S}'(\mathbb{R}_x^d)$. The results of the last section yield explicit formulas for u_ε and thereby an evaluation of the limit.

By Theorem 4.6.4, there is an $F_\varepsilon \in C^\infty(\mathbb{R})$ with

$$u_\varepsilon(t, x) = \begin{cases} (F_\varepsilon(ct + r) - F_\varepsilon(ct - r))/r, & r = |x| \neq 0, \\ 2F_\varepsilon'(ct), & r = |x| = 0. \end{cases} \tag{4}$$

Setting $t = 0$ yields $F_\varepsilon(r) - F_\varepsilon(-r) = 0$, so F_ε is an even function of r. Computing $\partial_t u_\varepsilon(0, \cdot)$ yields

$$F_\varepsilon' = \frac{r j_\varepsilon(r)}{2c}. \tag{5}$$

Note that any solution of (5) is automatically even in r since $rj_\varepsilon(r)$ is odd. F is only determined modulo an additive constant, so we may take

$$F_\varepsilon(r) = \int_{-\infty}^{r} \frac{r j_\varepsilon(r)}{2c}\, dr. \tag{6}$$

To compute $\lim u_\varepsilon$ the following properties are important:

$$\operatorname{supp} u_\varepsilon \subset \{(t, x) : c|t| - \varepsilon < |x| < c|t| + \varepsilon\}, \tag{7}$$

$$(\operatorname{sgn} t) u_\varepsilon(t, x) \geq 0, \tag{8}$$

and

$$\int u_\varepsilon(t, x)\, dx = t. \tag{9}$$

PROOFS. Property (7) follows from formula (4.6.6).

To prove (8), note that for $t > 0$ and $x \in \mathbb{R}^d$

$$|x| u_\varepsilon(t, x) = F_\varepsilon(ct + r) - F_\varepsilon(ct - r) = \int_{ct-r}^{ct+r} \frac{r j_\varepsilon(r)}{2c}\, dr.$$

If $ct - r \geq 0$, this is nonnegative since $j \geq 0$. If $ct - r < 0$, the integral from $ct - r$ to $r - ct$ vanishes so

$$|x| u_\varepsilon = \int_{r-ct}^{r+ct} \frac{r j_\varepsilon(r)}{2c}\, dr \geq 0.$$

This proves (8) for $t > 0$. The solution of (4.6.1) is odd in t since $-u(-t, x)$ solves the same initial value problem, so (8) for $t < 0$ follows from the case $t > 0$.

For (9) note that since u_ε is smooth and supported in $|x| \leq c|t| + \varepsilon$, differentiation under the integral sign is justified to give

$$\partial_t^2 \int u_\varepsilon(t, x)\, dx = \int \partial_t^2 u_\varepsilon(t, x)\, dx = \int c^2 \Delta_x u_\varepsilon\, dx.$$

The last equality uses the equation $\square u = 0$. Integrating by parts shows that the last integral vanishes. Thus, $\int u_\varepsilon\, dx = at + b$ with $a, b \in \mathbb{C}$. Evaluating at $t = 0$ yields $b = 0$. Differentiating with respect to time yields $a = \int \partial_t u_\varepsilon\, dx$. Evaluating this at $t = 0$ shows that $a = 1$. □

We now compute the $\mathscr{S}'(\mathbb{R}^3)$ limit of the $u_\varepsilon(t)$. For $\psi \in \mathscr{S}(R^3)$, use polar coordinates to write, for $t > 0$,

$$\langle u_\varepsilon(t), \psi \rangle = \int_{S^2} \int_0^\infty u_\varepsilon(t, r) \psi(r\omega) r^2\, dr\, d\omega.$$

Here $d\omega$ is the element of surface area on the unit sphere S^2. Now u_ε vanishes if r is more that ε units form ct. This suggests writing

$$\langle u_\varepsilon(t), \psi \rangle = \int_{S^2} \int_0^\infty u_\varepsilon(t, r) \psi(ct\omega) r^2\, dr\, d\omega + \text{error},$$

$$\text{error} \equiv \iint u_\varepsilon(t, r)(\psi(r\omega) - \psi(ct\omega)) r^2\, dr\, d\omega.$$

In the support of u_ε, $|r\omega - ct\omega| < \varepsilon$, so

$$|\text{error}| \leq \varepsilon \|\nabla_x \psi\|_{L^\infty} \iint |u_\varepsilon| r^2\, dr\, d\omega = \varepsilon \|\nabla_x \psi\|_{L^\infty} \int u_\varepsilon\, dx = \varepsilon \|\nabla_x \psi\|_{L^\infty} t.$$

The integral with $\psi(ct\omega)$ is computed explicitly using Fubini's Theorem

$$\int_{S^2}\int_0^\infty u_\varepsilon(t, r)\psi(ct\omega)r^2\,dr\,d\omega = \int_0^\infty u_\varepsilon(t, r)r^2\,dr \int_{S^2} \psi(ct\omega)\,d\omega. \tag{10}$$

The last term is the integral of ψ over the surface of the sphere of radius ct, on which the element of surface area is given by $d\sigma = (ct)^2\,d\omega$. The integral of u_ε in (10) is equal to $t/4\pi$, since it is equal to $(4\pi)^{-1}$ times the integral of u over $\mathbb{R}^3$. Thus

$$\langle u_\varepsilon(t), \psi\rangle = \frac{t}{4\pi}\frac{1}{(ct)^2}\int_{|x|=ct} \psi\,d\sigma + O(\varepsilon).$$

This completes the proof of the following important formula when $t > 0$.

Theorem 1. *For $d = 3$ and $t \neq 0$*

$$K_W(t) = \frac{1}{4\pi c^2 t}\,d\sigma_{|x|=c|t|}. \tag{11}$$

PROOF. We have just proved the identity for $t > 0$. It follows for $t < 0$ since both sides are odd functions of t. □

Formula (4.2.6) shows that for $f, g \in \mathscr{S}(\mathbb{R}^3)$, the solution to the initial value problem for the wave equation is given by

$$u(t, x) = \frac{1}{4\pi c^2 t}\int_{|y-x|=ct} g(y)\,d\sigma(y) + \frac{\partial}{\partial t}\left(\frac{1}{4\pi c^2 t}\int_{|x-y|=ct} f(y)\,d\sigma(y)\right). \tag{12}$$

The rest of this section is devoted to studying this formula.

The Cauchy data of K_W are supported at $\{0\}$, so the finite propagation speed (Problem 3.7.5) implies that

$$\operatorname{supp} K_W(t) \cup \operatorname{supp}\frac{\partial K_W(t)}{\partial t} \subset \{|x| \le c|t|\}.$$

This is clearly visible in formula (11).

Since K_W is compactly supported, formula (5) extends immediately to all $f, g \in C^\infty(\mathbb{R}^{1+3})$, the corresponding solution belonging to $C^\infty(\mathbb{R}^{1+3})$. The existence part of this assertion, but not the formula for the solution, is valid in all dimensions (Problem 3.7.4).

(i) **Huygens' Principle.** *If $u \in C^\infty(\mathbb{R}^{1+3})$ satisfies $\Box_{1+3}u = 0$ and $\Xi(t) \equiv \{\operatorname{supp} u(t) \cup \operatorname{supp} u_t(t)\}$, then*

$$\Xi(t) \subset \{(t, x)\colon \operatorname{dist}(x, \Xi(0)) = c|t|\}.$$

The set $\Xi(t)$ is the set occupied by the wave at time t. The result is an immediate consequence of formula (12). It expresses the fact that signals travel with speed no less than 1, in addition to the already known fact that they propagate no faster than 1. The lower bound on the speed is most clearly seen for data supported in $|x| < \rho$. In that case, the solution is supported in $-\rho < |x| - c|t| < \rho$. When $t > 0$, this is the region between the two light cones $|x| - ct = \pm\rho$ sketched in Figure 4.6.2 There is a hole in the support of radius $c|t| - \rho$ corresponding to the fact that signals cannot travel slowly.

In the next section we show that the conclusion of (i) is not correct when $d = 2$. It holds precisely when d is odd and greater than 1. For $d = 1$ it is nearly correct, an analogous result is correct for the support of $\nabla_{t,x}u$.

(ii) *The result* (i) *is true for any* $u \in \mathscr{D}'(\mathbb{R}^{1+3})$ *which satisfies* $\Box u = 0$.

SKETCH OF PROOF. Let j_ε be a standard approximate identity in $\mathbb{R}^{1+3}$ and apply (i) to $j_\varepsilon * u$. Then pass to the limit $\varepsilon \to 0$. □

The assertion (i) can be expressed in another way in terms of domains of influence and determinancy.

(iii) **Sharp Domain of Influence/Dependence.** *Suppose that* $u \in C(\mathbb{R} : H^s(\mathbb{R}^3))$ *satisfies* $\Box u = 0$. *Then values of the Cauchy data in an open set* $\mathcal{O}$ *influence the solution only on* $\{(t, x)\colon \operatorname{dist}(x, \mathcal{O}) = c|t|\}$. *The values of the solution u in an open set U in space–time depend only on the values of the Cauchy data in* $\{x \in \mathbb{R}^3\colon \exists(t, y) \in U, \operatorname{dist}(x, y) = c|t|\}$.

(iv) **Monotonicity.** *If* $u \in C^\infty(\mathbb{R}^{1+3})$ *satisfies* $\Box u = 0$ *and*

$$u(0, \cdot) = 0, \qquad u_t(0, \cdot) \geq 0,$$

then $(\operatorname{sgn} t)u \geq 0$.

This is an immediate consequence of the formula. The analogous result is false when $d > 3$ (see Treves [Tr], Folland [Fo], Courant and Hilbert [CH], or Garabedian [Gara] for the propagator K_W when $d > 3$). Using Duhamel's formula together with an approximation argument as in (ii), one finds the same conclusion under the weaker hypotheses, $u \in C^2(\mathbb{R} : \mathscr{D}'(\mathbb{R}^3))$, $(\operatorname{sgn} t)\Box u \geq 0$, $u(0) = 0$, $u_t(0) \geq 0$. Here ≥ 0 is interpreted in the sense of distributions.

(v) *If u is a solution of the wave equation on* $\mathbb{R}^{1+3}$ *and the Cauchy data satisfy* $u(0) \in C^k$, $u_t(0) \in C^{k-1}$, *then* $u \in C^{k-1}(\mathbb{R}^{1+3})$.

We have seen that this loss of one derivative actually occurs for spherically symmetric solutions. The above result shows that one never looses more than one classical derivative when $d = 3$. If one measures regularity in the H^s sense there is no loss of derivatives in any dimension.

Problems

We have seen that the regularity $u, u_t \in C^\alpha \times C^{\alpha-1}$ is not propagated while $u, u_t \in H^s \times H^{s-1}$ is. The latter regularity is called *continuable*. One might think that the difference is that, in the second case, the norms are defined by integrals. Littman's Theorem asserts that norms based on L^p for $p \neq 2$ are not propagated by the wave operator in dimensions $d > 1$. This is the wave equation analogue of Problem 3.4.2. For $d = 3$, a simple proof is available.

Consider the solutions to $(\partial_t^2 - c^2\Delta)u = 0$ with initial data

$$u(0, \cdot) = 0, \qquad u_t(0, \cdot) = g.$$

Then the map $u_t(0) \mapsto u_t(t)$ is the Fourier multiplier $\sin(ct|\xi|)$. If $g \in L^2$, then $u_t \in C(\mathbb{R} : L^2)$. It is reasonable to ask whether $g \in L^p$ yields a solution with $u_t \in C(\mathbb{R} : L^p)$. The next problem shows that for $t \neq 0$, the set of g in L^p with the property that $u_t(t) \in L^p$ is a set of first category in L^p.

1. **Littman's Theorem for** $d = 3$. *Prove that for* $d = 3$, $p \neq 2$, *and* $t \neq 0$

$$\sup \frac{\|u_t(t)\|_{L^p}}{\|u_t(0)\|_{L^p}} = \infty,$$

the supremum over all $g \in \mathcal{S}(\mathbb{R}^d)\backslash 0$.

Hint. Consider rotationally symmetric g and the corresponding explicit spherical wave solutions. Solutions with small support at $t = 0$ spread over an annular region of much larger volume at time t. The general principle described in Problem 3.4.2 is in operation.

For $g \in \mathcal{S}(\mathbb{R}^d)$ the Radon transform of g is a function in $C^\infty(\mathbb{R} \times S^{d-1})$ defined by $h(s, \omega) \equiv \int_{x\cdot\omega=s} g\, d\sigma$. Thus h encodes the integrals of u over the hyperplanes of $\mathbb{R}^d$.

2. Let u be the solution of the wave equation with $u(0) = 0$, $u_t(0) = g \in \mathcal{S}(\mathbb{R}^3)$. Prove that $\lim_{t\to\infty} tu(t, y + ct\omega) = h(y\cdot\omega, \omega)/4\pi c^2$, the limit being uniform for $\omega \in S^{d-1}$ and y in compact subsets of $\mathbb{R}^d$.
Discussion. It follows that for $t \gg 1$, $u(t, x) \cong h(|x| - ct, x/|x|)/4\pi c^2 t$. If h did not depend on ω this would be an outgoing spherical wave as in §4.6. The general case is an outgoing wave whose cross section depends on the direction $\omega = x/|x|$. A similar formula holds for initial data, $u(0) = 0$, $u_t(0) \neq 0$ (exercise). The corresponding description of u for $t \gg 1$ is due to Friedlander [Fr]. The map $u \mapsto h$ gives the translation representation which is central to the Scattering Theory of Lax and Phillip [LP]. The Radon transform, h, also yields an elegant proof of Huygen's principle in all odd dimensions (see Folland [Fo]).
 Note that Theorem 4.3.6 shows that the asymptotic state for the heat equation is described by one scalar quantity, $\int u(0, x)\, dx$. For the Schrödinger equation, Theorem 4.4.2 shows that all of $\mathcal{F}u(0)$ is needed, that is, a function of three variables. For the wave equation we need $h(s, \omega)$, a function on $\mathbb{R} \times S^2$. Again there is a function of three variables. In the last two examples, the asymptotic states have as much variety as the initial data.
 Many partial differential equations have wave-like solutions. Hyperbolic equations are peculiar in having such a wide variety of waves. This is important for nondigital transmission of information. It is no accident that hearing and sight rely

on acoustic and light waves both governed by linear hyperbolic equations. Information in the nervous system is transmitted in digital form and the governing equations are parabolic.

§4.8. The Method of Descent

Knowing K_W in dimension d yields a formula for K_W in dimension d' for all $d' < d$. The method is called Hadamard's method of descent and rests on the simple observation that if one has a function of f of $x_1, \ldots, x_{d'}$ and $d > d'$, then there is a naturally defined function F of d variables obtained by ignoring the values of $x_{d'+1}, \ldots, x_d$, that is, $F(x_1, \ldots, x_d) \equiv f(x_1, \ldots, x_{d'})$. If $\pi: \mathbb{R}^d \to \mathbb{R}^{d'}$ is the canonical projection $x \mapsto x_1, \ldots, x_{d'}$, then this relation is simply that $F = f \circ \pi$.

Proposition 1. *If $d > d'$ and $f, g \in C^\infty(\mathbb{R}^{d'})$ and $F, G \in C^\infty(\mathbb{R}^d)$ are defined by $F \equiv f \circ \pi$, $G = g \circ \pi$, then the unique solution $w \in C^\infty(\mathbb{R}^{1+d})$ to*

$$\Box_{1+d} w = 0, \qquad w(0, \cdot) = F, \qquad w_t(0, \cdot) = G, \tag{1}$$

and the solution u to

$$\Box_{1+d'} u = 0, \qquad u(0, \cdot) = f, \qquad u_t(0, \cdot) = g, \tag{2}$$

are related by $w(t, x_1, \ldots, x_d) = u(t, x_1, \ldots, x_{d'})$, that is, $w = u(t, \pi(x))$.

PROOF. Define $w \in C^\infty(\mathbb{R}^{1+d})$ by $w(t, x) = u(t, \pi(x))$. Then w solves the initial value problem (1) and uniqueness of solutions completes the proof. □

Take $d = 3$, $d' = 2$, and $f = 0$. Then

$$u(t, \cdot) = K_{d=2}(t) * g. \tag{3}$$

Since $K(t, x) = K(t, -x)$, formula (3) for $(x_1, x_2) = 0$ yields

$$u(t, 0) = \langle K_{d=2}(t), g \rangle. \tag{4}$$

On the other hand,

$$u(t, 0) = w(t, 0) = \langle K_{d=3}(t), g \rangle$$
$$= \frac{1}{4\pi t c^2} \int_{|x_1, x_2, x_3| = c|t|} g(x_1, x_2)\, d\sigma.$$

As g is independent of x_3, the latter integral is twice the integral over the hemisphere in $x_3 > 0$. On the hemisphere, x_1, x_2 can be chosen as coordinates with $x_3 = ((ct)^2 - x_1^2 - x_2^2)^{1/2}$. The unit upward normal is $n = (x_1, x_2, x_3)/c|t|$. The element of surface area is

$$\frac{1}{n_3}\, dx_1\, dx_2 = \frac{c|t|}{\sqrt{(ct)^2 - x_1^2 - x_2^2}}\, dx_1\, dx_2.$$

Thus

$$u(t, 0) = \frac{2c|t|}{4\pi c^2 t} \int \frac{g(x_1, x_2)}{\sqrt{(ct)^2 - x_1^2 - x_2^2}} \, dx_1 \, dx_2.$$

Comparing with (4) we have proved the following theorem.

Theorem 2. *For $d = 2$ and $t \neq 0$*

$$K_W(t, x) = \begin{cases} \dfrac{\operatorname{sgn} t}{2\pi c\sqrt{(ct)^2 - x_1^2 - x_2^2}}, & |x| < c|t|, \\ 0, & |x| \geq c|t|. \end{cases} \tag{5}$$

K_W is a smooth function except when $|x| = |ct|$. There K diverges like $(c|t| - |x|)^{-1/2}$. Thus, for $t \neq 0$, $K_W(t, \cdot) \in L^1(\mathbb{R}_x^2)$. This is more regular than K_W in dimension $d = 3$, where K is a measure, and less regular than in $d = 1$, where K_W is a bounded function. Measured in the scale $H^s(\mathbb{R}^d)$, $K_W(t, \cdot) \in H^{-d/2+1-\varepsilon}(\mathbb{R}_x^d)$ for all $\varepsilon > 0$. K_W loses one-half a derivative for each dimension.

As in the previous section, finite speed is reflected in the fact that $K_W(t)$ is supported in $|x| \leq c|t|$. However, for $d = 2$, $(\operatorname{sgn} t)K > 0$ on $|x| < c|t|$, the support fills the interior of the forward and backward light cones. Thus one cannot improve upon the general inclusion

$$\operatorname{supp} u \subset \{(t, x) \colon \exists y \in \operatorname{supp} u(0) \cup \operatorname{supp} u_t(0), |x - y| \leq c|t|\}, \tag{6}$$

which follows from the finite speed of propagation. For example, if $g \geq 0$ and $f = 0$, then $(\operatorname{sgn} t)u$ is strictly positive on the interior of the set on the right-hand side of (6).

For an initial disturbance supported near the origin, the solution in $d = 2$ decays like $1/|t|$ as $|t| \to \infty$ in $\{(t, x) \colon |x| < (c - \varepsilon)|t|\}$. For $d = 3$ a similar initial disturbance leads to a wave which, for t large, vanishes identically in such a shrunken forward light cone. This is a manifestation of the fact that for $d = 3$, K is supported on the surface of the light cone while when $d = 2$ the support fills the entire insides of the cone. We say that Huygens' principle is valid when $d = 3$ and not when $d = 2$. I can offer no persuasive physical intuition to explain the failure of Huygens' principle in dimension 2. The principle is true for odd $d \geq 3$ and not for other dimensions. Even for $d \geq 3$ and odd, the Huygens' principle is destroyed if the wave operator is slightly perturbed, say to $\square + \varepsilon$. The historical association with Huygens is unconvincing.

There is a strong vestige of Huygens' principle, namely, singularities propagate with speed exactly equal to c.

Corollary 3. *Suppose that $f, g \in H^s \times H^{s-1}$ and that u is the solution of $\square u = 0$ with Cauchy data equal to f, g. Let*

$$\Gamma \equiv \operatorname{sing\,supp}(f) \cup \operatorname{sing\,supp}(g),$$

then

$$\text{sing supp } u(t) \cup \text{sing supp } u_t(t) \subset \{x: (\exists y \in \Gamma), |x - y| = c|t|\}. \qquad (7)$$

PROOF. Write f and g as a sum of two terms, one in C^∞ and the other supported in the set of points at distance $\varepsilon/2$ from sing supp$(f) \cup$ sing supp(g). The contribution from the smooth part is smooth, so we may suppose that f and g are supported in such an $\varepsilon/2$ neighborhood of the singular support of the Cauchy data.

The key observation is that

$$\text{sing supp } K_W(t) \subset \{x: |x| = c|t|\}. \qquad (8)$$

This is valid in all dimensions though we have proved it only for $d = 1, 2, 3$. A general proof is outlined in Problem 2.

If $t \neq 0$ and x does not belong to the right-hand side of (7), choose $\varepsilon > 0$ so that the $\mathbb{R}^d$ ball of radius ε with center x is disjoint from the set on the right-hand side. Choose $\varphi \in C_0^\infty(|x| < c|t|)$ with φ equal to 1 at all points in the $c|t| - \varepsilon/2$ disk.

Write $u = K * g + \partial_t K * f$ and $u_t = \partial_t K * g + K * \Delta f$. Write $K(t)$ as a sum $\varphi K + (1 - \varphi)K$. Then $\varphi K(t)$ and $\varphi \partial_t K(t)$ are smooth and compactly supported so the contributions from φK are smooth on $\mathbb{R}^d$. The contributions from $(1 - \varphi)K$ are supported in an ε neighborhood of the right-hand side of (7). It follows that the singularities of u belong to this ε neighborhood. Since this is true for all ε, the theorem is proved. □

The sharp propagation speed (7) for singular supports is sometimes called the *generalized Huygens principle*. It is valid in all dimensions (problem 2). The natural generalization is true for variable coefficient hyperbolic equations as well. The generalized Huygens principle is of much wider utility than the strict Huygens principle.

The fact that $(\text{sgn } t)K_W \geq 0$ when $d = 1, 2, 3$ shows that $\square_{1+1}$, $\square_{1+2}$, and $\square_{1+3}$ share the monotonicity property. On the other hand, if $d > 3$, it is no longer true that $(\text{sgn } t)K_W \geq 0$.

PROBLEMS

The singular support can be criticized because it treats discontinuities in a function on an equal footing with discontinuities in the ten millionth derivative. This weakness is overcome by introducing the H^s singular support as follows.

If $x \in \Omega \subset \mathbb{R}^d$ and $u \in \mathscr{D}'(\Omega)$ we say that u is H^s at x, and write $u \in H^s(x)$ if there is a $\varphi \in C_0^\infty(\Omega)$ with φ identically equal to 1 on a neighborhood of x and $\varphi u \in H^s(\mathbb{R}^d)$. Note that this makes sense since $\varphi u \in \mathscr{E}'(\Omega)$ so extends naturally to an element of $\mathscr{E}'(\mathbb{R}^d) \subset \mathscr{S}'(\mathbb{R}^d)$ with support in Ω.

Definition. For $u \in \mathscr{D}'(\Omega)$ and $s \in \mathbb{R}$

$$\text{sing supp}_s\, u \equiv \{x \in \Omega: u \text{ is not } H^s \text{ at } x\}.$$

Sing supp_s is closed and increasing with s and

$$\text{sing supp}(u) \supset \bigcup_s \text{sing supp}_s\, u. \tag{9}$$

Since the singular support is closed, it follows that

$$\text{sing supp}(u) \supset \text{cl}\left(\bigcup_s \text{sing supp}_s\, u\right). \tag{10}$$

1. Construct an example where the inclusion (9) is strict. Show that the inclusion (10) is an equality.

In the next problem you will prove (8) using the method of nonstationary phase described at the end of §4.4. The strategy is as follows. Use the representation

$$K_W(t) = \mathscr{F}^{-1}\left(\frac{\sin(ct|\xi|)}{c|\xi|}\right),$$

and write $\sin(\theta)$ as $(e^{i\theta} - e^{-i\theta})/2i$ to show that formally

$$2icK_W(t, x) = \sum_{\pm} \int \frac{\pm e^{i\varphi^{\pm}(t,x,\xi)}}{|\xi|}\, d\xi, \qquad \varphi^{\pm} = x\xi \pm ct|\xi|.$$

Fix t, x with $|x| \neq c|t|$. Then the phase is nonstationary, that is, $\nabla_\xi \varphi \neq 0$. As in §4.4, construct first-order partial differential operators $L^{\pm}$ with $L^{\pm} e^{i\varphi^{\pm}} = e^{i\varphi^{\pm}}$. Dropping the $\pm$, K_W is a sum of terms of the form

$$\int (L^m e^{i\varphi})\left(\frac{1}{|\xi|}\right) d\xi = \int e^{i\varphi}(L')^m\left(\frac{1}{|\xi|}\right) d\xi.$$

For $m = m(k)$ large the resulting expression can be shown to be C^k on a neighborhood of t, x.

2. Fill in the details in the above proof as follows:
   (i) Choose $\chi \in C_0^\infty(\mathbb{R}^d_\xi)$ with χ identically equal to 1 on a neighborhood of $\xi = 0$. Write $\mathscr{F}K_W = \chi\mathscr{F}K_W + \psi\mathscr{F}K_W$ with $\psi \equiv 1 - \chi$. Prove that the first term has smooth inverse transform.
   (ii) For the second term show that

$$\mathscr{F}^{-1}(\psi\mathscr{F}K_W) = (1 - \Delta_x)^d \mathscr{F}^{-1}((1 + |\xi|^2)^{-d}\psi\mathscr{F}K_W),$$

   so it suffices to show that $\mathscr{F}^{-1}((1 + |\xi|^2)^{-d}\psi\mathscr{F}K_W)$ is smooth near t, x. Show that this inverse Fourier transform is an absolutely convergent integral.
   (iii) Analyze the absolutely convergent integral following the nonstationary phase ideas sketched above.

3. Suppose $f, g \in \bigcup (H^s(\mathbb{R}^d) \times H^{s-1}\mathbb{R}^d))$ and u is the solution of $\Box_{1+d} u = 0$ with Cauchy data equal to f, g. Let

$$\Gamma_\sigma \equiv \text{sing supp}_\sigma f \cup \text{sing supp}_{\sigma-1}\, g.$$

Prove that

$$\text{sing supp}_\sigma\, u(t) \cup \text{sing supp}_{\sigma-1}\, u_t(t) \subset \{x : (\exists y \in \Gamma_\sigma), |x - y| = c|t|\}. \tag{11}$$

Hint. Use the result of Problem 2.

§4.9. Radiation Problems

Consider the radiation from an antenna which begins to radiate at a time t_0. One finds an initial value problem

$$\square_{1+d} u = F, \qquad u = F = 0 \qquad \text{for} \quad t < t_0. \tag{1}$$

If $F \in L^1_{\text{loc}}([0, \infty[: H^s(\mathbb{R}^d))$ for some $s \in \mathbb{R}$, Duhamel's formula yields

$$u(t) = \int_{-\infty}^{t} R(t-\sigma)F(\sigma)\, d\sigma = \int_{-\infty}^{t} K_{\mathrm{W}}(t-\sigma) * F(\sigma)\, d\sigma, \tag{2}$$

where K_{W} is given by formulas (4.5.3), (4.8.5), and (4.7.11) for $d = 1, 2, 3$. For $t > \sigma$ we have

$$d = 1, \qquad R(t-\sigma)F(\sigma) = \frac{1}{2c} \int_{|x-y| \le c(t-\sigma)} F(\sigma, y)\, dy,$$

$$d = 2, \qquad R(t-\sigma)F(\sigma)$$

$$= \frac{1}{2\pi c} \iint_{|x-y| \le c(t-\sigma)} F(\sigma, y)(c^2(t-\sigma)^2 - |x-y|^2)^{-1/2}\, dy,$$

$$d = 3, \qquad R(t-\sigma)F(\sigma) = (4\pi c^2)^{-1} \iint_{|x-y| = c(t-\sigma)} F(\sigma, y) \frac{1}{t-\sigma}\, d\Sigma(y).$$

Using the expression for $d = 1$ in (2) yields

$$u(t, x) = \frac{1}{2c} \iint_{|x-y| \le c(t-\sigma)} F(\sigma, y)\, d\sigma\, dy. \tag{3}$$

The double integral is over the backward light cone from (t, x). For $d = 2$ we find

$$u(t, x) = \frac{1}{2\pi c} \iiint_{|x-y| \le c(t-\sigma)} \frac{F(\sigma, y)}{(c^2(t-\sigma)^2 - |x-y|^2)^{1/2}}\, d\sigma\, dy. \tag{4}$$

Here we have a triple integral over the solid backward light cone dropped from (t, x). For $d = 3$ we find

$$u(t, x) = (4\pi c^2)^{-1} \iiint_{|x-y| = c(t-\sigma)} \frac{F(\sigma, y)}{(t-\sigma)} \frac{dA}{(1+c^2)^{-1}}, \tag{5}$$

where dA is the element of three-dimensional area on the surface of the backward light cone from (t, x). The Pythagorean factor at the end comes from the relation $dA = (1 + c^2)^{1/2}\, d\sigma\, d\Sigma$ which is explained in Figure 4.9.1.

Typically, an observer is at a distance which is large compared to the dimensions of an antenna. This implies that $|x - y|$ varies little for x an observation point and $y \in \text{supp}(F)$.

With this in mind, suppose that $\text{supp}(F) \subset \{|y| < r\}$, $r \ll |x|$. Let

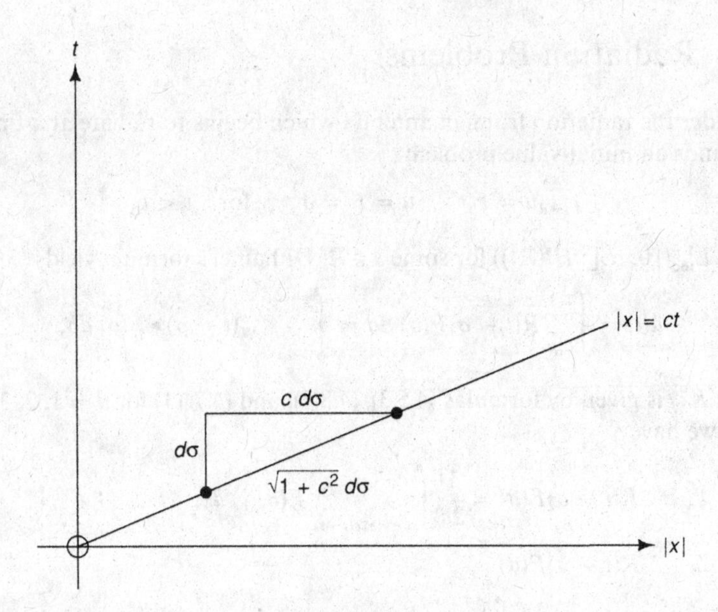

Figure 4.9.1

$f(t) \equiv \int F(t, y)\, dy$. We obtain the following approximations:

$$d = 1, \qquad u(t, x) \cong \frac{1}{2c} \int_{-\infty}^{t-|x|/c} f(\sigma)\, d\sigma,$$

$$d = 2, \qquad u(t, x) \cong \frac{1}{2\pi c} \int_{-\infty}^{t-|x|/c} f(\sigma)(c^2(t-\sigma)^2 - |x|^2)^{-1/2}\, d\sigma,$$

$$d = 3, \qquad u(t, x) \cong \frac{f(t - |x|/c)}{4\pi c|x|}.$$

This sort of approximation is common. One has a function $\Phi(x)$ supported near 0 and one must approximate $\langle \Phi, \psi \rangle$ when ψ varies little over supp Φ. Replacing ψ by its Taylor polynomial of degree N yields

$$\langle \Phi, \psi \rangle \sim \sum_{|\alpha| \le N} \partial^\alpha \psi(0) \left\langle \Phi, \frac{x^\alpha}{\alpha!} \right\rangle.$$

The approximations above correspond to $N = 0$. From the point of view of distribution theory, this amounts to replacing Φ by

$$\sum \left\langle \Phi, \frac{x^\alpha}{\alpha!} \right\rangle \partial^\alpha \delta.$$

This is called the *multipole approximation* to Φ and the coefficients of $\partial^\alpha \delta$, the *multipole moments*. From this point of view, the approximate solution of the radiation problem comes from approximating $F(t, y)$ by

$$F(t, y) \sim \int F(t, y)\, dy\, \delta(y) = f(t)\delta(y).$$

Thus it is not surprising that the approximations are exact solutions to

$$\square_{1+d} u = f(t)\delta(y), \qquad u(t, \cdot) = 0 \qquad \text{and} \qquad f(t) = 0 \qquad \text{for} \quad t < t_0. \tag{6}$$

This is called the *radiation problem with point source of strength* $f(t)$. If $f \in C^\infty(\mathbb{R})$

$$f(t)\delta \in C^\infty(\mathbb{R} : H^s) \qquad \text{for all} \quad s < -d/2.$$

The general theory implies unique solvability of the radiation problem with

$$u \in C^\infty(\mathbb{R} : H^{s+1}), \qquad \text{for all} \quad s < -d/2.$$

We next verify that the exact solution of the radiation problem (6) is equal to the approximation determined above. In all d

$$u(t) = \int_{-\infty}^{t} K(t-\sigma) f(\sigma)\delta \, d\sigma.$$

The values of the integrand for $d = 1, 2, 3$ are

$$d = 1, \qquad K(t-\sigma)f(\sigma)\delta = \frac{f(\sigma)}{2c} \chi_{|x| \le c(t-\sigma)},$$

$$d = 2, \qquad K(t-\sigma)f(\sigma)\delta = \frac{f(\sigma)}{2\pi c}(c^2(t-\sigma)^2 - |x|^2)^{-1/2} \chi_{|x| \le c(t-\sigma)},$$

$$d = 3, \qquad K(t-\sigma)f(\sigma)\delta = \frac{f(\sigma)}{4\pi c|x|} \, d\Sigma_{|x| = c(t-\sigma)}.$$

Integrating $d\sigma$ from $-\infty$ to t gives the solution to (6). For $d = 1$, and t, x fixed,

$$K(t-\sigma)f(\sigma)\delta = \begin{cases} \dfrac{f(\sigma)}{2c} & \text{if } |x| \le c(t-\sigma), \\ 0 & \text{otherwise.} \end{cases}$$

Integrating yields

$$u(t, x) = \int_{-\infty}^{t-|x|/c} f(\sigma) \frac{d\sigma}{2c}. \tag{7}$$

Similarly, for $d = 2$

$$K(t-\sigma)f(\sigma)\delta = \begin{cases} \dfrac{f(\sigma)}{2\pi c(c^2(t-\sigma)^2 - |x|^2)^{1/2}} & \text{if } |x| \le c(t-\sigma), \\ 0 & \text{otherwise,} \end{cases}$$

$$u(t, x) = \int_{-\infty}^{t-|x|/c} \frac{f(\sigma)}{2\pi c(c^2(t-\sigma)^2 - |x|^2)^{1/2}} \, d\sigma. \tag{8}$$

Both cases give the approximation formulas from our first calculations.

Both computations are formal since they deal with point values. We know that $\sigma \mapsto K(t-\sigma)f(\sigma)\delta$ belongs to $C(\mathbb{R}: H^s)$ for $s < -d/2$. Thus the integral is a Riemann integral of an H^s valued function. We have performed natural *pointwise* integrals. To show that the distribution $u(t)$ is given by the function $u(t, \cdot)$ in formula (7) or (8), one must show that for $\psi \in C_0^\infty(\mathbb{R}^d)$

$$\int u(t, x)\psi(x)\, dx = \int_{-\infty}^{t} \langle K(t-\sigma)f(\sigma)\delta, \psi\rangle\, d\sigma.$$

This is not difficult to do. The justification for $d = 1, 2$ is analogous to, but simpler than, the case $d = 3$ which follows. When $d = 3$ we have for all $\psi \in C_0^\infty(\mathbb{R}^3)$

$$\langle K(t-\sigma)f(\sigma)\delta, \psi\rangle = \int \frac{\psi(x)f(\sigma)}{4\pi c^2(t-\sigma)}\, d\Sigma_{|x|=c(t-\sigma)}.$$

Thus the formula for u as an integral yields

$$\langle u(t), \psi\rangle = \int_{-\infty}^{t} \left(\int \frac{\psi(x)}{4\pi c^2(t-\sigma)} f(\sigma)\, d\Sigma_{|x|=c(t-\sigma)} \right) d\sigma.$$

Introduce spherical coordinates, $x = r\omega$, $r \in]0, \infty[$, $\omega \in S^2$, $dx = r^2\, dr\, d\omega$, to find

$$\int \frac{\psi(x)}{4\pi c^2(t-\sigma)} f(\sigma)\, d\Sigma_{|x|=c(t-\sigma)} = \int \frac{\psi(r\omega)}{4\pi c r} f\left(t - \frac{r}{c}\right) r^2\, d\omega,$$

where we have used the fact that $r = c(t-\sigma)$ in the region of integration. Then $\sigma = t \leftrightarrow r = 0$ and $\sigma = -\infty \leftrightarrow r = \infty$, so

$$\begin{aligned} \langle u, \psi\rangle &= \int_0^\infty \int \frac{\psi(r\omega)}{4\pi c r} f\left(t - \frac{r}{c}\right) r^2\, d\omega\, dr \\ &= \left\langle \frac{f(t - r/c)}{4\pi c r}, \psi \right\rangle, \end{aligned}$$

which proves the expected result, $u = f(t - |x|/c)/4\pi c|x|$.

An observer at a fixed position x

> Must differentiate the observed field to measure f if $d = 1$.
> Observes an average of f weighted heavily at $t - |x|/c$ if $d = 2$.
> Observes $f(t - |x|/c)$ exactly with $1/|x|$ decay if $d = 3$.

Happily for us, our environment is three-dimensional and electromagnetic or acoustic signals transmitted from an antenna are received without any need of decoding. This makes the technology of radio, television, radar, and sonar much simpler than they would be if d were any dimension other than 3. To the question, "Why is space-time four-dimensional?", this yields the facetious reply, "For the sake of better TV reception".

PROBLEMS

1. The digital signal 010100... is transmitted via the source function $f = \chi_{]1,2[\cup]4,5[}(t)$. Let $x = (20, 0, \dots, 0) \in \mathbb{R}^d$ and sketch roughly the form of the signal received at $\mathbb{R}_t \times \{x\}$ when $d = 1, 2, 3$.

2. If $f \in C^k(\mathbb{R})$, estimate the smoothness of the solution u to the radiation problem (6) at points $x \neq 0$. *Hints.* Using the Sobolev regularity $f(t)\delta \in C^k(\mathbb{R} : H^s)$ for $s < -d/2$ gives a lousy estimate. Use the formulas of this section. The answer depends on d.

CHAPTER 5

The Dirichlet Problem

§5.1. Introduction

This chapter is devoted to studying boundary value problems for second-order elliptic equations. The variational (also known as Hilbert space) approach to the Dirichlet problem is emphasized. Maximum principles are discussed in §5.10 and §5.11, which are independent of the preceding sections and are essential reading along with §5.1, §5.2, and §5.3. Sections 5.8 and 5.9 address the technically difficult question of the regularity of the weak solution constructed in §5.3. Sections 5.4–5.7 treat a variety of topics which rely solely on the H^1 variational approach to the Dirichlet problem. It is interesting that this straightforward argument propels one so far.

Elliptic boundary value problems arise in a striking number of distinct settings in science and geometry. For a first example, consider the flow of heat in a bounded regular subset $\Omega \subset \mathbb{R}^d$. The dimensions $d \leq 3$ are the most important in practice. The temperature $v(t, x)$ satisfies the heat equation

$$v_t = \nu \Delta_x v \qquad \text{in} \quad \mathbb{R}_t \times \Omega_x \tag{1}$$

with $\nu > 0$.

Suppose that a fixed-temperature distribution at the boundary is maintained by a heating and refrigeration system. Then there is a function g on $\partial\Omega$ such that

$$v(t, x) = g(x) \qquad \text{for all} \quad x \in \partial\Omega. \tag{2}$$

Heat flowing from hot to cold will smooth out irregularities in an initial temperature distribution $v(0, \cdot)$. Given the fact that the boundary temperatures are kept at a steady state, it is plausible that the system will evolve toward an equilibrium state $u(x)$, that is,

$$\lim_{t\to\infty} v(t, x) - u(x) = 0.$$

Put another way, as $\lambda \to \infty$ the functions $v_\lambda(t, x) \equiv v(t + \lambda, x)$ converge to $u(x)$ on $[0, \infty[\times \Omega$. Since each of the v_λ satisfies the boundary value problem (1), (2), one expects that u is also a solution. Since u does not depend on time this yields

$$\Delta u = 0 \quad \text{in } \Omega \qquad \text{and} \qquad u = g \quad \text{on } \partial\Omega. \tag{3}$$

This boundary value problem for u is the classical Dirichlet problem. It is not clear that we have extracted all the necessary physical information to determine the asymptotic state of the heat equation. Thus it is reasonable to expect existence of solutions to (3), but there may be many such solutions. Note also that the boundary condition gives u and not $\partial u/\partial \nu$, while Cauchy data for the second-order operator Δ consists of both functions. This might lead one to believe that u is not completely specified. If there were many solutions, that would show that additional properties would have to be given to determine which is the equilibrium state to which v converges. In fact there is uniqueness, so that there are no additional physical constraints on the state. The first indication that (3) is uniquely solvable comes from explicitly solvable problems with exceptionally simple geometry.

The very simplest example is the case $\Omega \equiv]a, b[$, an open interval in $\mathbb{R}^1$. The Dirichlet problem is then

$$\left(\frac{d}{dx}\right)^2 u = 0 \quad \text{in }]a, b[, \qquad u(a) = g(a) \quad \text{and} \quad u(b) = g(b).$$

The unique solution is the linear function $u(x) = g(a) + (x - a)(g(b) - g(a))/(b - a)$. More complicated examples are given in Problems 1, 2, and 5.

An entirely different problem leading to (3) is the construction of a conformal mapping, $\varphi(z)$ from a simply connected domain, $\Omega \subset \mathbb{C} \approx \mathbb{R}^2$, to the disc $\{|z| < 1\}$. Translating Ω if necessary, suppose that $0 \in \Omega$. Using a Möbius transformation in the disc we see that if such a φ exists, then there is one with $\varphi(0) = 0$. Then $\varphi(z)/z$ is nonzero, so can be written as e^{h+ik} with h and k real and harmonic. Then

$$\log|\varphi(z)| = \log|z| + h(z).$$

Thus h is a solution of the Dirichlet problem

$$\Delta h = 0 \quad \text{in } \Omega \qquad \text{and} \qquad h = -\log|z| \quad \text{on } \partial\Omega.$$

Riemann's idea for a proof of existence for this Dirichlet problem is described in the next sections.

A third example is the problem of finding the equilibrium position of a membrane stretched over a domain $\Omega \subset \mathbb{R}^2$, whose height above points in the boundary of Ω is known. This problem occurs if the membrane is attached to a fixed support surrounding Ω.

Seek the equilibrium position as a minimum of the potential energy function. If $u(x)$ is the height of the membrane above the point $x \in \Omega$, then a reasonable candidate for the potential energy is the surface area of the surface $z = u(x)$ since energy is stored in the membrane when it is stretched, that is, when its area is increased. This leads to the following nonlinear variational problem called *Plateau's problem*

$$\text{minimize} \quad \int_\Omega (1 + |\nabla u|^2)^{1/2}\, dx\, dy \tag{4}$$

the minimum taken over all functions u with $u|_{\partial\Omega} = g$, the given height of the membrane at the boundary. The functional to be minimized in (4) is the surface area. If one includes gravitational effects there is an additional contribution to the potential energy leading to the variational problem

$$\underset{u=g \text{ on } \partial\Omega}{\text{minimize}} \quad \int_\Omega (1 + |\nabla u|^2)^{1/2} + cu\, dx\, dy, \tag{5}$$

where c is constant.

The minimum principle leads to a differential equation for u in the standard fashion. Denote by J the functional to be minimized. Then if $\varphi \in C^1(\bar{\Omega})$, $\varphi|_{\partial\Omega} = 0$, then $u + \varepsilon\varphi$ is a competing function. If J is minimized at u, then $J(u + \varepsilon\varphi)$ is minimized at $\varepsilon = 0$. Thus $(d/d\varepsilon)J(u + \varepsilon\varphi)|_{\varepsilon=0} = 0$. This equation is called the *Euler* or *Euler–Lagrange equation.* For many problems it is a partial differential equation. In the present case, a straightforward computation yields

$$\left(\frac{d}{d\varepsilon}\right) J(u + \varepsilon\varphi)|_{\varepsilon=0} = \int_\Omega \sum (\partial_{x_i}\varphi) \frac{(\partial_{x_i} u)}{(1 + |\nabla u|^2)^{1/2}} + c\varphi\, dx.$$

Integrate by parts to move the derivative from φ to the u terms. Since φ vanishes at $\partial\Omega$, the boundary term vanishes. This yields

$$\int_\Omega \varphi \left[\sum \partial_{x_i} \frac{(\partial_{x_i} u)}{(1 + |\nabla u|^2)^{1/2}} + c \right] dx = 0.$$

This can vanish for all φ with $\varphi|_{\partial\Omega} = 0$ if and only if the function in square brackets vanishes. Thus, for u we find the boundary value problem

$$\sum \partial_{x_i} \frac{(\partial_{x_i} u)}{(1 + |\nabla u|^2)^{1/2}} + c = 0 \quad \text{in } \Omega, \qquad u = g \quad \text{on } \partial\Omega. \tag{6}$$

The differential equation in (6) asserts that the mean curvature of the surface $z = u(x)$ is equal to $-c/d$. If $c \equiv 0$, the surface is a *minimal surface*, the name coming from the fact that it minimizes area as in (4).

A connection between the nonlinear problem (6) with $c \equiv 0$ and the classical Dirichlet problem is that the latter is a good approximation of the former when the surface is nearly horizontal. To see this, note that for nearly flat surfaces, the derivatives of u are small. If we drop higher-order terms in these derivatives, the boundary value problem (6) becomes the classical Dirichlet problem $\Delta u = 0$ in Ω, and $u|_{\partial\Omega} = g$. Essentially the same observation is that Δ is the linearization of the differential equation in (6) at the solution $u = 0$. A third way to view this approximation is to note that Taylor expansion of the integrand in (4) about $|\nabla u| = 0$ shows that

$$\int_\Omega (1 + |\nabla u|^2)^{1/2}\, dx\, dy = |\Omega| + \int_\Omega |\nabla u|^2\, dx\, dy + O(|\nabla u|^4).$$

Dropping the higher-order terms is essentially the same approximation as above so should lead to the classical Dirichlet problem. Dropping those terms yields the minimum principle

$$\underset{u=g \text{ on } \partial\Omega}{\text{minimize}} \int_\Omega |\nabla u|^2\, dx\, dy.$$

Thus, one should not be surprised to find that this minimization problem is equivalent to the classical Dirichlet problem. Exploiting this, or similar, minimum principles is called the *variational approach to the Dirichlet problem.* It is the path that we follow.

In the problems, we will study some examples where the Dirichlet problem is solvable more or less explicitly. These examples are important since they illustrate in concrete cases the general principles to be proved later, and they also serve as a testing ground for conjectures.

Problems

The simplest multi-dimensional Dirichlet problems are those where the domain Ω has spherical symmetry. For example, if Ω is the region between two balls, $\Omega = \{x : r_1 < |x| < r_2\}$ and the boundary data are spherically symmetric. The Dirichlet problem becomes

$$\Delta u = 0 \quad \text{in } \Omega, \qquad u(x) = g(r_j) \quad \text{for } |x| = r_j, \qquad j = 1, 2.$$

1. Find the unique rotationally symmetric solution to the above annular Dirichlet problem. *Hint.* First solve Problem 4.6.1. Then match the boundary data.

Next turn to the case $\Omega = \{x \in \mathbb{R}^2 : |x| < 1\}$. Use polar coordinates, $u = u(r, \theta)$ with u periodic in θ with period 2π. Fourier series expansion in θ yields

$$u(r, \theta) = \sum_{n \in \mathbb{Z}} u_n(r) e^{in\theta},$$

with coefficients rapidly decreasing in n if u is smooth. The boundary condition at $r = 1$ implies that

$$u_n(1) = g_n, \qquad \text{where} \quad g(\theta) = \sum g_n e^{in\theta}$$

with g_n rapidly decreasing if $g \in C^\infty(\partial\Omega)$.

2. Show that for boundary data $g \in C^\infty(\partial\Omega)$, the Dirichlet problem in the disc has exactly one solution $u \in C^\infty(\text{cl}(\Omega))$. *Hint.* Show that the $u_n(r)$ are uniquely and explicitly determined and the resulting series can be differentiated termwise.

3. In the spirit of Chapter 4, we analyze the explicit formula in Problem 2.
   (i) Use the formula to show that $u(0, 0) = \int g(\theta)\, d\theta/2\pi$. This is the case $d = 2$ of the *mean value property*, Corollary 5.10.3.
   (ii) For $0 < r < 1$, define a smooth function $K_r(\theta)$ on the unit circle by
   $$K_r(\theta) = \sum_{n \in \mathbb{Z}} r^{|n|} e^{in\theta}.$$
   Show that the formula for u is equivalent to $u(r, \cdot) = K_r * g$, the convolution performed in S^1.
   (iii) Use the formula for the sum of a geometric series to find an explicit expression for K_r.

   DISCUSSION. The function K is called the *Poisson kernel.*

4. Consider the Dirichlet problem in $\Omega \equiv \{x \in \mathbb{R}^d : |x| < R\}$ for the equation of constant mean curvature equal to H,
   $$\sum \partial_{x_i} \frac{(\partial_{x_i} u)}{(1 + |\nabla u|^2)^{1/2}} = dH, \qquad u = 0 \quad \text{on } \partial\Omega.$$
   (i) If $0 \leq H < 1/R$, use spheres to find an explicit solution.
   (ii) If $H > 1/R$, show that there can be no solution in $C^2(\text{cl}(\Omega))$ as follows (see Finn, [Fi] for a discussion of this result of Bernstein and related results): (1) Integrate the equation over Ω. The right-hand side is equal to $d\omega_d H R^d$ where ω_d is the d volume of the unit ball in $\mathbb{R}^d$. (2) Perform an integration by parts to express the left-hand side as a boundary integral and show that that integral is less than or equal to $d\omega_d R^{d-1}$.

   DISCUSSION. The linearization of this boundary value problem at $H = 0$, $u = 0$, is the classical Dirichlet problem. Thus, for H small, one expects the two problems to behave similarly. For large u the nonlinear aspects dominate leading to nonsolvability.

5. Let $\Omega \equiv \{x \in \mathbb{R}^d : 0 < x_d < 1\}$. Use the Fourier transform with respect to the variables $(x_1, \ldots, x_{d-1}) \equiv x'$ to prove that for $g_0, g_1 \in \mathcal{S}(\mathbb{R}^{d-1}_{x'})$ there is exactly one solution $u \in C^\infty([0, 1] : \mathcal{S}(\mathbb{R}^d_{x'}))$ to the Dirichlet problem
   $$\Delta u = 0 \quad \text{in } \Omega, \qquad u|_{x_d=0} = g_0 \quad \text{and} \quad u|_{x_d=1} = g_1.$$
   Find an explicit expression for the solution.

   DISCUSSION. This example is sometimes used to explain why one boundary condition is sufficient, in contrast to the two functions which comprise the Cauchy data. The reasoning is that there are really two functions, one for the bottom of the boundary and one for the top. However, considering the square region $0 < x_i < 1$ for $i = 1, 2$, one finds four boundary functions corresponding to top, bottom, and two sides. The region $0 < x_i < 1$ for $i = 1, 2, \ldots, k$ yields $2k$ boundary functions. Thus, this explanation should not be taken too seriously.

   The fact that the boundary conditions are given on a boundary which surrounds Ω, in contrast to the Cauchy problem where the initial plane is on one side of $\{t > 0\}$, is at least part of the explanation.

§5.2. Dirichlet's Principle

In the last section a variational equivalent of the Dirichlet problem was motivated by linearizing the principle of minimum potential energy for a nonlinear membrane. We begin this section by giving an independent motivation based on an analogy between the heat equation and *gradient flows* in finite dimensions. These are the flows of ordinary differential equations

$$\dot{x} = -\text{grad}\,\Phi(x). \tag{1}$$

The integral curves of this equation move in the direction of most rapid decrease of Φ, in particular, Φ decreases along orbits. Precisely

$$\frac{d\Phi(x(t))}{dt} = \langle \text{grad}\,\Phi(x(t)), \dot{x}(t)\rangle = -|\text{grad}\,\Phi(x(t))|^2. \tag{2}$$

Flowing along such integral curves is a reasonable way to seek minima for Φ. This is called the *method of steepest descent*. The flow is also a centerpiece in Morse Theory. The ordinary differential equation (1) is equivalent to $\langle \dot{x}, \psi\rangle = \langle -\nabla\Phi, \psi\rangle$ for all $\psi \in \mathbb{R}^d$. This in turn is equivalent to

$$\frac{d}{dt}\langle x(t), \psi\rangle = \frac{d}{d\varepsilon} - \Phi(x(t) + \varepsilon\psi)|_{\varepsilon=0}, \qquad \text{for all} \quad \psi \in \mathbb{R}^d. \tag{3}$$

The heat equation with Dirichlet boundary condition

$$v_t = \nu\Delta v \quad \text{on } [0, \infty[\times \Omega, \qquad v(t, x) = g(x) \quad \text{for } x \in \partial\Omega, \tag{4}$$

has an analogous structure. Note that since the heat equation does not satisfy the Hadamard–Petrowsky condition for backward evolution we only expect a solution in $t \geq 0$. Suppose that Ω is a nice boounded open subset of $\mathbb{R}^d$. The role of the function Φ is played by the Dirichlet integral

$$J(w) \equiv \frac{\nu}{2}\int_\Omega |\nabla_x w|^2\, dx.$$

Begin by considering J as a functional defined on the set of $w \in C^1(\bar{\Omega})$ whose restrictions to $\partial\Omega$ are equal to g. Call such w *admissible*. If w is admissible and $\varphi \in C^1(\bar{\Omega})$ with $\varphi|_{\partial\Omega} = 0$, then $w + \varepsilon\varphi$ is admissible and computing as in the last section yields

$$\frac{d}{d\varepsilon}J(w + \varepsilon\varphi)|_{\varepsilon=0} = \nu\int_\Omega \nabla\varphi\cdot\nabla w\, dx = -\nu\int_\Omega \varphi\Delta w\, dx + \nu\int_{\partial\Omega} \varphi\frac{\partial w}{\partial n}\, d\sigma. \tag{5}$$

The boundary term in the integration by parts vanishes since φ is equal to zero on $\partial\Omega$. Thus for solutions of (4)

$$\frac{d}{dt}\int_\Omega \varphi v\, dx = \int_\Omega \varphi v_t\, dx = \int_\Omega \varphi\nu\Delta v\, dx = \frac{d}{d\varepsilon} - J(v + \varepsilon\varphi)|_{\varepsilon=0}. \tag{6}$$

The analogy with (3) is clear.

This analogy suggests that $J(v(t))$ is a decreasing function of time. To verify this, differentiate to find

$$\frac{d}{dt} J(v(t)) = v \int_{\Omega} \nabla v_t \cdot \nabla v \, dx = -v \int_{\Omega} v_t \Delta v \, dx + v \int_{\partial\Omega} v_t \frac{\partial v}{\partial n} \, d\sigma.$$

The boundary term in the integration by parts vanishes since $v_t = \partial_t g(x) = 0$ on $\partial\Omega$. Using the differential equation yields

$$\frac{d}{dt} J(v(t)) = -v^2 \int_{\Omega} (\Delta v)^2 \, dx \le 0.$$

For the gradient system (1) on $\mathbb{R}^d$, it is easy to show that if $|\Phi(x)| \to \infty$ as $|x| \to \infty$, then as t tends to infinity, orbits tend to critical points of Φ. Recall that critical or stationary points are points $x \in \mathbb{R}^d$ such that grad $\Phi(x) = 0$. By analogy, it is reasonable to expect that the solution of the heat equation tends to a critical point of the functional J. Since J is strictly convex there is only one such critical point, a global minimum (Proposition 1(iii)). In the previous section, we argued that the asymptotic state for v is the solution of the Dirichlet problem. These arguments suggest two things:

(1) the approach to equilibrium for the heat equation can be studied using this gradient structure; and
(2) the solution of the Dirichlet problem minimizes the functional J taken over all functions equal to g at $\partial\Omega$.

Both of these ideas are correct. We pursue the second.

Proposition 1 (Dirichlet's Principle). *Suppose that* $u \in C^2(\bar{\Omega})$ *and* $u|_{\partial\Omega} = g$. *Then the following are equivalent:*

(i) $\Delta u = 0$ *in* Ω.
(ii) u *is a critical point of* J *in the sense that*

$$\frac{d}{d\varepsilon} J(u + \varepsilon\varphi)|_{\varepsilon=0} = 0 \qquad \text{for all} \quad \varphi \in C^2(\bar{\Omega}) \quad \text{with} \quad \varphi|_{\partial\Omega} = 0.$$

(iii) u *minimizes* J *in the sense that* $J(u) \le J(w)$ *for all* $w \in C^2(\bar{\Omega})$ *with* $w|_{\partial\Omega} = g$.

Proof. The equivalence (i) $\Leftrightarrow$ (ii) is an immediate consequence of (5).

To implication (ii) $\Rightarrow$ (iii) is proved by using (ii) with $\varphi \equiv w - u$ to show that $\int \nabla u \nabla (w - u) \, dx = 0$. Then since J is quadratic

$$J(w) = J(u + (w - u)) = J(u) + 2 \int \nabla u \nabla (w - u) \, dx + J(w - u).$$

The middle term vanishes and the last is nonnegative so $J(w) \ge J(u)$.

Conversely, assuming (iii), the function $J(u + \varepsilon\varphi)$ has a minimum at $\varepsilon = 0$ for any φ as in (ii). Thus $(d/d\varepsilon) J(u + \varepsilon\varphi)|_{\varepsilon=0} = 0$ proving (ii). □

Riemann concluded that the Dirichlet problem was solvable, reasoning that J is nonnegative and so must have a minimum value. Choosing a function u with $J(u) = \min(J)$ solves the problem.

Brought up as we are, on the rigorous analysis of the end of the nineteenth century, the flaw in this proof is apparent. A function which is bounded below has an infimum, but there is no guarantee that the infimum is a minimum. There may be no point where the infimum is achieved.

The first rigorous proofs that the Dirichlet problem is solvable followed other lines. Poincaré's method of balayage, Perron's method of subharmonic functions, and Neumann's method of integral equations are described in many texts (e.g. [CH]). Hilbert, as a part of his study of the Calculus of Variations, showed that Riemann's original strategy is valid. This is the path that we take.

For technical reasons, it is easier to treat the inhomogeneous differential equation with homogeneous boundary conditions rather than the other way around. Thus we will solve

$$\Delta u = f \quad \text{in } \Omega \qquad \text{and} \qquad u = 0 \quad \text{on } \partial\Omega. \tag{7}$$

If we know how to solve this problem for smooth data, and are given a smooth g on $\partial\Omega$, then, to find a harmonic function with boundary values equal to g, simply choose a $G \in C^\infty(\bar{\Omega})$ with $G|_{\partial\Omega} = g$, let $f \equiv -\Delta G$, and solve (7). The sum $u + G$ does the trick.

A variational formulation of problem (7) is given in the next proposition whose proof parallels that of Proposition 1.

Proposition 2 (Dirichlet's Principle). *Suppose that $u \in C^1(\bar{\Omega})$, $u|_{\partial\Omega} = 0$, and $f \in L^2(\Omega)$. Then the following are equivalent:*

(i) *$\Delta u = f$ in $\mathscr{D}'(\Omega)$.*

(ii) *$J(u) \le J(w)$ for all $w \in C^1(\bar{\Omega})$ with $w|_{\partial\Omega} = 0$, where J is the functional defined by*

$$J(w) \equiv \int |\nabla w(x)|^2 + 2w(x)f(x)\,dx. \tag{8}$$

(iii) *u is a critical point of J in the sense that*

$$\frac{d}{d\varepsilon} J(u + \varepsilon\varphi)|_{\varepsilon=0} = 0 \qquad \textit{for all} \quad \varphi \in C^1(\bar{\Omega}) \quad \textit{with} \quad \varphi|_{\partial\Omega} = 0. \tag{9}$$

Computing the left-hand side of (9) yields

$$\frac{d}{d\varepsilon} J(u + \varepsilon\varphi)|_{\varepsilon=0} = 2 \int_\Omega \nabla\varphi \cdot \nabla u + fu\,dx.$$

When the right-hand side vanishes for all φ as in (9), we say that u is a *variational solution of the Dirichlet problem.* This formulation is the starting point of the *Galerkin method* in numerical analysis.

PROBLEMS

We give a third motivation for Dirichlet's principle following an analogy with classical mechanical systems with damping.

In the presence of damping, a vibrating membrane tends to an equilibrium state as t tends to $+\infty$. A mechanical analogue is the system of ordinary differential equations

$$m\ddot{x} + a\dot{x} = -\operatorname{grad} \Phi(x), \qquad m, a \in]0, \infty[. \tag{10}$$

The energy for solutions is a decreasing function of time

$$\frac{d}{dt}\left(\frac{m|\dot{x}|^2}{2} + \Phi(x)\right) = \dot{x}(m\ddot{x} + \operatorname{grad} \Phi(x)) = -a|\dot{x}|^2 \leq 0,$$

where Φ is the potential energy. If Φ tends to $+\infty$ as $|x|$ tends to infinity, then as $t \to \infty$, each orbit converges to a critical point of Φ.

The differential equation (10) is equivalent to

$$m\left(\frac{d}{dt}\right)^2 \langle x, \varphi\rangle + a\frac{d}{dt}\langle x, \varphi\rangle = -\frac{d}{d\varepsilon}\Phi(x + \varepsilon\varphi)|_{\varepsilon=0} \qquad \text{for all} \quad \varphi \in \mathbb{R}^d. \tag{11}$$

The damped wave equation on Ω is

$$\Box v + av_t = 0 \quad \text{on } [0, \infty[\times \Omega, \qquad v|_{\partial\Omega} = g(x). \tag{12}$$

1. Show that $v \in C^2([0, \infty[\times \bar{\Omega})$ with $v|_{\partial\Omega} = g$ satisfies (12) if and only if for all φ vanishing at $\partial\Omega$,

$$\left(\frac{d}{dt}\right)^2 \int_\Omega v\varphi \, dx + a\frac{d}{dt}\int_\Omega v\varphi \, dx = -\frac{d}{d\varepsilon}J(v + \varepsilon\varphi)|_{\varepsilon=0},$$

where $J(w) \equiv \int |\nabla w|^2 \, dx/2$. By analogy with the mechanics case, find and prove a law of energy decay.

DISCUSSION. These computations suggest that one can study the approach to equilibrium by analogy with the finite-dimensional case. If one supposes that there is approach to equilibrium, that is, $v_\lambda \equiv v(t + \lambda)$ approaches a limit $u(x)$ as $\lambda \to \infty$, then, as in §5.1, u is a solution of the Dirichlet problem (5.1.3). This is a third path leading to the idea that one should look for solutions of the Dirichlet problem among the critical points of J.

2. Prove an analogue of Proposition 2 relating the solution of the boundary value problem

$$\Delta u + cu = f \quad \text{in } \Omega, \qquad u = 0 \quad \text{on } \partial\Omega, \tag{13}$$

and extrema of the functional

$$J(w) \equiv \int_\Omega |\nabla w|^2 - cw^2 + 2fw \, dx.$$

Here $C \leq 0$ is a nonpositive real number.

3. For Ω fixed show that there is a $c_0 > 0$, so that for all $c > c_0$ the functional J in Problem 2 is not bounded below, that is, the infimum of J over $C_0^\infty(\Omega)$ is equal to minus infinity.

DISCUSSION. This shows that the mere existence of a variational formulation is not a panacea. Using the results of §5.7, one can show that the largest c_0 with this property is the largest eigenvalue of Δ on Ω with Dirichlet boundary conditions.

§5.3. The Direct Method of the Calculus of Variations

Proposition 5.2.2 shows that to solve the Dirichlet problem (5.2.7), it is reasonable to look for a function u which minimizes the functional

$$J(w) \equiv \int_{\Omega} |\nabla w|^2 + 2wf\,dx, \tag{1}$$

in the class of functions vanishing at the boundary of Ω. We suppose that $f \in \operatorname{Re} L^2(\Omega)$. The strategy we follow is a standard procedure called the *direct method of the calculus of variations.* It was used by Hilbert not only to solve the Dirichlet problem, but also to prove the existence of length minimizing geodesics on complete Riemannian manifolds and for a variety of other minimization problems. The method consists of five steps.

Step 1. Show that $\inf J > -\infty$.

Step 2. Choose a *minimizing sequence* u_n. That is, choose a sequence u_n such that $\lim J(u_n) = \inf J$. Sometimes one can arrange additional special properties.

Step 3. Derive estimates for the u_n. Usually, the key fact is that for n large $J(u_n) \le \varepsilon + \inf J$.

Step 4. Based on the estimates in Step 3, extract a subsequence of the u_n which converges. This step is a compactness argument. The topology is dictated by the estimates. The better the estimates, the stronger the convergence, and the easier is the next step.

Step 5. Show that if u is the limit of the subsequence, then $J(u)$ is equal to $\inf J$. This step is often achieved by showing that J is lower semicontinuous with respect to the convergence in Step 4.

In the present case, Steps 1 and 3 are performed simultaneously and represent the heart of the analysis. The functional J has two terms, one of which is nonnegative. To show that J is bounded below amounts to showing that the other term cannot by very negative without the positive term being just as positive. The set of functions over which we are minimizing is taken, provisionally, as the set of $\varphi \in C^1(\bar{\Omega})$ with $\varphi|_{\partial\Omega} = 0$. The class will be enlarged (completed) for Step 4. The second term in J is estimated using the Schwartz inequality

$$\left| \int 2\varphi f\,dx \right| \le 2\|\varphi\|_{L^2(\Omega)} \|f\|_{L^2(\Omega)}.$$

To bound this in terms of the nonnegative part of J, we use a lower bound for the latter.

For any $\omega \in S^{d-1}$, the width of Ω in the direction of ω (see Figure 5.3.1) is equal to

$$w(\omega) \equiv \sup_{x \in \Omega} \langle x, \omega \rangle - \inf_{x \in \Omega} \langle x, \omega \rangle.$$

The function w is a positive continuous function on the compact sphere S^{d-1}.

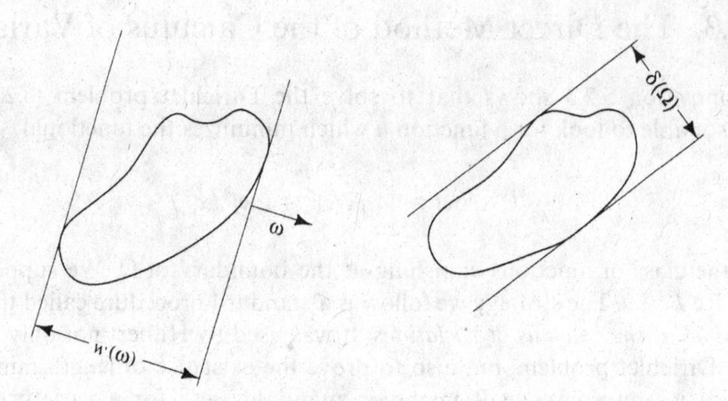

Figure 5.3.1

The minimum width of Ω, denoted $\delta = \delta(\Omega)$, is defined as

$$\delta(\Omega) \equiv \min\{w(\omega): \omega \in S^{d-1}\} > 0.$$

If the minimum is attained at a direction ω, then the domain $\bar{\Omega}$ is contained between and touching two parallel hyperplanes in $\mathbb{R}^d$ with normals equal to ω and at a distance δ from each other.

Theorem 1. *If $\varphi \in C^1(\bar{\Omega})$ with $\varphi|_{\partial\Omega} = 0$, then $\|\varphi\|_{L^2(\Omega)} \le \delta(\Omega)\|\nabla\varphi\|_{L^2(\Omega)}$.*

Assuming this for the moment, we show that J is bounded from below. Estimate

$$2\|\varphi\|_{L^2(\Omega)}\|f\|_{L^2(\Omega)} = 2\sqrt{\varepsilon}\|\varphi\|_{L^2(\Omega)}\frac{1}{\sqrt{\varepsilon}}\|f\|_{L^2(\Omega)} \le \varepsilon\|\varphi\|^2_{L^2(\Omega)} + \frac{1}{\varepsilon}\|f\|^2_{L^2(\Omega)}.$$

This trick of estimating a product as a small factor times the square of the first factor, plus a large factor times the square of the second is sometimes called the *Peter–Paul inequality* and is very useful. Theorem 1 then yields

$$2\|\varphi\|_{L^2(\Omega)}\|f\|_{L^2(\Omega)} \le \varepsilon\delta^2\|\nabla\varphi\|^2_{L^2(\Omega)} + \frac{1}{\varepsilon}\|f\|^2_{L^2(\Omega)}.$$

Choosing $\varepsilon = \delta^{-2}/2$, we find that

$$J(\varphi) \ge \tfrac{1}{2}\|\nabla\varphi\|^2_{L^2(\Omega)} - 2\delta^2\|f\|^2_{L^2(\Omega)}. \tag{2}$$

In particular, $\inf J > -\infty$.

Proof of Theorem 1. It suffices to prove the theorem for real valued φ, since the complex case follows upon applying the real result to $\mathrm{Re}(\varphi)$ and $\mathrm{Im}(\varphi)$ and adding the squares of the results.

Choose Cartesian coordinates such that $\Omega \subset \{-a < x_1 < a\}$ with $a = \delta/2$. Let $x \equiv (x_1, x')$, $x' \equiv (x_2, \ldots, x_d)$. Extend φ and $\nabla\varphi$ to vanish outside Ω. Then for any x' and $-a \le x_1 \le 0$

$$\varphi^2(x_1, x') = \int_{-a}^{x_1} \partial_1(\varphi^2)\,dx_1 = 2\int_{-a}^{x_1} \varphi\partial_1\varphi\,dx_1$$
$$\le 2\left(\int_{-a}^{0} \varphi^2\,dx_1 \int_{-a}^{0} (\partial_1\varphi)^2\,dx_1\right)^{1/2}$$

The right-hand side is independent of x_1. Integrate dx_1 from $-a$ to 0 to find

$$\int_{-a}^{0} \varphi^2\,dx_1 \le 2a\left(\int_{-a}^{0} \varphi^2\,dx_1 \int_{-a}^{0} (\partial_1\varphi)^2\,dx_1\right)^{1/2}$$

Square to find

$$\int_{-a}^{0} \varphi^2\,dx_1 \le (2a)^2 \int_{-a}^{0} (\partial_1\varphi)^2\,dx_1.$$

Integrating dx' yields

$$\|\varphi\|^2_{L^2(]-a,0[\times\mathbb{R}^{d-1})} \le (2a)^2\|\partial_1\varphi\|^2_{L^2(]-a,0[\times\mathbb{R}^{d-1})}.$$

Adding this to the corresponding result for $0 \le x_1 \le a$, and noting that $2a = \delta$, proves the theorem. □

For $\Omega =]0, \delta[$, using Fourier sine series shows that

$$\|\varphi\|_{L^2(]0,\delta[)} \le \frac{\delta}{\pi}\left\|\frac{d\varphi}{dx_1}\right\|_{L^2(]0,\delta[)}$$

with equality holding if and only if φ is a multiple of $\sin(\pi x/\delta)$. Using this in place of the Fundamental Theorem of Calculus and the Schwartz inequality in the proof above improves the constant in Theorem 1 to δ/π.

Having proved that J is bounded below, let $i \equiv \inf(J)$, the infimum taken over those $\varphi \in \mathrm{Re}\, C^1(\bar{\Omega})$ with $\varphi|_{\partial\Omega} = 0$. Choose φ_n from this set with $J(\varphi_n) \to i$, corresponding to Step 2 of the direct method.

Step 3 is to find estimates for the φ_n. Inequality (2) implies that

$$\limsup \|\nabla\varphi_n\|^2_{L^2(\Omega)} \le 2i + 4\delta^2\|f\|^2_{L^2(\Omega)}. \tag{3}$$

Theorem 1 yields

$$\limsup \|\varphi_n\|_{L^2(\Omega)} \le \delta(2i + 4\delta^2\|f\|_{L^2(\Omega)})^{1/2}. \tag{4}$$

These estimates show that the derivatives of order less than or equal to 1 of the minimizing sequence are bounded in $L^2(\Omega)$. This bound recalls the norm in H^1, and the next lemma shows that the φ_n is naturally a bounded sequence in $H^1(\mathbb{R}^d)$.

Lemma 2. *Suppose that $\Omega \subset \mathbf{R}^d$ is a bounded open set and $\phi \in C^1(\bar{\Omega})$. Define a distribution $T \in \mathscr{D}'(\mathbf{R}^d)$ by extending ϕ by zero, that is $\langle T, \psi\rangle := \int_\Omega \phi\psi\,dx$.*

(i) *If* $\phi|_{\partial\Omega} = 0$, *then* $T \in H^1(\mathbf{R}^d)$, *in fact T belongs to the closure of* $C_0^\infty(\Omega)$ *in* $H^1(\mathbf{R}^d)$.
(ii) *Conversely, if* Ω *is sufficiently regular that integration by parts as in* (1.7.3) *is valid, then* $T \in H^1(\Omega)$ *implies that* $\phi|_{\partial\Omega} = 0$.

PROOF. (i) Let $K(\varepsilon) \equiv \{x \in \Omega\colon \operatorname{dist}(x, \partial\Omega) \geq \varepsilon\}$ be the standard exhaustion of Ω. Choose $j \in C_0^\infty(|x| < \frac{1}{2})$ with $j \geq 0$ and $\int j\,dx = 1$, and define approximate delta functions $j_\varepsilon(x) \equiv \varepsilon^{-d} j(x/\varepsilon)$. Let $\chi_\varepsilon \equiv j_\varepsilon * \chi_{K(\varepsilon/2)}$. Then $\chi_\varepsilon \in C_0^\infty(\Omega)$, $0 \leq \chi \leq 1$, $\chi = 1$ on $K(\varepsilon)$, and for $0 < \varepsilon < 1$,

$$|\nabla \chi_\varepsilon| \leq \|\nabla j_\varepsilon\|_{L^1(\mathbb{R}^d)} \|\chi_{K(\varepsilon/2)}\|_{L^\infty(\mathbb{R}^d)} \leq \frac{\|\nabla j\|_{L^1(\mathbb{R}^d)}}{\varepsilon}.$$

Let $\varphi_\varepsilon \equiv \chi_\varepsilon \varphi \in C_0^1(\Omega)$. Then φ_ε converges to T in $L^2(\mathbb{R}^d)$ and we next show that $\nabla\varphi_\varepsilon$ converges to $\nabla T \in L^2(\mathbb{R}^d)$. This is equivalent to $\nabla\varphi_\varepsilon \to \nabla\varphi$ in $L^2(\Omega)$.

Now $\nabla\varphi_\varepsilon = \chi_\varepsilon \nabla\varphi + \varphi\nabla\chi_\varepsilon$. Lebesgue's Dominated Convergence Theorem implies that the first term converges to $\nabla\varphi$ in $L^2(\Omega)$. For the second term, note that in the support of $\nabla\chi_\varepsilon$, φ is no larger than ε times the sup norm of $\nabla\varphi$. Thus the product $\varphi\nabla\chi_\varepsilon$ is bounded independent of ε. Since the product vanishes outside $\Omega \backslash K(\varepsilon)$ and the measure of $\Omega\backslash K(\varepsilon)$ tends to zero, the second term tends to zero in $L^2(\Omega)$.

As $\eta \to 0$, the function $j_\eta * (\chi_\varepsilon\varphi) \in C_0^\infty(\Omega)$ converges to $\chi_\varepsilon\varphi$ in $H^1(\mathbb{R}^d)$.

(ii) The definition of distribution derivative yields

$$\langle \partial_j T, \psi\rangle = -\int_\Omega \phi\partial_j\psi\,dx = \int_\Omega (\partial_j\phi)\psi\,dx - \int_{\partial\Omega} n_j\phi\,d\sigma.$$

Thus $\partial_j T + n_j\phi\,d\sigma \in L^2(\mathbf{R}^d)$. If $T \in H^1(\mathbf{R}^d)$ then $\partial_j T \in L^2(\mathbf{R}^d)$ and it follows that $n_j\phi\,d\sigma \in L^2(\mathbf{R}^d)$ which can hold only if $n_j\phi = 0$. Multiply by n_j and sum to find that $\phi|_{\partial\Omega} = 0$. □

The lemma shows that the estimates (3), (4) for the minimizing sequence can be interpreted as saying that $\{\varphi_n\}$ is a bounded sequence in $H^1(\mathbb{R}^d)$. It also shows that we may replace our minimizing sequence with another, still denoted φ_n, with $\varphi_n \in C_0^\infty(\Omega)$. This suggests the introduction of the following space which should be thought of as the set of elements in $H^1(\mathbb{R}^d)$ which vanish on $\partial\Omega$ and on the exterior of Ω.

Definition. $\mathring{H}^1(\Omega)$ is the closure in $H^1(\mathbb{R}^d)$ of $C_0^\infty(\Omega)$.

Elements of $\mathring{H}^1(\Omega)$ belong to $H^1(\mathbb{R}^d)$ and have support in $\overline{\Omega}$.

EXAMPLE. If $\Omega =]0, 1[\subset \mathbb{R}$, then Sobolev's Theorem implies that $H^1(\mathbb{R}) \subset C(\mathbb{R})$. The elements of $\mathring{H}^1(\Omega)$ are continuous on $[0, 1]$ and, as uniform limits of elements of $C_0^\infty(]0, 1[)$, must vanish at the endpoints. In this simple case one sees immediately that membership in $\mathring{H}^1$ implies that homogeneous Dirichlet boundary conditions are satisfied.

The closed subspace $\mathring{H}^1(\Omega)$ of $H^1(\mathbb{R}^d)$ is a Hilbert space in the $H^1(\mathbb{R}^d)$ norm. Since $H^1(\mathbb{R}^d)$ is separable so is $\mathring{H}^1(\Omega)$. Theorem 1 implies that on $\mathring{H}^1(\Omega)$, $(\int_\Omega |\nabla u|^2\, dx)^{1/2}$ is a norm equivalent to the $H^1(\mathbb{R}^d)$ norm. Because of its close relation to the functional J, this is the norm we will use for $\mathring{H}^1(\Omega)$. The functional J is continuous from $\mathring{H}^1(\Omega)$ to $\mathbb{R}$. In particular, the infimum of J on Re $\mathring{H}^1(\Omega)$ is equal to its infimum on Re $C_0^\infty(\Omega)$. For regular Ω, Lemma 2 shows that $\mathring{H}^1(\Omega) \cap C^1(\bar{\Omega})$ coincides with those C^1 functions which vanish at the boundary.

The compactness required in Step 4 of the direct method is provided by the fact that a bounded sequence in a Hilbert space has a weakly convergent subsequence. We recall some of the basic results concerning weak convergence in Hilbert spaces. Let $\mathscr{H}$ denote a Hilbert space. A sequence h_n in $\mathscr{H}$ *converges weakly* to a limit h if and only if for all $k \in \mathscr{H}$, $(h_n, k) \to (h, k)$, where $(\cdot, \cdot)$ denotes the scalar product in $\mathscr{H}$. Weak convergence is denoted $h_n \rightharpoonup h$.

Weak convergence is equivalent to the pointwise convergence of the continuous linear functionals, $l_n(\cdot) \equiv (\cdot, h_n)$ to $l(\cdot) \equiv (\cdot, h)$. The Uniform Boundedness Principle shows that for weakly convergent sequences, $\{l_n\}$ and therefore h_n are bounded independent of n.

If $h_n \rightharpoonup h$, then $\|h\| \le \liminf \|h_n\|$. That is, $\|\cdot\|$ is lower semicontinuous with respect to weak convergence. The proof is simple. One need only consider $h \ne 0$. Then

$$\|h\|^2 = (h, h) = \lim (h_n, h) \le \liminf |(h_n, h)| \le \liminf \|h_n\|\, \|h\|.$$

Dividing by $\|h\|$ yields the result.

Every bounded sequence in a separable Hilbert space has a weakly convergent subsequence (Problem 2). Therefore the minimizing sequence has a subsequence, still denoted φ_n, which converges weakly in $\mathring{H}^1(\Omega)$ to a real valued limit u.

The final step in the direct method is to show that $J(u) = i$. Write $J(w) = \|w\|^2 + 2\int wf\, dx$. The lower semicontinuity of norm with respect to weak convergence shows that $\|u\|^2 \le \liminf \|\varphi_n\|^2$. For the second term, note that the map $w \mapsto 2\int fw\, dx$ is a continuous linear functional on $\mathring{H}^1(\Omega)$. Weak convergence of the φ_n is equivalent to the convergence $\ell(\varphi_n) \to \ell(u)$ for all $\ell \in \mathring{H}^1(\Omega)'$. In particular,

$$2 \int_\Omega \varphi_n f\, dx \to 2 \int_\Omega uf\, dx.$$

Thus $J(u) \le \liminf J(\varphi_n) = i$.

This completes the proof of all but the uniqueness part of the next theorem.

Theorem 3. *Suppose that Ω is an open subset of $\mathbb{R}^d$ which is contained between a pair of parallel hypersurfaces. If $f \in \operatorname{Re} L^2(\Omega)$, then the functional*

$J(w) = \int_\Omega |\nabla w|^2 + 2wf\,dx$ is continuous and bounded below on $\operatorname{Re} \mathring{H}^1(\Omega)$. It achieves its minimum value at one and only one $u \in \operatorname{Re} \mathring{H}^1(\Omega)$.

PROOF. The existence of a minimizer is proved above. If u_1 and u_2 are both minimizers, we prove equality by a convexity argument. In the next computation $\|\cdot\|$ denotes the norm $(\int |\nabla u|^2\,dx)^{1/2}$ on $\mathring{H}^1(\Omega)$

$$\left\|\frac{u_1 + u_2}{2}\right\|^2 = \frac{(\|u_1\|^2 + \|u_2\|^2)}{4} + \tfrac{1}{2}\int_\Omega \nabla u_1 \cdot \nabla u_2\,dx,$$

$$\tfrac{1}{2}\int_\Omega \nabla u_1 \cdot \nabla u_2\,dx \le \tfrac{1}{2}\|u_1\|\,\|u_2\| \le \frac{\|u_1\|^2 + \|u_2\|^2}{4}. \tag{5}$$

Thus $J((u_1 + u_2)/2) \le (J(u_1) + J(u_2))/2$. Since $J(u_j) = i$ for $j = 1, 2$, and i is the minimum value of J, we must have equality in the inequalities in (5).

Equality in the first implies that $\nabla u_1 = a\nabla u_2$ or $\nabla u_2 = a\nabla u_1$ with $a \in \mathbb{R}_+$. From the second we conclude that $\|u_1\| = \|u_2\|$. Recall that the norm is the L^2 norm of the gradient. Thus $a = 1$ and therefore $\|u_1 - u_2\| = \|\nabla u_1 - \nabla u_2\|_{L^2(\Omega)} = 0$. □

The uniqueness proof above shows that J is a strictly convex function on $\mathring{H}^1(\Omega)$.

It is natural that we had to enlarge the class of admissible functions in order to produce a minimum. This is entirely analogous to enlarging the rationals to the reals in order to gain completeness. Here the set of C^1 functions vanishing at the boundary was completed in a norm which was suggested by the estimates for the minimizing sequence. Whenever one admits more candidates into competition there is a danger that there may be too many solutions. The uniqueness in Theorem 3 is therefore reassuring.

To show that u solves the Dirichlet problem we would like to show that in some sense $\Delta u = f$ and $u|_{\partial\Omega} = 0$. The first is not hard to justify. The second is hidden in the fact that $u \in \mathring{H}^1(\Omega)$. The latter relation will be examined in more detail in §5.5. For the differential equation, we have a result analogous to Proposition 5.2.2.

Proposition 4. *Suppose that $u \in \mathscr{D}'(\Omega)$, $f \in L^2(\Omega)$, and $\nabla u \in L^2(\Omega)$. Then the following are equivalent:*

(i) *$\Delta u = f$ in the sense of $\mathscr{D}'(\Omega)$.*

(ii) *For all $\psi \in \mathscr{D}(\Omega)$*

$$\int_\Omega \nabla u \cdot \nabla \psi + f\psi\,dx = 0.$$

(iii) *The identity in* (ii) *holds for all $\psi \in \mathring{H}^1(\Omega)$.*

(iv) *For all $\psi \in \mathring{H}^1(\Omega)$, $dJ(u + \varepsilon\psi)/d\varepsilon|_{\varepsilon=0} = 0$.*

Proof. To show that (i)$\Leftrightarrow$(ii) note that for any $\psi \in \mathscr{D}(\Omega)$, $\langle -\Delta u, \psi \rangle = \sum \langle \partial_j u, \partial_j \psi \rangle$ from the definition of distribution derivative. Since $\partial_j u \in L^2$, we find that

$$\langle f - \Delta u, \psi \rangle = \int_\Omega \nabla u \cdot \nabla \psi + f\psi \, dx,$$

which yields the equivalence of (i) and (ii).

It is obvious that (iii) $\Rightarrow$ (ii). That (ii) $\Rightarrow$ (iii) follows from the fact that the integral in (ii) is continuous on $\mathring{H}^1(\Omega)$, and $\mathscr{D}(\Omega)$ is dense in $\mathring{H}^1(\Omega)$.

The equivalence of (iii) and (iv) follows from the fact that the derivative in (iv) is equal to the integral in (iii). □

The direct method of the calculus of variations usually yields solutions which satisfy the differential equations in a sense weaker than the classical sense. Historically, this was one of the motivations for the idea of distribution derivatives.

Corollary 5. *If $f \in \operatorname{Re} L^2(\Omega)$, then there is exactly one $u \in \operatorname{Re} \mathring{H}^1(\Omega)$ such that $\Delta u = f$.*

Proof. Proposition 4 shows that the minimizer of Theorem 3 is such a solution. Conversely, if u_1 and u_2 are solutions, then the difference $w \equiv u_1 - u_2$ belongs to $\mathring{H}^1(\Omega)$, and part (iii) of Proposition 4 shows that $\int \nabla w \cdot \nabla \psi \, dx = 0$ for all $\psi \in \mathring{H}^1(\Omega)$. Take $\psi = w$ to find that $\|w\| = 0$ so $w = 0$. □

This ends an important first step in studying the Dirichlet problem. The notion of solution has been extended and for this notion there is unique solvability. It is natural to ask whether one gets classical solutions when the data are sufficiently regular. For example, if Ω is regular and $f \in C^\infty(\bar{\Omega})$ we will show that $u \in C^\infty(\bar{\Omega})$ and is a solution in the classical sense. The next sections are devoted to showing just how much can be done in the $\mathring{H}^1(\Omega)$ context. The proof of the Regularity Theorems will be given in §5.8 and §5.9. The main results assert that u has two more derivatives than f at any point $x \in \bar{\Omega}$ when the derivatives are measured in the sense of H^s.

Definition. If $u \in \mathscr{D}'(\Omega)$, $s \in \mathbb{N}$, and $x \in \bar{\Omega}$, then we say that $u \in H^s(x)$ if there is an $r > 0$ so that for all $|\alpha| \le s$, $D^\alpha u \in L^2(\Omega \cap B_r(x))$.

The Interior Elliptic Regularity Theorem 5.9.2 asserts that if $x \in \Omega$ and $f \in H^s(x)$, then $u \in H^{s+2}(x)$. The Boundary Regularity Theorem 5.9.3 asserts that the same conclusion is true for $x \in \partial\Omega$, provided that $\bar{\Omega} \hookrightarrow \mathbb{R}^d$ is a smooth embedded manifold with boundary. In either case, the Sobolev Embedding Theorem allows us to conclude that u is C^k on an $\bar{\Omega}$ neighborhood of x provided that $k < s - d/2$. The C^∞ regularity asserted in the previous paragraph then follows.

PROBLEMS

1. Use the steps from the Dircet Method of the Calculus of Variations to prove the following abstract result:

Theorem. *Suppose that H is a separable Hilbert space and $J: H \to \mathbb{R}$ satisfies:*
(i) $\inf J > -\infty$.
(ii) *J is sequentially lower semicontinuous with respect to weak convergence in H.*
(iii) $\lim_{\|w\| \to \infty} J(w) = \infty$.
Then there is a $u \in H$ such that $J(u) \le J(w)$ for all $w \in H$.

DISCUSSION. To apply this result to the J in (1), one needs to verify (i), (ii), (iii) which is the heart of the argument presented in this section.

The weak compactness of the unit ball in Hilbert space is sometimes proved as a corollary to the Banach–Alouglu Theorem asserting the weak-star compactness of the unit ball in the dual of a Banach space. The latter result is proved using Tychonov's Theorem and therefore uses a strong version of the axiom of choice. The next problem presents a proof which uses minimal set-theoretic subtlety.

2. Suppose that $\{h_n\}$ is a bounded sequence in the separable Hilbert space $\mathscr{H}$. Prove that there is a weakly convergent subsequence by carrying out the following steps. Choose an orthonormal basis $e_1, e_2, \ldots$ for $\mathscr{H}$.
(i) Prove that there is a subsequence $\{h_{n_m}\}$ with the property that for all j, $\lim(e_j, h_{n_m})$ exists. *Hint.* Cantor diagonal process. Call the subsequence k_m, and the limits a_j.
(ii) Prove that $\sum |a_j|^2$ converges. *Hint.* Show that the sum of the first N terms is the limit of $(k_m, \sum_1^N a_j e_j)$ as $m \to \infty$. Estimate the limit using the Schwartz inequality.
Since $\{a_j\} \in l^2$, define $h \equiv \sum a_j e_j \in \mathscr{H}$.
(iii) Prove that $k_m \rightharpoonup h$.

§5.4. Variations on the Theme

In this section the proof in §5.3 is generalized in several directions. The changes are small and the domain of applicability of the method is shown to be very wide. Among the extensions considered are more general second-order operators, more general boundary conditions including the Neumann problem, and the Laplace–Beltrami operator on Riemannian manifolds.

First, the method is extended to symmetric divergence form operators generalizing Δ. The key observation here is that the Dirichlet integral, $\int |\nabla w|^2\, dx$, can be replaced by any other quadratic form which is equivalent to the norm in $\mathring{H}^1(\Omega)$. Toward that end, let

$$a(v, v) \equiv \int_\Omega \sum a_{ij}(x)\partial_i v \partial_j v + a(x)v^2\, dx. \tag{1}$$

Assume that

$$a_{ij} = a_{ji} \text{ and } a \text{ all belong to } L^\infty(\Omega : \mathbb{R}), \tag{2}$$

$$(\exists \mu > 0), (\forall \xi \in \mathbb{R}^d, x \in \Omega), \qquad \sum a_{ij}(x)\xi_i\xi_j \geq \mu|\xi|^2. \tag{3}$$

The constant μ in (3) is called an *ellipticity constant.*

$$a(u, u)^{1/2} \text{ is equivalent to the norm in } \operatorname{Re} \mathring{H}^1(\Omega). \tag{4}$$

Hypothesis (4) is equivalent to the existence of a $c > 0$ such that

$$a(u, u) \geq c\left(\int_\Omega |\nabla u|^2 + u^2\, dx\right) \qquad \text{for all} \quad u \in \operatorname{Re} C_0^\infty(\Omega). \tag{5}$$

Condition (3) is necessary for the validity of (4) and then $\mu \geq c$ (Problem 1). When inequality (5) is satisfied, $a(u, u)$ is called *coercive.*

Proposition 1. *Suppose that* (2) *and* (3) *are satisfied.*

(i) *If* $\inf\{a(x): x \in \Omega\} > 0$, *then* (4) *is satisfied.*
(ii) *If* Ω *lies between two parallel hyperplanes at distance* $\delta < \infty$, *then* (4) *is satisfied so long as* $\inf\{a(x): x \in \Omega\} > -\mu/\delta^2$.

PROOF. The first assertion is immediate. For the second, choose $\mu' < \mu$ so that $\inf a \geq -\mu'/\delta^2$. Then for all $u \in C_0^\infty(\Omega)$

$$\int_\Omega a(x)|u|^2\, dx \geq (\inf a)\|u\|_{L^2(\Omega)}^2 \geq \frac{\mu'}{\delta^2}\|u\|_{L^2(\Omega)}^2 \geq -\mu'\|\nabla u\|_{L^2(\Omega)}^2,$$

the last inequality using Theorem 5.3.1. Then $a(u, u) \geq (\mu - \mu')\|\nabla u\|^2$ which, together with Theorem 5.3.1, proves the desired estimate (5). □

EXAMPLE. If a_{ij} and a and Ω are fixed, then for sufficiently small r, $\Omega' \equiv B_r(x) \subset \Omega$ satisfies hypothesis (4).

What was needed of the term $2\int wf\, dx$ in J is that the linear map $w \mapsto 2\int wf\, dx$ was continuous from $\operatorname{Re} \mathring{H}^1$ to $\mathbb{R}$. This was used in showing that J was bounded from below, and in showing that J was lower semicontinuous with respect to weak convergence in $\mathring{H}^1$. Suppose that

$$l: \operatorname{Re} \mathring{H}^1(\Omega) \to \mathbb{R} \text{ is a linear and continuous map.} \tag{6}$$

Equivalently, $l \in \mathscr{D}'(\Omega)$ and

$$(\exists c), (\forall \varphi \in \operatorname{Re} C_0^\infty(\Omega)), \qquad |l(\varphi)| \leq c\|\varphi\|_{\mathring{H}^1(\Omega)}. \tag{6'}$$

Clearly (6) implies (6′). Conversely, if (6′) is satisfied, then since C_0^∞ in dense in $\mathring{H}^1$, l has a unique extension satisfying (6).

The proof of the last section yields the following result.

Theorem 2. *Under hypotheses* (1), (2), (4), *and* (6) *the functional* $J(v) \equiv a(v, v) + 2l(v)$ *is a strictly convex continuous function from* $\operatorname{Re} \mathring{H}^1(\Omega)$ *to* $\mathbb{R}$. *J is bounded below and achieves its infimum at a unique* $u \in \operatorname{Re} \mathring{H}^1$. *The minimizing u is* *characterized by*

$$(\forall \varphi \in \operatorname{Re} \mathring{H}^1(\Omega)), \qquad a(u, \varphi) + l(\varphi) = 0. \tag{7}$$

The above theorem is valid in an arbitrary open set. No regularity or boundedness is required.

The left-hand side of (7) is equal to $dJ(u + \varepsilon\varphi)/d\varepsilon|_{\varepsilon=0}$. Equations derived by setting such directional derivatives equal to zero at stationary points of a functional J are often called *Euler equations.* The interpretation in terms of differential equations is the following exact analogue of Proposition 5.3.3.

Proposition 3. *The minimizer u in Theorem* 1 *is the unique $u \in \operatorname{Re} \mathring{H}^1(\Omega)$ such that*

$$Lu \equiv \sum \partial_i(a_{ij}(x)\partial_j u) - a(x)u = l \quad \text{in } \mathscr{D}'(\Omega). \tag{8}$$

Note that for $u \in \mathring{H}^1(\Omega)$, $a_{ij}\partial_j u$ is square integrable on Ω so the distribution derivative $\partial_i(a_{ij}\partial_j u)$ is meaningful. On the other hand, the individual terms in the expression $a_{ij}\partial_{ij}u + (\partial_i a_{ij})\partial_j u$ from the product rule do not have a simple interpretation unless $\partial_i a_{ij}$ is square integrable. It is wise to leave L in the divergence form (8). The next proposition shows that L is not only well defined but continuous.

Proposition 4. *If $Q(x, D)u \equiv \sum \partial_j(q_j(x)u)$ with $q_j \in L^\infty(\Omega)$, then Q is a continuous map from $L^2(\Omega)$ to the dual of $\mathring{H}^1(\Omega)$.*

Proof. For $u \in L^2(\Omega)$, $\psi \in C_0^\infty(\Omega)$ compute

$$\langle Qu, \psi\rangle = -\sum \langle q_j(x)u, \partial_j\psi\rangle = -\sum \int q_j(x)u(x)\partial_j\psi(x)\,dx$$

$$|\langle Q\varphi, \psi\rangle| \le \sum \|\varphi\|_{L^2}\|q_j\|_{L^\infty}\|\partial_j\psi\|_{L^2} \le c\|\varphi\|_{L^2}\|\psi\|_{\mathring{H}^1}.$$

Since the set of ψ are dense in $\mathring{H}^1(\Omega)$, the result follows. □

This proposition implies that L is a continuous map of $\mathring{H}^1(\Omega)$ to $\operatorname{Re} \mathring{H}^1(\Omega)'$ whenever (2) is satisfied.

Corollary 5. *The map L in* (8) *is an isomorphism from* $\operatorname{Re} \mathring{H}^1(\Omega)$ *to the dual* $\operatorname{Re} \mathring{H}^1(\Omega)'$.

Proof. We have established that L is a continuous bijection between these spaces. To prove that the inverse is continuous is equivalent to finding an estimate for the $\mathring{H}^1(\Omega)$ norm of the solution of $Lu = l$ in terms of the $\mathring{H}^1(\Omega)'$ norm of l. To do this, take $\varphi = u$ in the Euler equation (5) to find that

$$a(u, u) \le c\|u\|_{\mathring{H}^1}\|l\|_{(\mathring{H}^1)'}.$$

Since $a(u, u)$ is equivalent to the square of the $\mathring{H}^1$ norm, this shows that

$$\|u\|_{\mathring{H}^1} \leq c\|f\|_{(\mathring{H}^1)'}, \tag{9}$$

which is equivalent to the continuity of the inverse of L. □

One could have concluded that L^{-1} is continuous by applying the Open Mapping Theorem. It is a general principle that if one has done so much analysis that an inequality follows by applying the Baire Category Theorem, then one has actually proved the inequality earlier. In our case, it is the lower bound (5.2.2) which is an alias for (9).

It is also useful to recall the derivation of (9) since one step is hidden in Proposition 3. What one does is apply $Lu = l$ to u. An integration by parts and (5) yields the estimate. Briefly, the estimate is proved by multiplying the equation by u, then integrating by parts. This is an example of the energy method with multiplier u.

In analogy with the Lax duality proved in Problem 3.5.1, we make the

Definition. $H^{-1}(\Omega)$ is the dual of $\mathring{H}^1(\Omega)$. That is, $H^{-1}(\Omega)$ is the set of distributions $l \in \mathscr{D}'(\Omega)$ satisfying (6′) for all $\varphi \in C_0^\infty(\Omega)$.

The Riesz Representation Theorem gives an isometry between $\mathring{H}^1(\Omega)$ and $H^{-1}(\Omega)$. In particular, $H^{-1}(\Omega)$ is a separable Hilbert space.

Under the hypotheses which imply coerciveness, Corollary 5 asserts that L: Re $\mathring{H}^1 \to$ Re H^{-1} is an isomorphism. Estimate (9) takes the elegant form

$$\|u\|_{\mathring{H}^1(\Omega)} \leq c\|Lu\|_{H^{-1}(\Omega)}. \tag{10}$$

Such an estimate giving a gain of two derivatives is typical of elliptic equations (recall Proposition 2.4.5 and contrast Problem 1.1.4).

The technique for the Dirichlet problem also solves a variety of other elliptic boundary value problems which arise in applications. As an example, conside heat flow in Ω with a constant source $f(x)$ and insulated boundary. The equations of motion take the form

$$v_t = \nu\Delta v + f(x), \qquad t, x \in \mathbb{R}_t \times \Omega, \tag{11}$$

$$\frac{\partial v}{\partial n} = 0 \quad \text{at } \partial\Omega \qquad \text{and} \qquad v(0, \cdot) \text{ given}. \tag{12}$$

The *Neumann boundary condition* (12) asserts that there is no flow of heat through the boundary. This is the meaning of perfect insulation. One finds, as $t \to \infty$, $v(t, x) \to u(x)$, a solution to the *Neumann problem*

$$\nu\Delta u = f \quad \text{in } \Omega, \qquad \frac{\partial u}{\partial n} = 0 \quad \text{at } \partial\Omega. \tag{13}$$

These conditions are insufficient to determine u. Any two solutions differ by a constant function. The constants are the elements of the null space of Δ with Neumann boundary conditions. To complete the determination of u use the conservation of energy

$$\int v(t, x)\, dx \quad \text{is independent of time} \tag{14}$$

for solution of (11), (12). To prove this, integrate the equation over Ω and note that $\int_\Omega \Delta v\, dx = \int_{\partial\Omega} \partial v/\partial n\, d\sigma = 0$, thanks to (12). Thus $\partial_t \int v\, dx = \int v_t\, dx = 0$ proving (14). Therefore, $\int_\Omega u(x)\, dx = \int_\Omega v(0, x)\, dx$. This additional condition together with (13) identifies u. Here the initial condition contributes to the determination of the steady state u. This is in contrast to heat flow in a domain whose boundary is kept at a time-independent temperature.

Corresponding to the fact that there is nonuniqueness of solutions to (13) there is also nonexistence. Integrating the equation over Ω yields

$$\int_\Omega f(x)\, dx = \nu \int_\Omega \Delta u\, dx = \nu \int_{\partial\Omega} \frac{\partial u}{\partial n}\, d\sigma = 0,$$

which is a necessary condition on f for solvability.

If one attempted to solve (11) by Laplace transform in time

$$\mathscr{L}v(\tau) \equiv \int_0^\infty e^{-\tau t} v(t)\, dt,$$

one finds the boundary value problem

$$(\tau - \nu\Delta)\mathscr{L}v = f(x) \quad \text{in } \Omega, \qquad \frac{\partial(\mathscr{L}v)}{\partial n} = 0 \quad \text{at } \partial\Omega.$$

For $\tau \in]0, \infty[$ (more generally $\tau \in \mathbb{C}\backslash]-\infty, 0]$), this is uniquely solvable. The key is to show the solvability of the Neumann problem

$$(a(x) - \nu\Delta)u = 0 \quad \text{in } \Omega, \qquad \frac{\partial u}{\partial n} = 0 \quad \text{on } \partial\Omega, \tag{15}$$

where $\nu > 0$ and $a \in L^\infty(\Omega : \mathbb{R})$ and $\inf a(x) \equiv \alpha > 0$. The solution is obtained by the variational method with a basic Hilbert space different from $\mathring{H}^1(\Omega)$.

Definition. If $\Omega \subset \mathbb{R}^d$ is open, then the Sobolev space $H^1(\Omega)$ is the set of distributions $u \in \mathscr{D}'(\Omega)$ such that for $|\beta| \le 1$, $\partial^\beta u \in L^2(\Omega)$.

Proposition 6. *$H^1(\Omega)$ is a separable Hilbert space with norm given by*

$$\|u\|^2_{H^1(\Omega)} \equiv \int_\Omega |\nabla u|^2 + |u|^2\, dx.$$

PROOF. To verify completeness suppose that u_n is a Cauchy sequence. Then, for all $|\beta| \le 1$, $\partial^\beta u$ is a Cauchy sequence in $L^2(\Omega)$. By completeness of L^2, $\partial^\beta u_n$ converges in $L^2(\Omega)$ to a limit g_β for all $|\beta| \le 1$.

Since L^2 convergence implies convergence in the sense of distributions and ∂ is continuous for distribution convergence, $\partial^\beta g_0 = \partial^\beta \lim u_n = \lim \partial^\beta u_n = g_\beta \in L^2(\Omega)$. Thus g_0 belongs to $H^1(\Omega)$ and by construction u_n converges to g_0 in $H^1(\Omega)$ proving completeness.

The map $u \mapsto (u, \partial_1 u, \ldots, \partial_d u)$ is an isometry from $H^1(\Omega)$ into $L^2(\Omega)^{d+1}$. The separability of the target implies the separability of the source. □

The Neumann problem (15) is solved by minimizing

$$J(u) \equiv \int_\Omega \nu|\nabla u|^2 + a(x)u^2 - 2f(x)u(x)\,dx, \tag{16}$$

where the minimum is taken over all $u \in \operatorname{Re} H^1(\Omega)$. The key fact in constructing a minimum is that the bilinear form

$$a(u, u) \equiv \int_\Omega \nu|\nabla u|^2 + a(x)u^2\,dx$$

defines a norm equivalent to the $H^1(\Omega)$ norm. This is the requisite coerciveness.

Even a trusting reader should be skeptical that this minimum solves the boundary value problem since the boundary condition appears nowhere in the description. If the minimum is achieved at v, the Euler equations show that

$$\int_\Omega \nu\nabla\varphi\cdot\nabla v + a(x)\varphi v - \varphi f\,dx = 0 \qquad \text{for all} \quad \varphi \in H^1(\Omega). \tag{17}$$

Choosing $\varphi \in C_0^\infty(\Omega)$ shows that the differential equation is satisfied in the sense of distributions. However, the boundary condition $\partial v/\partial n = 0$ is not hidden in the condition that $v \in H^1(\Omega)$. For example, the restriction of an arbitrary member of $C^\infty(\mathbb{R}^d)$ to Ω lies in the space $H^1(\Omega)$ and in no sense satisfies homogeneous Neumann conditions.

However, the equation of variation, (17), holds for a large class of test functions φ. For the Dirichlet problem, Proposition 5.3.4 showed that the equation of variation was satisfied if and only if it was satisfied for all $\varphi \in C_0^\infty(\Omega)$. This is not the case for the Neumann problem. To see the difference suppose that $\bar\Omega$ is smooth and $v \in C^2(\bar\Omega)$ satisfies (17). An integration by parts yields

$$\int_\Omega \varphi(-\nu\Delta v + a(x)v - \varphi f)\,dx + \int_{\partial\Omega} \frac{\nu\varphi\partial v}{\partial n}\,d\sigma = 0 \qquad \text{for all} \quad \varphi \in H^1(\Omega). \tag{18}$$

For $\varphi \in C_0^\infty(\Omega)$ the boundary term vanishes and we recover the differential equation, $(a - \nu\Delta)v = f$. This implies that the integral over Ω vanishes for all $\varphi \in H^1(\Omega)$. Thus the boundary integral vanishes for all $\varphi \in H^1(\Omega)$, in particular, for all $\varphi \in C^\infty(\bar\Omega)$. Then $\varphi|_{\partial\Omega}$ is an arbitrary smooth function on the boundary of Ω and it follows that $\partial v/\partial n = 0$ on the boundary. For irregular sets Ω or nonsmooth v the equation of variation gives a weak sense to the homogeneous Neumann boundary condition which is still strong enough to imply uniqueness. When the data are regular the solutions are regular and satisfy the boundary condition in the classical sense. Elliptic regularity of the sort $f \in H^s(x) \Rightarrow u \in H^{s+2}(x)$ can be proved by the same methods as those employed for the Dirichlet problem in §5.8 and §5.9. Boundary conditions which are hidden in the equation of variation, rather than being imposed directly on the admissible functions in a variational argument, are called *natural boundary conditions.*

Another example amenable to analysis as above is the Laplace–Beltrami operator. Suppose that M is a compact smooth *Riemannian manifold.* We use the standard notation $T_x(M)$ for the tangent space to M at x. It is identified with equivalence classes of curves $\gamma(t)$ with $\gamma(0) = x$ with $\gamma_1 \sim \gamma_2$ if and only if their first derivatives are equal in any coordinate system. The dual space to $T_x(M)$ is denoted $T_x^*(M)$ and is the fiber of the cotangent bundle $T^*(M)$. For a $\varphi \in C^\infty(M : \mathbb{R})$ the exterior derivative $d\varphi$ is a section of the cotangent bundle. The Riemannian metric on $T_x(M)$ induces a correspondence between $T_x(M)$ and $T_x^*(M)$ and thereby a metric on $T_x^*(M)$. Thus if $\omega \in T_x^*(M)$, its length $|\omega|$ is well defined as is the scalar product $\langle \omega, \xi \rangle$ of two elements of $T_x^*(M)$. The Laplace–Beltrami operator Δ_M is the differential operator associated to the Dirichlet integral $D(\varphi) \equiv \int_M |d\varphi|^2 \, dV$, where $|d\varphi(x)|$ is the length of $d\varphi(x)$ and dV is the Riemannian volume element on M. Thus, for $\varphi \in C_0^\infty(M)$, $\Delta_M \varphi \in C_0^\infty(M)$ is defined by the relation

$$\int_M \psi \Delta_M \varphi \, dv = -\int_M \langle d\psi(x), d\varphi(x) \rangle \, dv = \tfrac{1}{2}(d/d\varepsilon) D(\varphi + \varepsilon\psi)|_{\varepsilon=0}.$$

Computing in local coordinates shows that Δ_M is a second-order differential operator like those considered in this section. $H^1(M)$ is defined as the set of elements of $u \in L^2(M)$ such that in any local coordinates, $\partial u/\partial x_i \in L^2_{\text{loc}}$. If ψ_j is a finite partition of unity subordinate to a coordinate atlas (U_j, β_j), then $H^1(M)$ is a Hilbert space with norm

$$\|u\|^2_{H^1(M)} \equiv \sum_j \sum_{|\alpha| \le 1} \|\partial^\alpha((\psi_j u) \circ \beta_j)\|^2_{L^2(\mathbb{R}^d)}.$$

Then the functional

$$\int_M u(x)^2 + |du(x)|^2 \, dV$$

is coercive, that is, defines a norm equivalent to the norm on $H^1(M)$, which shows that $1 - \Delta_M$ is an isomorphism of $H^1(M)$ to $H^1(M)'$. If $\Omega \subset M$ is a bounded open set and $\mathring{H}^1(\Omega)$ is defined to be the closure of $C_0^\infty(\Omega)$ in $H^1(M)$, then the functional $D(u)$ is coercive on $\mathring{H}^1(\Omega)$ and one finds that Δ_M is an isomorphism of $\mathring{H}^1(\Omega)$ to $\mathring{H}^1(\Omega)'$. If $f \in C^\infty(M)$ and $x \in M$, we can take Ω to be a small neighborhood of x and thereby solve the equation $\Delta_M u = f$ on a neighborhood of x. In §5.8 we will show that $u \in C^\infty(\Omega)$. This local solvability of Δ_M is a crucial step in the construction of isothermal coordinates on Riemannian 2-manifolds. This in turn is how they are given a complex structure and are identified as *Riemann surfaces.*

In the same vein, the *Hodge Theory* is associated with the quadratic form

$$\int_M |d\omega(x)|^2 + |d^*\omega(x)| \, dv,$$

where $d\colon \Lambda^k(M) \to \Lambda^{k+1}(M)$ is the exterior derivative and $d^*\colon \Lambda^{k+1}(M) \to \Lambda^k(M)$ (and therefore $\Lambda^k \to \Lambda^{k-1}$) is the transposed differential operator based on the natural L^2 scalar products in $\Lambda^k(M)$ and $\Lambda^{k+1}(M)$. The associated second-order differential operator from Λ^k to itself is equal to $d^*d + dd^*$ and is called the *Hodge Laplacian*. The difficult part of the classical Hodge Theorem is an Elliptic Regularity Theorem which can be proved as in §5.8.

PROBLEMS

1. Prove that if inequality (5) holds, then (4) must be satisfied and $c \le \inf\{\sum a_{ij}(x)\xi_i\xi_j\colon x \in \Omega \text{ and } \xi \in \mathbb{R}^d \text{ with } |\xi| = 1\}$. *Hint.* Consider highly oscillatory test functions $e^{i\lambda x\xi}$ with $\lambda \to \infty$ localized by a cutoff $\chi(n(x - \underline{x}))$, with $n = n(\lambda)$ tending to infinity.

The existence theorems of the last two sections can be proved using the Riesz Representation Theorem for elements of the dual of a Hibert space. That theorem is proved by a variational argument, and one can view what we have done as repeating the proof.

This is not a wasted exercise since there are many other problems which can be attacked by the Direct Method of the Calculus of Variations, including many nonlinear problems. Problem 2 describes the Riesz Representation Theorem proof. We take the opportunity to give a generalization to complex equations and solutions. The hypotheses on the coefficients are

$$a_{ij} = \bar{a}_{ji} \quad \text{and} \quad a \ge 0 \quad \text{belong to } L^\infty(\Omega). \tag{19}$$

In this case

$$a(v, v) \equiv \int_\Omega \sum a_{ij}(x)\partial_i v\overline{\partial_j v} + a(x)|v|^2\,dx, \tag{20}$$

and we assume coercivity, that is

$$a(v, v)^{1/2} \text{ is equivalent to the norm in } \mathring{H}^1(\Omega). \tag{21}$$

The Riesz Representation Theorem then shows that every continuous linear functional on $\mathring{H}^1(\Omega)$ is of the form $a(\cdot, u)$ for appropriate $u \in \mathring{H}^1(\Omega)$.

As in Problem 1, (21) can hold only if there is a $\mu > 0$ such that

$$\sum a_{ij}(x)\xi_i\bar{\xi}_j \ge \mu|\xi|^2, \qquad \text{for all} \quad x \in \Omega \text{ and } \xi \in \mathbb{C}^d. \tag{22}$$

If the a_{ij} are real and satisfy (3), then this holds since in that case one has

$$\sum a_{ij}(x)\xi_i\bar{\xi}_j = \sum a_{ij}(x)((\text{Re}\,\xi_i)(\text{Re}\,\xi_j) + (\text{Im}\,\xi_i)(\text{Im}\,\xi_j)). \tag{23}$$

In the real case, if L is a second order and elliptic, then either L or $-L$ satisfies (3). In the complex case, the analogous result is not true, for example, $L = (\partial_x + i\partial_y)^2$ is elliptic, but neither L nor $-L$ is positive in the sense of (22).

2. Use the Riesz Representation Theorem to show that if L satisfies (19), (20), (21), then L is an isomorphism of $\mathring{H}^1(\Omega)$ to $H^{-1}(\Omega)$. *Warning.* Beware of complex conjugates.
3. Prove
   (i) If $\Omega =]0, b[$, then $l \in H^{-1}(\Omega)$ if and only if there is a $u \in L^2(\Omega)$ such that $l = du/dx$ (derivative in the sense of distributions).

(ii) Let $\Gamma = \text{kernel}(d/dx)$ be the linear subspace of constant functions in L^2. Prove that the map $u \mapsto du/dx$ is an isomorphism of $L^2(]0, b[)/\Gamma$ to $H^{-1}(]0, b[)$.

4. Prove that if $\Omega \subset \mathbb{R}^d$ is bounded and open, then $l \in H^{-1}$ if and only if there are functions $f_j \in L^2(\Omega)$ such that $u = \sum \partial_j f_j$. Is this an isomorphism of $L^2(\Omega)^d$/constants to $H^{-1}(\Omega)$?

5. $C_0^\infty(\Omega) \hookrightarrow H^{-1}(\Omega)$ by the usual identification $\varphi \mapsto l$

$$\langle l, \psi \rangle \equiv \int_\Omega \psi(x)\varphi(x)\, dx \qquad \text{for all} \quad \psi \in \mathring{H}^1(\Omega).$$

Prove that $C_0^\infty(\Omega)$ is dense in $H^{-1}(\Omega)$.

6. Show that the natural boundary condition associated to the variational problem

$$\text{minimize over } v \in \text{Re } H^1(\Omega): \int_\Omega \sum a_{ij}(x)\partial_i v \partial_j v + v^2\, dx$$

is

$$\sum a_{ij}(x) n_j(x) \partial_j v(x) = 0, \qquad (n_1, \ldots, n_d) = \text{outward conormal to } \partial\Omega. \tag{24}$$

Discussion. The derivative on the left is called the *normal derivative* associated to the operator with principal symbol $\sum a_{ij}\partial_{ij}$. It arises in a variety of problems in geometry and mechanics.

The next problem treats the so-called *third boundary value problem* which has the Robin boundary condition

$$\frac{\partial u}{\partial n} + \alpha(x)u = 0.$$

One origin is the study of heat flow in a region Ω surrounded by a temperature bath at temperature T. The outward heat flux at the boundary is proportional to the difference between the temperature at the boundary and T

$$-\frac{\partial v}{\partial n} = \alpha(v - T), \qquad \alpha > 0.$$

Thus $v - T$ satisfies the heat equation in Ω and a homogeneous Robin condition at the boundary. Equilibrium solutions with a source f satisfy

$$\Delta u = f \quad \text{in } \Omega \qquad \text{and} \qquad \frac{\partial u}{\partial n} + \alpha u = 0 \quad \text{on } \partial\Omega. \tag{25}$$

This is equivalent to the variational problem

$$\text{minimize} \quad J(u) \equiv \int_\Omega |\nabla u|^2 + 2uf\, dx + \int_{\partial\Omega} \alpha u^2\, d\sigma, \qquad u \in H^1(\Omega).$$

The fact that $u|_{\partial\Omega}$ makes sense and is square integrable for $u \in H^1(\Omega)$ is called a *Trace Theorem.* A model is Theorem 5.5.3.

7. Prove that if Ω is regular and $u \in C^2(\bar{\Omega})$ satisfies $J(u) \le J(w)$ for all $w \in C^2(\bar{\Omega})$, then u is a solution of (25).

Discussion. The Robin condition is a natural boundary condition associated to the functional J on $H^1(\Omega)$. The variational approach to a variety of elliptic boundary value problems can be found in the books of Agmon [A] and Lions [Lio].

§5.5. $\mathring{H}^1$ and the Dirichlet Boundary Condition

Our solutions of the Dirichlet problem satisfy the differential equation in the sense of distributions and the homogeneous Dirichlet boundary condition in the sense that $u \in \mathring{H}^1(\Omega)$. In this section we examine more closely the structure of $\mathring{H}^1$. In the process, we prove several ways in which the condition $u|_{\partial\Omega} = 0$ is satisfied.

Theorem 1. *If $\bar{\Omega} \hookrightarrow \mathbb{R}^d$ is a compact smooth submanifold with boundary, then $\mathring{H}^1(\Omega)$ is equal to the set of all $u \in H^1(\mathbb{R}^d)$ with support in $\bar{\Omega}$.*

PROOF. If $u \in \mathring{H}^1(\Omega)$, choose $u_n \in C_0^\infty(\Omega)$ with $u_n \to u$ in $H^1(\mathbb{R}^d)$. Then u_n vanishes on $\mathbb{R}^d \backslash \bar{\Omega}$ and u_n converges to u in $\mathscr{D}'(\mathbb{R}^d)$. It follows that u vanishes on $\mathbb{R}^d \backslash \bar{\Omega}$, so supp $u \subset \bar{\Omega}$.

Conversely, if $u \in H^1(\mathbb{R}^d)$ with $\text{supp}(u) \subset \bar{\Omega}$ and $\varepsilon > 0$, we will construct $u_\varepsilon \in H^1(\mathbb{R}^d)$ with $\text{supp}(u_\varepsilon) \subset \Omega$ and $\|u - u_\varepsilon\|_{H^1(\mathbb{R}^d)} < \varepsilon$. Then, for η sufficiently small, $j_\eta * u_\varepsilon \in C_0^\infty(\Omega)$ and $\|u - j_\eta * u_\varepsilon\|_{H^1(\mathbb{R}^d)} < \varepsilon$.

To construct u_ε we push u inward using the flow of a vector field. Choose a smooth compactly supported vector field V on $\mathbb{R}^d$ which is transverse to $\partial\Omega$ and points out of Ω at points of $\partial\Omega$. Let Φ_t be the flow generated by the vector field V. Since V is compactly supported the flow is globally defined and Φ_t is equal to the identity map outside supp V.

The transversality hypothesis implies that there is a $c > 0$ such that for $0 \le t \le 1$, $\Phi_{-t}(\bar{\Omega}) \subset \{x \in \Omega: \text{dist}(x, \partial\Omega) \ge ct\}$. The function u_ε is defined to be $u \circ \Phi_t$ for t small positive so $\text{supp}(u \circ \Phi_t) \subset \Phi_{-t}(\bar{\Omega})$. It suffices to show that $u \circ \Phi_t$ converges to u in $H^1(\mathbb{R}^d)$.

Lemma 2. *Suppose that V is a smooth compactly supported vector field on $\mathbb{R}^d$ and Φ_t is the flow generated by V. For any $w \in L^2(\mathbb{R}^d)$, $w \circ \Phi_t \to w$ in $L^2(\mathbb{R}^d)$ as $t \to 0$.*

PROOF. For w belonging to the dense subset $C_0(\mathbb{R}^d) \subset L^2(\mathbb{R}^d)$, the conclusion is immediate. In fact, one has uniform convergence and support contained in a compact set independent of $t \in [-1, 1]$.

The maps $w \to w \circ \Phi_t$ are linear maps. To prove the lemma it is sufficient to prove that for $|t| \le 1$ the maps are uniformly bounded in $\text{Hom}(L^2(\mathbb{R}^d))$. Compute for $w \in C_0(\mathbb{R}^d)$ using the change of variable $y = \Phi_t(x)$

$$\int_{\mathbb{R}^d} |w \circ \Phi_t(x)|^2 \, dx = \int_{\mathbb{R}^d} |w(y)|^2 \det D\Phi_{-t}(y) \, dy,$$

where we note that $\det D\Phi_t > 0$ since $\det D\Phi_t$ is continuous in t, nonvanishing, and equal to 1 at $t = 0$. The uniform boundedness is a consequence of the fact that $\det D\Phi_{-t}(y)$ is bounded independent of $y \in \mathbb{R}^d$ and $t \in [-1.1]$. □

For diffeomorphisms Φ of $\mathbb{R}^d$, which equal the identity outside a compact set, the map $\mathscr{S}(\mathbb{R}^d) \ni u \to u \circ \Phi$ extends uniquely to a sequentially continuous map of $\mathscr{S}'(\mathbb{R}^d)$ to itself. The transpose of the map is given by $v \to (\det D\Phi)^{-1}(v \circ \Phi^{-1})$ and takes $\mathscr{S}$ to itself. Thus, for $u \in \mathscr{S}'(\mathbb{R}^d)$, $u \circ \Phi$ is given by $\langle u \circ \Phi, \psi \rangle = \langle u, (\det D\Phi)^{-1}(\psi \circ \Phi^{-1})\rangle$. The chain rule for the partial derivatives of $u \circ \Phi$ follows by continuity from its validity on the sequentially dense subset $\mathscr{S} \subset \mathscr{S}'$. Therefore

$$\partial_j(u \circ \Phi) = \sum ((\partial_k u) \circ \Phi) \frac{\partial \Phi_k}{\partial x_j}, \qquad \Phi = (\Phi_1, \ldots, \Phi_d), \tag{1}$$

so $\|u \circ \Phi\|_{H^1} \leq c\|u\|_{H^1}$ and it follows that $\circ\, \Phi$ maps $H^1(\mathbb{R}^d)$ to itself. Lemma 2 together with formula (1) imply that $u \circ \Phi_t$ converges to u in $H^1(\mathbb{R}^d)$ as $t \to 0$. This completes the proof of Theorem 1. □

When $d = 1$, the elements of $H^1(\mathbb{R}^d)$ are continuous functions. Thus supp $u \subset \bar{\Omega}$ implies that $u|_{\partial\Omega} = 0$. When $d > 1$, u does not have well-defined values at points. However, $u|_{\partial\Omega}$ is a well-defined element of $\mathscr{D}'(\partial\Omega)$ and for $u \in \mathring{H}^1(\Omega)$ this distribution vanishes.

Theorem 3. *Suppose that $\Sigma \subset \mathbb{R}^d$ is an smooth embedded compact $d - 1$ dimensional manifold. Then the map $C_0^\infty(\Omega) \ni \varphi \to \varphi|_\Sigma \equiv \gamma(\varphi)$ has a unique continuous extension to a linear map of $H^1(\mathbb{R}^d)$ to $L^2(\Sigma)$. In addition, there is a constant $c = c(\Sigma)$ such that for all $u \in H^1(\mathbb{R}^d)$.*

$$\|\gamma(u)\|^2_{L^2(\Sigma)} \leq c(\|u\|_{L^2(\mathbb{R}^d)}\|\nabla u\|_{L^2(\mathbb{R}^d)} + \|u\|^2_{L^2(\mathbb{R}^d)}). \tag{2}$$

The norm in $L^2(\Sigma)$ is that with respect to any smooth volume element, for example, the volume element $d\sigma$ induced by the Euclidean metric in $\mathbb{R}^d$.

For $u \in H^1(\mathbb{R}^d)$ the value of $\gamma(u)$ is called the *trace of u on Σ* and is also denoted $u|_\Sigma$. A more careful estimate shows that $u|_\Sigma$ belongs to the fractional Sobolev space $H^{1/2}(\Sigma)$ (see Hormander, [H2, Vol. I]).

Proof of Theorem 3. Since $C_0^\infty(\mathbb{R}^d)$ is dense in $H^1(\mathbb{R}^d)$, it suffices to prove the estimate (2) for elements ψ of $C_0^\infty(\mathbb{R}^d)$.

Choose a compactly supported smooth vector field, V, which is transverse to Σ. Let Φ_t be the flow generated by V. Then there is an $\varepsilon > 0$ such that the map η: $(t, \sigma) \mapsto \Phi_t(\sigma)$ is a diffeomorphism from $]-\varepsilon, \varepsilon[\times \Sigma$ to an $\mathbb{R}^d$ neighborhood, $\mathscr{U}$, of Σ. The Euclidean volume element in $\mathscr{U}$ is equal to $w(t, \sigma)\, dt\, d\sigma$ with w smooth on $]-\varepsilon, \varepsilon[\times \Sigma$.

Choose $\chi(t) \in C_0^\infty(]-\varepsilon, \varepsilon[)$ with $\chi(0) = 1$. Integrate $\partial_t(\chi(|\psi \circ \eta|^2)$ over $-\varepsilon \leq t \leq 0$ to find

$$|\psi|^2(0, \sigma) = \int \chi_t |\psi|^2\, dt + 2\,\text{Re} \int \chi\psi\bar{\psi}_t\, dt.$$

Integrate $d\sigma$ and then apply the Schwartz inequality in the second term to prove (2) for ψ. □

The next result shows that the traces at $\partial\Omega$ of elements of $\mathring{H}^1(\Omega)$ vanish. In this sense they satisfy the Dirichlet boundary condition.

Corollary 4. *If Ω is as in Theorem 1 and $u \in \mathring{H}^1(\Omega)$, then $u|_{\partial\Omega} = 0$.*

PROOF. Choose $u_n \in C_0^\infty(\Omega)$ converging to u in $H^1(\mathbb{R}^d)$. Then $u_n|_{\partial\Omega}$ converges to $u|_{\partial\Omega}$ in $L^2(\Omega)$. The result follows since $u_n|_{\partial\Omega} = 0$. □

The traces defined in the previous theorem enter in the formula for integration by parts.

Theorem 5. *Suppose that Ω is as in Theorem 1. Then, for all $u, v \in H^1(\mathbb{R}^d)$ and $1 \le j \le d$,*

$$\int_\Omega u\partial_j v \, dx = -\int_\Omega v\partial_j u \, dx + \int_{\partial\Omega} uv \, dx_1 \wedge \cdots \wedge \widehat{dx_j} \wedge \ldots dx_d. \tag{3}$$

The integrand of the last term is $u|_{\partial\Omega} v|_{\partial\Omega} \in L^1(\partial\Omega)$. The $\widehat{dx_j}$ means that this factor is omitted. The formula is often written noticing that $dx_1 \wedge \cdots \wedge \widehat{dx_j} \wedge \ldots dx_d = n_j \, d\sigma$, where $d\sigma$ is the element of Euclidean area on $\partial\Omega$ and $(n_1, \ldots, n_d)$ is the Euclidean outward unit normal to $\partial\Omega$.

PROOF. Formula (3) is true for u, v in $C_0^\infty(\mathbb{R}^d)$ which is a dense subset of $H^1(\mathbb{R}^d)$. Thanks to Theorem 4, both sides of (3) define continuous bilinear forms on $H^1(\mathbb{R}^d) \times H^1(\mathbb{R}^d)$. The identity follows since continuous functions which are equal on a dense subset are everywhere equal. □

The next result expresses a sense in which u vanishes at the boundary without passing through the intermediary of traces at the boundary. It asserts that the average value of $|u|^2$ in a band of width ε about $\partial\Omega$ tends to zero with ε.

Theorem 6. *Suppose that Ω is as in Theorem 1 and $\mathcal{T}_\varepsilon \equiv \{x \in \Omega: \operatorname{dist}(x, \partial\Omega) < \varepsilon\}$. Then there is a constant c and $\varepsilon_0 > 0$ so that, for all $u \in \mathring{H}^1(\Omega)$ and $\varepsilon < \varepsilon_0$,*

$$\int_{\mathcal{T}_\varepsilon} |u|^2 \, dx \le c\varepsilon^2 \|\nabla u\|^2_{L^2(\mathcal{T}_\varepsilon)}. \tag{4}$$

PROOF. Introduce $\Psi: \mathbb{R} \times \partial\Omega \to \mathbb{R}^d$ by $\Psi(t, \sigma) = \sigma + tn(\sigma)$ where $n(\sigma)$ is the unit outward normal to $\partial\Omega$. Then there is an $\varepsilon_0 > 0$ so that Ψ is a diffeomorphism of $]-\varepsilon_0, \varepsilon_0[\times \partial\Omega$ to a neighborhood of $\partial\Omega$ and t is the distance of $\Psi(t, \sigma)$ to $\partial\Omega$.

It is sufficient to prove (4) for φ in the dense subset $C_0^\infty(\Omega)$. For $t \in [-\varepsilon_0/2, 0]$ and such φ integrate $\partial_t(\varphi \circ \Psi)$ over $[t, 0]$ to find

$$|\varphi(t, \sigma)| = \left| \int_t^0 \varphi_t \, dt \right| \le |t|^{1/2} \left(\int_t^0 |\varphi_t|^2 \, dt \right)^{1/2}, \tag{5}$$

where the last estimate uses the Schwartz inequality.

Square (5) and integrate $d\sigma$ to find

$$\int_{\partial\Omega} |\varphi(t, \sigma)|^2 \, d\sigma \leq c|t| \, \|\nabla\varphi \circ \Psi\|^2_{L^2(]-a,0[\times\nabla\Omega)}.$$

For any $0 < \varepsilon \leq \varepsilon_0/2$, integrate dt over $[-\varepsilon, 0]$ to find

$$\int_{-\varepsilon}^{0} \int_{\partial\Omega} |\varphi(t, \sigma)|^2 \, dt \, d\sigma \leq c'\varepsilon^2 \, \|\nabla\varphi \circ \Psi\|^2_{L^2(]-\varepsilon,0[\times\partial\Omega)} \tag{6}$$

and the theorem follows. □

Since the volume of $\mathcal{T}_\varepsilon$ is of order ε and the integral of u^2 over $\mathcal{T}_\varepsilon$ is $o(\varepsilon^2)$, the theorem shows that the average of $|u|^2$ over $\mathcal{T}_\varepsilon$ is $o(\varepsilon)$, which is best possible (Problem 2).

Corollary 7. *With Ω and $\mathcal{T}_\varepsilon$ as above and $u \in \mathring{H}^1(\Omega)$*

$$\frac{1}{\mathrm{Vol}(\mathcal{T}_\varepsilon)} \int_{\mathcal{T}_\varepsilon} |u|^2 \, dx = o(\varepsilon) \qquad \text{as} \quad \varepsilon \to 0+. \tag{7}$$

PROBLEMS

1. Suppose that $\bar{\Omega} \hookrightarrow \mathbb{R}^d$ is a smooth compact submanifold with boundary and $u \in C^\infty(\mathbb{R}^d)$. Let $v \equiv u\chi_\Omega \in L^2(\Omega)$. Prove that $v \in H^1(\mathbb{R}^d)$ if and only if $u|_{\partial\Omega} = 0$. In this case, show that $v \in \mathring{H}^1(\Omega)$.
   DISCUSSION. This is an H^1 version of Lemma 5.3.2.

2. Show that the $o(\varepsilon)$ in Corollary 7 cannot be strengthened to $O(\varepsilon^\alpha)$ for any $\alpha > 1$.
   Hint. Consider $\Omega =]0, 1[\subset \mathbb{R}$ and $u = x^{1/2}(\ln x)^p$ near $x = 0$ with p suitably chosen.

3. Let t, σ be the coordinates on a tubular neighborhood of $\partial\Omega$ as in the proof of Theorem 3. Prove that there is an $\varepsilon > 0$ such that for $u \in H^1(\mathbb{R}^d)$ the map $t \mapsto (u \circ \Psi)(t, \cdot)$ is continuous on $[-\varepsilon, \varepsilon]$ with values in $L^2(\partial\Omega)$.
   DISCUSSION. This result shows that the traces of u on the hypersurfaces "parallel" to $\partial\Omega$ depend continuously on the distance to $\partial\Omega$.

§5.6. The Fredholm Alternative

In §5.3 and §5.4 we proved unique solvability of the Dirichlet problem for coercive divergence form operators, $\sum \partial_i a_{ij}(x)\partial_j + a(x)$. In this section we show how this can be used to study operators which are not in divergence form, for example, $\Delta + \partial/\partial x_1$, and operators which are not coercive, for example, $\Delta + \lambda$ for λ large positive. The final results assert that the null spaces are finite dimensional and the ranges are equal to the annihilators of the null spaces of the transposed operators.

Suppose that Ω is a bounded open set and consider a second-order operator

$$L(x, \partial) = \sum a_{ij}(x)\partial_i\partial_j + \sum a_j(x)\partial_j + a(x), \tag{1}$$

$$a_{ij}, a_j, a \text{ belong to } C^\infty(\Omega: \mathbb{R}) \text{ and } a_{ij} = a_{ji}, \tag{2}$$

$$(\exists \mu > 0), (\forall x \in \Omega, \xi \in \mathbb{R}^d), \qquad \sum a_{ij}(x)\xi_i\xi_j \geq \mu|\xi|^2, \tag{3}$$

$$(\forall \alpha \in \mathbb{N}^d), \qquad \partial^\alpha(a_{ij}, a_j, a) \in L^\infty(\Omega). \tag{4}$$

The *transposed operator* is

$$L^t(x, \partial)v \equiv \sum \partial_i\partial_j(a_{ij}(x)v) + \sum -\partial_j(a_j(x)v) + a(x)v. \tag{5}$$

Note that to form L^t the coefficients must be differentiable. L is of divergence form precisely when $L = L^t$, that is, when L is *symmetric*. For all $\varphi, \psi \in C_0^\infty(\Omega)$ we have Green's identity

$$\int_\Omega \varphi L\psi - \psi L^t\varphi \, dx = 0.$$

Both L and L^t are continuous maps of $\mathring{H}^1(\Omega)$ to $H^{-1}(\Omega)$ and Green's identity extends by continuity to all $\varphi, \psi \in \mathring{H}^1(\Omega)$ if one interprets the integral of $L\psi$ times φ as the value of $L\psi \in \mathring{H}^1(\Omega)'$ at $\varphi \in \mathring{H}^1$ and similarly for the second term. Thus

$$\langle L\psi, \varphi \rangle = \langle L^t\varphi, \psi \rangle \qquad \text{for all} \quad \varphi, \psi \in \mathring{H}^1(\Omega). \tag{6}$$

In particular, if φ belongs to the kernel of L^t, then φ is annihilated by every element of the range of L. The goal of this section is to show that this is an exact description, that is, the range of L is equal to the set of elements of $H^{-1}(\Omega)$ which annihilate the kernel of L^t. At the same time, we show that kernel(L) and kernel(L^t) are finite dimensional with the same dimension. In particular, L is surjective if and only if it is injective. The Elliptic Regularity Theorems of §5.9 show that the kernels belong to $C^\infty(\Omega)$ and that when Ω is regular the kernels lie in $C^\infty(\bar{\Omega})$.

EXAMPLES. 1. $\Omega =]0, \pi[\subset \mathbb{R}$ and $L = d^2/dx^2 - \lambda$. Here $L = L^t$ and the kernel is empty except when $\lambda = -n^2$, $n = 1, 2, \ldots$. In that case the kernel is spanned by $\sin(nx)$. The operator is an isomorphism of $\mathring{H}^1$ to H^{-1} except for those values of λ. The operator satisfies the coerciveness inequality (5.4.5) if and only if $\lambda > -1$, so one has an isomorphism in many noncoercive cases. When $\lambda = -n^2$, the range of L is the set of elements f of $H^{-1}(\Omega)$ such that $\langle f, \sin(nx) \rangle = 0$. In this case solutions are nonunique, if u is a solution then so is $u + c(\sin(nx))$ for any $c \in \mathbb{C}$. The eigenvectors $\sin(nx)$ for the eigenvalues $\lambda = -n^2$ form the basis for Fourier sine series. Note that the eigenfunctions are determined by the eigenvalue equation $d^2u/\partial x^2 = \lambda u$ and the boundary condition $u|_{\partial\Omega} = 0$. Analogous eigenfunction expansions for general symmetric L are discussed in §5.7.

2. $L = \Delta + \sum a_j(x)\partial_j$. In §5.11 we use the maximum principle to show that the kernel of L is empty. This implies that the kernel of L^t is empty and we find that L is an isomorphism for any real valued a_j.

One of the advantages of considering inhomogeneous equations with homogeneous boundary conditions is that the Fredholm alternative has a simple description in this context.

The proof of the basic result is not hard. The idea is to consider L as a perturbation of the symmetric operator

$$L_\sigma \equiv -1 + \sum \partial_i a_{ij}(x)\partial_j.$$

Then L_σ is equal to its own transpose and is coercive. Thus L_σ is an isomorphism of $\mathring{H}^1(\Omega)$ to $H^{-1}(\Omega)$. In addition

$$L = L_\sigma + Q(x, \partial),$$

where Q is a differential operator of order 1 with smooth coefficients. Since Q is lower order, it is in a sense small compared to L_σ.

To solve the Dirichlet problem

$$Lu = f \in H^{-1}(\Omega) \qquad \text{with} \quad u \in \mathring{H}^1(\Omega), \tag{7}$$

we take as a first approximation the solution with L replaced by L_σ. That is, write $u = v + w$ where $w \in \mathring{H}^1(\Omega)$ and $L_\sigma w = f$. The equation for $v \in \mathring{H}^1(\Omega)$ is then $(L_\sigma + Q)(v + w) = f$ which simplifies to $(L_\sigma + Q)v = -Qw$. Applying $(L_\sigma)^{-1}$ shows that this equation in $H^{-1}(\Omega)$ is equivalent to an equation in $\mathring{H}^1(\Omega)$

$$(I + (L_\sigma)^{-1}Q)v = -(L_\sigma)^{-1}Q(L_\sigma)^{-1}f. \tag{8}$$

To summarize, $u \in \mathring{H}^1(\Omega)$ solves (2) if and only if $u - (L_\sigma)^{-1}f \equiv v \in \mathring{H}^1(\Omega)$ solves (8).

Turning to equation (8) note that Proposition 5.4.5 shows that Q is a continuous map of $L^2(\Omega)$ to $H^{-1}(\Omega)$, so $(L_\sigma)^{-1}Q$ is a bounded linear map of $L^2(\Omega)$ to $\mathring{H}^1(\Omega)$. If $v \in L^2(\Omega)$ solves (8), then $v = -(L_\sigma)^{-1}Q(v + w)$ belongs to $\mathring{H}^1(\Omega)$. Thus it suffices to study solutions v to (8) in the space $L^2(\Omega)$.

The key observation is that $(L_\sigma)^{-1}Q$ is a smoothing operator. It maps L^2 to $\mathring{H}^1$ so it gains one derivative. This implies that it is a compact operator on $L^2(\Omega)$ thanks to the following criterion of Rellich. Recall that a linear map from one Banach space to another is called compact if the image of any bounded set is precompact.

Theorem 1 (Rellich Compactness Theorem). *If $s > t$, $c \in \mathbb{R}$, and K is a compact subset of $\mathbb{R}^d$, then*

$$\{u \in H^t(\mathbb{R}^d) : \|u\|_{H^s(\Omega)} \le c \text{ and } \operatorname{supp}(u) \subset K\} \tag{9}$$

is precompact in $H^t(\mathbb{R}^d)$.

Example. The case $s = 1$, $t = 0$, and $K = \bar{\Omega}$ shows that bounded sets in $\mathring{H}^1(\Omega)$ are precompact in $L^2(\Omega)$. This shows that bounded linear maps (e.g. $(L_\sigma)^{-1}Q$) from $L^2(\Omega)$ to $\mathring{H}^1(\Omega)$ are compact operators from $L^2(\Omega)$ to itself.

Proof of Theorem 1. Suppose that u_n is a sequence in the set (9). We must show that u_n has a subsequence which converges in $H^t(\mathbb{R}^d)$. Since the u_n are bounded in $H^s(\mathbb{R}^d)$, we may pass to a subsequence $v_n = u_{k(n)}$ such that v_n converges weakly in $H^s(\mathbb{R}^d)$ to a limit v.

Choose a $\varphi \in C_0^\infty(\mathbb{R}^d)$ with φ identically equal to 1 on a neighborhood of K. Then $\varphi v_n = v_n$ and compute

$$\mathscr{F} v_n(\xi) = \langle \varphi v_n, (2\pi)^{-d/2} e^{-ix\xi} \rangle = (2\pi)^{-d/2} \langle v_n, \varphi e^{-ix\xi} \rangle.$$

Now $\varphi e^{-ix\xi}$ belongs to $H^{-s}(\mathbb{R}^d)$ with norm bounded by a multiple of $\langle \xi \rangle^{|s|}$ (exercise). Thus $|\hat{v}_n(\xi)| \le c \langle \xi \rangle^{|s|}$ and the weak convergence of the v_n implies that $\hat{v}_n(\xi) \to \hat{v}(\xi)$ for all $\xi \in \mathbb{R}^d$. Lebesgue's Dominated Convergence Theorem implies that for any $R > 0$

$$\int_{|\xi| \le R} |\hat{v}_n(\xi) - \hat{v}(\xi)|^2 \langle \xi \rangle^{2t} \, d\xi \to 0.$$

On the other hand,

$$\int_{|\xi| > R} |\hat{v}_n(\xi) - \hat{v}(\xi)|^2 \langle \xi \rangle^{2t} \, d\xi \le \int_{|\xi| > R} |\hat{v}_n(\xi) - \hat{v}(\xi)|^2 \langle \xi \rangle^{2s} \langle \xi \rangle^{2(t-s)} \, d\xi$$
$$\le (1 + R^2)^{(t-s)/2} \sup \|v_n - v\|^2_{H^s(\mathbb{R}^d)},$$

since $\langle \xi \rangle^{2(t-s)} \le (1 + R^2)^{(t-s)/2}$ when $|\xi| \ge R$. The sup is bounded thanks to the weak convergence $v_n \rightharpoonup v$.

Given $\varepsilon > 0$ we may choose $R > 0$ so that the integral over $|\xi| > R$ is smaller than $\varepsilon/2$ for all n. Then choose N so that for $n > N$ the integral over $|\xi| \le R$ is less than $\varepsilon/2$. Then for $n > N$ the $H^t(\mathbb{R}^d)$ norm of $v_n - v$ is less than or equal to ε. This proves that v_n converges to v in $H^t(\mathbb{R}^d)$. □

Theorem 2 (Fredholm Alternative for the Dirichlet Problem). *The mapping* $L: \mathring{H}^1(\Omega) \to H^{-1}(\Omega)$ *has finite-dimensional kernel and closed range equal to the annihilator of kernel*(L^t). *Moreover, kernel*(L) *and kernel*(L^t) *have the same dimension.*

It follows that the index of L, which is equal to the difference of the codimension of the range of L and the dimension of kernel of L, is equal to zero.

Proof of Theorem 2. Let $K \equiv (L_\sigma)^{-1} Q$ so that K is a compact operator on $L^2(\Omega)$. The strategy is simply to study the equation $(I + K)v = g$ and use the fact (see (8)) that $Lu = f$ is equivalent to $(I + K)v = g$ with $g \equiv -K(L_\sigma)^{-1} f$ and $u \equiv v + (L_\sigma)^{-1} f$.

For $I + K$ we use the Fredholm Alternative for Compact Operators which was invented by Fredholm to solve the Dirichlet problem by the method of integral equations. Thus there is justice in using the result to solve the Dirichlet problem, albeit by a different method.

Theorem 3 (Fredholm/Riesz Theory of Compact Operators). *If B is a Banach space and $K: B \to B$ is a compact linear transformation then:*

(i) *the kernel of $I + K$ is finite dimensional;*
(ii) *the range of $I + K$ is equal to the annihilator of kernel$(I + K^t)$;*
(iii) *$I + K$ and $I + K^t$ have kernels of equal dimension.*

Note that K^t is a map of B' to itself. This result is proved in most functional analysis texts, for example, Reed and Simon [RS] or Riesz and Sz-Nagy [RSzN].

To apply this result we must identify the transpose of $K \equiv L_\sigma^{-1} Q$. Formally, compute

$$K^t = (L_\sigma^{-1} Q)^t = Q^t (L_\sigma^{-1})^t = Q^t (L_\sigma^t)^{-1} = Q^t L_\sigma^{-1}, \tag{10}$$

the last since L_σ is equal to its transpose. As L_σ and Q are unbounded operators and there are at least two dualities involved in this computation (that between $\mathring{H}^1$ and H^{-1} and between $L^2(\Omega)$ and itself) we present a careful derivation.

Lemma 4. *Identify the dual of $L^2(\Omega)$ with $L^2(\Omega)$ by the mapping $L^2(\Omega) \ni \varphi \mapsto M(\varphi) \in L^2(\Omega)'$ by $M(\varphi)(\psi) \equiv \int_\Omega \psi\varphi \, dx$ Then $K^t = MQ^t(L_\sigma)^{-1}M^{-1}$.*

Note that the right-hand side is a continuous map of $L^2(\Omega)'$ to itself, as it should be. Note also that the identification of L^2 with its dual is linear rather than antilinear as is the case with the standard Riesz Representation Theorem. The difference is a complex conjugate in the integral. The formula for K^t is identical with that derived in (10) except that the identification M is made explicit.

Proof of Lemma 4. The goal is to show that for all $f \in L^2(\Omega)$ and $l \in L^2(\Omega)'$

$$l(Kf) = (MQ^t(L_\sigma)^{-1}M^{-1}l)(f). \tag{11}$$

It suffices to prove (11) for f, l chosen from dense subsets of $L^2(\Omega)$ and its dual. Suppose that $f \in C_0^\infty(\Omega)$ and $l = Mg$ with $g \in C_0^\infty(\Omega)$. Then the left-hand side is equal to $\int_\Omega gQ^t(L_\sigma)^{-1}f \, dx$.

Since both g and $(L_\sigma)^{-1}f$ belong to $\mathring{H}^1(\Omega)$ an integration by parts shows that the left-hand side is equal to $\int_\Omega (Qg)(L_\sigma)^{-1}f \, dx$.

Let $w \equiv (L_\sigma)^{-1}f$ and $v \equiv (L_\sigma)^{-1}Qg$, both in $\mathring{H}^1(\Omega)$. The integral is then equal to $\int_\Omega (L_\sigma v)((L_\sigma)^{-1}f) \, dx = \langle L_\sigma v, (L_\sigma)^{-1}f\rangle$. Green's identity (6) for $L_\sigma = L_\sigma^t$ shows that this is equal to

$$\langle v, L_\sigma(L_\sigma)^{-1}f\rangle = \langle v, f\rangle = \langle (L_\sigma)^{-1}Qg, f\rangle = \int f(L_\sigma)^{-1} \, Qg \, dx. \tag{12}$$

The last expression in (12) is equal to the right-hand side of (11). □

Returning to the proof of Theorem 2 we identify the kernels of $I + K$ and $I + K^t$. The equivalence of $Lu = f$ and $(I + K)v = g$ from the first paragraph of the proof of Theorem 2 shows that $\ker(I + K) = \ker(L)$.

Next note that $l \in \ker(I + K^t)$ if and only if $l = Mh$ with $(I + Q^t L_\sigma^{-1})h = 0$. Let $v \equiv L_\sigma^{-1} h$, to see that this holds if and only if $(L_\sigma + Q^t)v = 0$. The operator on the left is equal to L^t so we have shown that the mapping $L_\sigma^{-1}M^{-1}$ is an isomorphism from $\ker(I + K^t)$ to $\ker(L^t)$.

Since $\dim \ker(I + K^t) = \dim \ker(I + K) < \infty$ the above identifications show that $\ker(L)$ and $\ker(L^t)$ have equal finite dimensions.

Finally, note that $f \in Rg(L)$ if and only if $-KL_\sigma^{-1} f$ is in the range of $I + K$ hence if and only if $KL_\sigma^{-1} f$ is annihilated by $\ker(I + K^t)$. By the above calculation $\ker(I + K^t) = ML_\sigma(\ker L^t)$. Thus f is in the range of L if and only if

$$\int_\Omega (KL_\sigma^{-1} f)(L_\sigma v)\, dx = 0 \qquad \text{for all} \quad v \in \ker(L^t). \tag{13}$$

Note that $L_\sigma v = -Q^t v$ since $v \in \ker(L^t)$ and therefore that $L_\sigma v \in L^2$. Since $K = L_\sigma^{-1} Q$, Green's identity (6) for L_σ shows that (13) holds if and only if

$$\int_\Omega (QL_\sigma^{-1} f) v\, dx = 0 \qquad \text{for all} \quad v \in \ker(L^t).$$

Since $v \in \mathring{H}^1(\Omega)$, an integration by parts yields

$$\int_\Omega (QL_\sigma^{-1} f) v\, dx = \int_\Omega L_\sigma^{-1} f Q^t v\, dx = -\int_\Omega L_\sigma^{-1} f L_\sigma v\, dx$$

the last equality since $v \in \ker(L^t)$. Relation (6) shows that the right-hand side is equal to $-\int_\Omega f v\, dx$. Thus f is in the range of L if and only if this integral vanishes for all $v \in \ker(L^t)$ which is the desired result. □

EXAMPLE. For ρ sufficiently large and positive, the operators $L - \rho$ and $L^t - \rho$ are isomorphisms of $\mathring{H}^1(\Omega)$ to $H^{-1}(\Omega)$.

Thanks to the Fredholm Alternative, it suffices to prove that $\ker(L - \rho) = \{0\}$. Since L is real, it suffices to show that if u is a real-valued element in $\ker(L - \rho)$ then $u = 0$.

For such a u one has $\langle u, (\rho - L)u \rangle = 0$. Write $L = L_\sigma + Q$ as above. For the L_σ contribution estimate

$$\langle u, (\rho - L_\sigma)u \rangle \geq \mu \|\nabla u\|_{L^2(\Omega)}^2 + \rho \|u\|_{L^2(\Omega)}^2.$$

For the Q contribution, the Schwartz inequality shows that

$$|\langle Qu, u \rangle| \leq c \|u\|_{L^2(\Omega)} (\|u\|_{L^2(\Omega)} + \|\nabla u\|_{L^2(\Omega)}).$$

The Peter–Paul inequality yields

$$|\langle Qu, u \rangle| \leq (\mu/2) \|\nabla u\|_{L^2(\Omega)}^2 + c' \|u\|_{L^2(\Omega)}^2.$$

Combining the two estimates yields

$$\langle u, (\rho - L)u \rangle \geq (\mu/2) \|\nabla u\|_{L^2(\Omega)}^2 + (\rho - c') \|u\|_{L^2(\Omega)}^2.$$

If $\rho \geq c'$ it follows that $(\rho - L)u = 0$ implies $u = 0$. □

The method of this section is to view L as a perturbation of L_σ which is invertible. A similar argument is a standard tool in considering nonlinear perturbations of L. As an example suppose that Ω is open and that L: Re $\mathring{H}^1(\Omega) \to$ Re $H^{-1}(\Omega)$ is an isomorphism. Suppose that $f \in C(\mathbb{R})$ is uniformly Lipshitzian so that

$$\Lambda \equiv \sup\left\{\left|\frac{f(a) - f(b)}{a - b}\right| : a \neq b\right\} < \infty.$$

Then $u \mapsto f(u)$ is a uniformly Lipshitz continuous map of Re $L^2(\Omega)$ to itself with Lipshitz constant Λ. The goal is to solve the *semilinear* ($\equiv$ linear in its highest-order terms) Dirichlet problem

$$Lu + f(u) = g \in \text{Re } H^{-1}(\Omega) \qquad \text{with} \quad u \in \text{Re } \mathring{H}^1(\Omega). \tag{14}$$

Apply L^{-1} to show that this is equivalent to the fixed point equation

$$\Gamma(u) = u, \qquad \Gamma(u) \equiv L^{-1}g - L^{-1}(f(u)). \tag{15}$$

Theorem 5. *The nonlinear Dirichlet problem* (14) *has exactly one solution provided that*

$$\Lambda \|L^{-1}\|_{\text{Re}\,L^2(\Omega) \to \text{Re}\,L^2(\Omega)} < 1. \tag{16}$$

The crucial hypothesis (16) says that f is sufficiently small. In particular, it is satisfied for $|\varepsilon| < \varepsilon_0$ if $f = \varepsilon F$ with F Lipshitzian.

PROOF. First note that Γ maps Re $L^2(\Omega)$ to Re $\mathring{H}^1(\Omega)$, so solving (14) is equivalent to finding a $u \in$ Re $L^2(\Omega)$ solving (15).

The map Γ from Re $L^2(\Omega)$ to istelf is Lipshitzian with Lipshitz constant dominated by the left-hand side of (16). By hypothesis this is strictly less than 1 so Banach's Contraction Mapping Theorem implies that Γ has a unique fixed point $u \in$ Re $L^2(\Omega)$. □

PROBLEMS

For the variational approach of §5.3 and §5.4, Ω could be bounded or unbounded. The key was coercivity. The compactness arguments of this section require Ω to be bounded. The first example provides counterexamples for unbounded Ω.

1. Suppose that $\Omega = \mathbb{R}^d$.
   (i) For $L = \Delta$ show that L: $\mathring{H}^1(\mathbb{R}^d) \to H^{-1}(\mathbb{R}^d)$ has kernel equal to $\{0\}$ and dense range which is not closed. In particular, the Fredholm alternative is violated.
   (ii) Show that $(\Delta - 1)^{-1}$ is not a compact operator on $L^2(\Omega)$ by exhibiting a bounded sequence $u_n \in L^2(\mathbb{R}^d)$ such that $(1 - \Delta)^{-1}u_n$ has no $L^2(\mathbb{R}^d)$-convergent subsequences.

   DISCUSSION. An argument like that in (ii) shows that the embedding $\mathring{H}^1(\mathbb{R}^d) \hookrightarrow L^2(\mathbb{R}^d)$ is not compact.

With two alterations, the methods of this section extend to the Neumann problem, and other boundary conditions. For the Dirichlet problem, we were able to integrate by parts with vanishing boundary contribution because the functions involved belong to $\mathring{H}^1(\Omega)$. For the Neumann problem, that is not the case. The replacement for integration by parts in $\mathring{H}^1(\Omega)$ is the variational form of the boundary value problem, for example, equation (5.4.17). The second change is the compactness criterion. For the Dirichlet problem, $(L_\sigma)^{-1}$ maps $H^{-1}(\Omega)$ into $\mathring{H}^1(\Omega) \hookrightarrow H^1(\mathbb{R}^d)$ so Theorem 1 applies directly. For the Neumann problem the range of $(L_\sigma)^{-1}$ is in $H^1(\Omega)$. It is not true that the inclusion map $H^1(\Omega) \hookrightarrow L^2(\Omega)$ is compact for arbitrary bounded open sets Ω. Fortunately, it is compact for moderately regular sets, for example, sets with Lipshitz boundaries.

Example. Let I_n be the open interval of diameter 20^{-n} and center at the point 2^{-n}, $n \geq 1$. Then the I_n are disjoint subsets of $]0, 1[$. Let $\Omega \equiv \bigcup I_n$. Then the inclusion $H^1(\Omega) \hookrightarrow L^2(\Omega)$ is not compact since if u_n is equal to $20^{n/2}$ times the characteristic function of I_n, then the u_n have $H^1(\Omega)$ and $L^2(\Omega)$ norms equal to 1 and they have no L^2-convergent subsequence.

A skeptical reader should think that this is because the set Ω is infinitely connected. A finitely connected bounded open set in $\mathbb{R}$ is a finite union of intervals and it is not hard to show that the inclusion is compact in that case. The next problem dashes the naive hope that this example inspires.

2. Find a bounded open connected subset $\Omega \subset \mathbb{R}^2$ such that the inclusion $H^1(\Omega) \hookrightarrow L^2(\Omega)$ is not compact. *Hint*. Think first of a disjoint union of open squares. Then connect them. The crux is to give a proof that the inclusion is not compact for the resulting set.

There is a good strategy for proving that $H^1(\Omega)$ is compactly included in $L^2(\Omega)$. One constructs a possibly nonlinear extension operator $E: H^1(\Omega) \to H^1(\mathbb{R}^d)$ with two properties

$$(\exists c), (\forall u \in H^1(\Omega)), \qquad \|Eu\|_{H^1(\mathbb{R}^d)} \leq c\|u\|_{H^1(\Omega)} \tag{17}$$

and

$$Eu = u \quad \text{on } \Omega. \tag{18}$$

If an extension operator exists it follows that $H^1(\Omega) \hookrightarrow L^2(\Omega)$ is compact. To prove this, choose $\psi \in C_0^\infty(\mathbb{R}^d)$ with ψ equal to 1 on a neighborhood of $\bar{\Omega}$. Then if $\{u_n\}$ is bounded in $H^1(\Omega)$, then ψ_{u_n} has an $L^2(\mathbb{R}^d)$-convergent subsequence by Theorem 1. The same subsequence is $L^2(\Omega)$ convergent which completes the proof.

If $\bar{\Omega} \hookrightarrow \mathbb{R}^d$ is a smooth submanifold with boundary, such extension operators are constructed in Theorem 5.9.6. A celebrated theorem of Calderon shows that extension operators exist for Lipshitz domains (see Agmon [A]).

§5.7. Eigenfunctions and the Method of Separation of Variables

In this section we will show that the natural unbounded operator L on $L^2(\Omega)$, defined by a real elliptic operator with Dirichlet boundary conditions, has adjoint equal to the unbounded operator defined by L^t with the same boundary condition. In particular, if $L = L^t$, the operator is self-adjoint. When Ω is bounded the spectrum is discrete and converges to $-\infty$. The resulting eigenfunction expansions generalize Fourier series and the eigenfunction expansions associated with regular Sturm–Liouville problems. They can be used to solve a variety of boundary value problems, justifying the *method of separation of variables.* In particular, we justify many of the heuristic ideas about heat flow and damped wave motion which motivated the variational approach to the Dirichlet problem. This section supposes some familiarity with elementary spectral theory. Good references are the first volume of Reed and Simon [RS] and the classic text of Riesz and Sz-Nagy [RSzN].

Suppose that Ω is a possibly unbounded open subset of $\mathbb{R}^d$ and that L and L^t are as in (5.6.1)–(5.6.5). For the sake of simplicity restrict attention to operators with real coefficients. The changes for complex coefficients are minimal. The ellipticity condition is (5.4.22) in the complex case.

Example. $L = \Delta - V(x)$ on $\Omega = \mathbb{R}^d$ is particularly important in quantum mechanics. For example, periodic V model crystalline structures while V which tend to zero at infinity lead to scattering theory.

Define an unbounded operator on $L^2(\Omega)$ to be the restriction of L, defined in the sense of distributions, to the domain

$$\mathscr{D}(L) \equiv \{u \in \mathring{H}^1(\Omega): Lu \in L^2(\Omega)\}. \tag{1}$$

In the same way, an operator L^t is the restriction of the differential operator L^t to the domain

$$\mathscr{D}(L^t) \equiv \{u \in \mathring{H}^1(\Omega): L^t u \in L^2(\Omega)\}. \tag{2}$$

Note that Dirichlet boundary conditions are hidden in both domains.

Theorem 1. *L and L^t are densely defined closed operators on $L^2(\Omega)$ and each is the adjoint of the other.*

Note that for elliptic operators with complex coefficients the adjoint would be the restriction of the operator

$$v \mapsto \sum \partial_i \partial_j (\bar{a}_{ij} v) - \sum \partial_j (\bar{a}_j v) + \bar{a} v.$$

Proof. The domains contain $C_0^\infty(\Omega)$ so are dense.

Since the adjoint of any densely defined operator is closed it suffices to show that the operators are adjoints of each other.

Since $(L^t)^t = L$, it is sufficient to show that the adjoint of L, denoted L^*, is equal to the operator L^t.

Since $(L - \lambda)^* = L^* - \lambda$ for any $\lambda \in \mathbb{R}$ it is sufficient to show that $(L - \lambda)^* = L^t - \lambda = (L - \lambda)^t$. Thus without loss of generality L may be replaced by $L - \lambda$.

Lemma 2. *If λ is sufficiently large and positive, then $L - \lambda$ is an isomorphism of $\mathring{H}^1(\Omega)$ to $H^{-1}(\Omega)$.*

This assertion is proved in the example following the proof of Theorem 5.6.2.

Thus, without loss of generality, we may suppose that L and L^t are isomorphisms from $\mathring{H}^1(\Omega)$ to $H^{-1}(\Omega)$. Then L is a bijection from $\mathscr{D}(L)$ to $L^2(\Omega)$.

A function v satisfies $L^*v = g$ if and only if for all $u \in \mathscr{D}(L)$, $(Lu, v) = (u, g)$ where $(\cdot, \cdot)$ denotes the scalar product in $L^2(\Omega)$. Taking $u \in C_0^\infty(\Omega)$ yields $L^t\bar{v} = \bar{g}$ in the sense of distributions. Since L^t is real we have $L^t v = g$ in the sense of distributions.

Choose $w \in \mathring{H}^1(\Omega)$ solving $L^t w = g$. Then a complex version of (5.5.6) shows that $(Lu, w) = (u, g)$. Thus $(Lu, w - v) = 0$ for all $u \in \mathscr{D}(L)$. Since L maps $\mathscr{D}(L)$ onto $L^2(\Omega)$, we conclude that $v = w$ so $v \in \mathscr{D}(L^t)$ and $L^t v = g$. The proof is complete. □

Corollary 3. *If $L = L^t$, then L with domain defined by Dirichlet boundary conditions as in* (1) *defines a self-adjoint operator on $L^2(\Omega)$. The spectrum of L is a subset of* $]-\infty, \sup a]$.

Proof. For $u \in \mathscr{D}(L)$, the identity (5.6.6) reads

$$(u, Lu) = \int_\Omega -a_{ij}\partial_i u \partial_j \bar{u} + a|u|^2 \, dx. \tag{3}$$

The terms in ∇u are nonpositive so the form (3) is bounded above by $\sup(a)$ times the $L^2(\Omega)$ norm of u. □

Theorem 4. *If Ω is bounded, then the spectrum of L is discrete in $\mathbb{R}$. In particular, in $L^2(\Omega)$ there is a complete orthonormal set of eigenfunctions.*

Proof. If $\rho > \sup a$, then $\ker(L - \rho) = \{0\}$ and the Fredholm alternative implies that $L - \rho$ is an isomorphism of $\mathring{H}^1(\Omega)$ to $H^{-1}(\Omega)$. The restriction of $(L - \rho)^{-1}$ to $L^2(\Omega)$ is the inverse of the operator $L - \rho$ with domain equal to $\mathscr{D}(L)$. This inverse maps bounded sets in $L^2(\Omega)$ to bounded sets in $\mathring{H}^1(\Omega)$, in particular, the inverse is a compact operator on $L^2(\Omega)$.

Thus $(L - \rho)^{-1}$ is a negative compact self-adjoint operator on $L^2(\Omega)$. Thus the spectrum of $(L - \rho)^{-1}$ is discrete except for possible accumulation at $\{0\}$.

Since the range of $(L - \rho)^{-1}$ is $\mathscr{D}(L)$, which is dense in $L^2(\Omega)$, the point 0 is not an eigenvalue. Thus 0 must be an accumulation point of eigenvalues, since L^2 is not finite dimensional. Label the eigenvalues as $\nu_1 \le \nu_2 \le \cdots \to 0$ with each eigenvalue repeated according to its multiplicity which must be finite.

Choose an orthonormal basis of eigenfunctions φ_j in $L^2(\Omega)$, $(L - \rho)^{-1}\varphi_j = \nu_j\varphi_j$. Since $\nu_j \ne 0$, the eigenvalue equation implies that φ_j belongs to Range $(L - \rho)^{-1} = \mathscr{D}(L) \subset \mathring{H}^1(\Omega)$. In addition

$$L\varphi_j = (L - \rho)\varphi_j + \rho\varphi_j = (L - \rho)((L - \rho)^{-1}\nu_j^{-1}\varphi_j) + \rho\varphi_j = (\rho + 1/\nu_j)\varphi_j.$$

Thus the eigenfunctions are also eigenfunctions of L with eigenvalues $\rho + 1/\nu_j \equiv \lambda_j$ converging to $-\infty$. Each eigenvalue has finite multiplicity. □

EXAMPLES. 1. If $\Omega =]0, \pi[$ and $L = d^2/dx^2$, then the normalized eigenfunctions are $(2/\pi)\sin(nx)$ for $n = 1, 2, \ldots$ with eigenvalues $\lambda_n = -n^2$. The associated eigenfunction expansions are called Fourier sine series.

2. Let $L = \Delta$ and the minimum width $\delta(\Omega) < \infty$. Then Theorem 5.3.1 shows that $\delta^2(L\varphi, \varphi) \le -(\varphi, \varphi)$ for all $\varphi \in \mathring{H}^1$ so $\lambda_j \le -\delta^{-2} < 0$ for all j.

For $u \in L^2(\Omega)$ the Fourier series $\sum \alpha_j \varphi_j$ converges to u in the L^2 norm and the $L^2(\Omega)$ norm of u is equal to $\sum |\alpha_j|^2$. Note that the Fourier coefficients $\alpha_j \equiv (u, \varphi_j) = \langle u, \bar{\varphi}_j \rangle$ are meaningful for any $u \in H^{-1}(\Omega)$ since $\bar{\varphi}_j$ belong to $\mathring{H}^1(\Omega)$. The next result complements the information from Bessel's identity.

Theorem 5.

(i) *If $u \in \mathring{H}^1(\Omega)$, then the eigenfunction expansion of u converges in $\mathring{H}^1(\Omega)$ and $(\sum (|\lambda_j| + 1)|\alpha_j|^2)^{1/2}$ is equivalent to the $\mathring{H}^1(\Omega)$ norm of u.*
(ii) *If $u \in H^{-1}(\Omega)$, then the eigenfunction expansion of u converges in $H^{-1}(\Omega)$ and $(\sum (|\lambda_j| + 1)^{-1}|\alpha_j|^2)^{1/2}$ is equivalent to the $H^{-1}(\Omega)$ norm of u.*
(iii) *If $k \in \mathbb{N}$ and $u \in \mathscr{D}(L^k)$, then the eigenfunction expansion of u converges in $\mathscr{D}(L^k)$ and $(\sum (|\lambda_j| + 1)^{2k}|\alpha_j|^2)^{1/2}$ is equivalent to the graph norm in $\mathscr{D}(L^k)$ of u.*

The third assertion is valid for any self-adjoint operator with discrete spectrum. It suggests that (i) and (ii) are the cases $k = \frac{1}{2}$ and $k = -\frac{1}{2}$, respectively. In fact, (i) proves that $\mathscr{D}(|L|^{1/2}) = \mathring{H}^1(\Omega)$.

It is interesting to note that the precise description of $\mathscr{D}(L)$ for regular Ω given in equation (5.9.15) was not known until about 1950. The characterization of $\mathscr{D}(L^k)$ in (5.9.15), together with the Sobolev Embedding Theorem, allow one to translate the convergence results above to uniform convergence. For example, one has $C^j(\Omega)$ convergence provided $u \in \mathscr{D}(L^k)$ with $j < 2k - d/2$. If Ω is regular this extends to $C^j(\bar{\Omega})$ convergence.

PROOF OF (i). Choose $\rho > \sup(a)$. For $u \in \mathscr{D}(L)$ the Spectral Theorem implies

$$((\rho - L)u, u) = \sum (\rho - \lambda_j)|\alpha_j|^2.$$

The right-hand side defines a norm equivalent to $(\sum (|\lambda_j| + 1)|\alpha_j|^2)^{1/2}$. On the other hand, (3) shows that the left-hand side defines a norm equivalent to the $\mathring{H}^1(\Omega)$ norm.

For $u \in \mathring{H}^1(\Omega)$ choose $u_n \in C_0^\infty(\Omega) \subset \mathscr{D}(L)$ converging to u in $\mathring{H}^1(\Omega)$. Then the Fourier coefficients $\alpha_j(u_n)$ of u_n converge to the corresponding coefficients of u. Fatou's Lemma implies that

$$\sum (\rho - \lambda_j)|\alpha_j(u)|^2 \le \liminf \sum (\rho - \lambda_j)|\alpha_j(u_n)|^2.$$

Since the sum on the right is equivalent to the square of the $\mathring{H}^1(\Omega)$ norm of u_n, the lim inf is dominated by a multiple of the $\mathring{H}^1$ norm of u, hence is finite.

Let $s_n \equiv \sum_{j \le n} \alpha_j(u)\varphi_j$ be the partial sums of the Fourier expansion of u. Then s_n converges to u in $L^2(\Omega)$. To show that s_n is a Cauchy sequence in $\mathring{H}^1(\Omega)$

notice that the s_n belong to $\mathscr{D}(L)$, so for $n \le m$

$$\|s_n - s_m\|^2_{\mathring{H}^1(\Omega)} = \sum_n^m (\rho - \lambda_j)|\alpha_j(u)|^2 \le c \sum_n^\infty (\rho - \lambda_j)|\alpha_j(u)|^2 = o(1)$$

as $n \to \infty$. It follows that the s_n converge to u in $\mathring{H}^1(\Omega)$ and

$$\|u\|^2_{\mathring{H}^1(\Omega)} = \lim \|s_n\|^2_{\mathring{H}^1(\Omega)} \approx \sum (\rho - \lambda_j)|\alpha_j(u)|^2. \qquad \square$$

PROOF OF (ii). For $u \in H^{-1}(\Omega)$ let v be the solution of $(\rho - L)v = u$. Then $\alpha_j(v) = (\rho - \lambda_j)^{-1}\alpha_j(u)$. Let s_n be the partial sum of the Fourier series of u and let v_n be the partial sum of the Fourier series of v. The result follows since v_n converges to v in $\mathring{H}^1$, so $s_n = (\rho - L)v_n$ converges to $u = (\rho - L)v$ in H^{-1} since $\rho - L$ is an isomorphism from $\mathring{H}^1$ to H^{-1}. $\square$

With L and Ω as in Theorem 5 consider the mixed initial boundary value problem

$$u_t = Lu \quad \text{in } [0, \infty[\times \Omega, \qquad u = 0 \quad \text{on } [0, \infty[\times \partial\Omega, \qquad \text{and} \quad u(0, \cdot) = f. \tag{4}$$

This is a parabolic equation in Ω which generalizes the heat equation. If φ_j is an eigenfunction of L with Dirichlet boundary conditions, then $u_j(t, x) \equiv \varphi_j(x)\exp(\lambda_j t)$ solves the boundary value problem with initial data φ_j. The eigenfunction expansion $f = \sum \alpha_j \varphi_j$ suggest $u = \sum \alpha_j u_j$ as the solution of the initial value problem. The solutions u_j are products of functions of t and a function of x. Seeking such product solutions is called the *method of separation of variables*. It is of very limited utility but when it works it is very informative.

The analysis of (4) is almost identical to the analysis in §3.6 with $\bigcap H^s(\mathbb{R}^d)$ replaced by $\bigcap \mathscr{D}(L^k)$ and $\mathscr{D}(L^k)$ playing the role of H^{2s}. Note that $f \in \bigcap \mathscr{D}(L^k)$ if and only if the Fourier, coefficients of f decay faster than $|\lambda_j|^{-N}$ for any N. It is not hard to show that $|\lambda_j|$ grows like $j^{2/d}$ (Problem 3), so that this is equivalent to decay faster than $|j|^{-N}$ for any N.

Theorem 6. (i) *If $f \in \bigcap \mathscr{D}(L^k)$, then there is one and only one $u \in \bigcap C^\infty([0, \infty[: \mathscr{D}(L^k))$ solving* (4). *The solution is given by the formula $u = \sum \alpha_j \varphi_j \exp(\lambda_j t)$ where $f = \sum \alpha_j \varphi_j$ is the eigenfunction expansion of f. The series converges in $C^\infty([0, \infty[: \mathscr{D}(L^k))$ for all k.*

The solution $u \in C([0, \infty[: \mathring{H}^1(\Omega))$, and it is in this sense that u satisfies the Dirichlet boundary condition. If Ω is nice, Theorem 5.9.3 shows that $\mathscr{D}(L^k) \subset H^{2k}(\Omega) \subset C^j(\bar{\Omega})$ if $2k > j - d/2$ and it follows that $u \in C^\infty([0, \infty[\times \bar{\Omega})$ and satisfies (4) in the classical sense.

PROOF. If u is a solution, let $u(t) = \sum c_j(t)\varphi_j$ be the eigenfunction expansion. Then

$$c_j' = (u_t, \varphi_j) = (Lu, \varphi_j) = (u, L\varphi_j) = (u, \lambda_j \varphi_j) = \lambda_j(u, \varphi_j) = \lambda_j c_j.$$

Therefore $c_j(t) = \text{const} \cdot \exp(\lambda_j t)$. Setting $t = 0$ shows that the constant must equal α_j. This proves uniqueness and the formula of the theorem.

Conversely, if u is defined by the formula, then $u \in \bigcap C^\infty([0, \infty[: \mathscr{D}(L^k))$, and differenting the formula for u shows u solves the initial value problem (4).

□

Define the evolution operator $S(t)$ from $\bigcap \mathscr{D}(L^k)$ to itself by

$$S(t)f \equiv \sum \alpha_j \exp(\lambda_j t) = u(t) \qquad \text{where} \quad f = \sum \alpha_j \varphi_j.$$

Theorem 7.

(i) *For $t \geq 0$, the operator S extends uniquely to a continuous map of $L^2(\Omega)$ to itself. For $f \in L^2(\Omega)$, the resulting function $u(t) \equiv S(t)f$ is called the generalized solution of* (4).

(ii) *For $t > 0$, the generalized solution belongs to $C^j(]0, \infty[: \mathscr{D}(L^k))$ for all j, $k \in \mathbb{N}$.*

(iii) *If $f \in \mathring{H}^1(\Omega)$, the solution belongs to $C([0, \infty[: \mathring{H}^1(\Omega)) \cap C^1([0, \infty[: H^{-1}(\Omega))$.*

(iv) *If $f \in \mathscr{D}(L^k)$ and $0 \leq j \leq k$, then the solution belongs to $C^j([0, \infty[: \mathscr{D}(L^{k-j}))$.*

This is a nearly immediate consequence of Theorem 5. The generalized solution is characterized by a variety of equivalent condition as in Theorem 3.6.3. For brevity we omit the discussion.

The strong regularity (ii) in $t > 0$ is the *smoothing property* of the heat equation in this context.

For unbounded Ω the operator L is self-adjoint and bounded above and the Spectral Theorem solves the initial value problem via the functional calculus, $u(t) = e^{tL}f$. This gives an analogue of Theorem 6. Theorem 7 is valid without modification.

Note that if $f \in L^2(\Omega)$ it need not satisfy the boundary condition at $t = 0$. For example, one could have $\Omega =]0, 1[$ and $f \equiv 1$. Since u is continuous with values in $\mathring{H}^1$ in $t > 0$, the boundary conditions will be satisfied for t positive. In the case $\Omega =]0, 1[$, $f = 1$, this shows that the generalized solution must be discontinuous at the corners (0, 0) and (0, 1).

Next consider $L = \Delta$ and the initial boundary value problem (5.1.1) and (5.1.2). Suppose that Ω and g are smooth. Then the argument following (5.2.7) shows that there is a $v \in \mathscr{D}(L)$ (actually $C^\infty(\bar{\Omega})$, see §5.9) solving (5.1.3). Then $w \equiv u - v$ satisfies

$$w_t = \Delta w \quad \text{in } [0, \infty[\times \Omega, \qquad w = 0 \quad \text{on } [0, \infty[\times \partial\Omega, \qquad w(0) \in L^2.$$

Thus $w = \sum \alpha_j \exp(\lambda_j t)$ with $\{\alpha_j\} \in l^2$. By Example 2 above we have $\lambda_j \leq -\delta^2$ and it follows that $w(t)$ converges exponentially fast to zero in in $\mathscr{D}(L^k)$ for all k. This proves that $u(t)$ converges exponentially to v as t tends to infinity. This shows that the heuristic argument at the start of §5.1 is correct.

One can study the Schrödinger equation, wave equation, and damped wave equation on Ω with Dirichlet condition on $\partial\Omega$ in the same way. This requires a simple blend of the above ideas with those of Chapter 3. The details are left to the interested reader. We remark that the damped wave equation associated to $L = \Delta$ with Dirichlet boundary conditions yields a justification for the discussion in Problem 5.2.1 (see Problem 5).

PROBLEMS

1. Suppose that L_1 and L_2 are self-adjoint second-order elliptic operators on bounded domains Ω_1 and Ω_2 in dimensions d_1 and d_2. Consider $L \equiv L_1(x, D_x) + L_2(y, D_y)$ on functions $u(x, y)$ in $\Omega_1 \times \Omega_2$ with Dirichlet boundary conditions on $\partial(\Omega_1 \times \Omega_2)$. Prove that the eigenfunctions of L are the products of the eigenfunctions of the L_i and that the eigenvalues are the sums of the eigenvalues of the L_i. *Hint*. It is easy to see that the products are eigenfunctions. You must show that these are all of them. Use completeness.

   DISCUSSION. This is another example of the *separation of variables*.

2. (i) Use Problem 1 and Example 2 following Theorem 4 to compute the eigenvalues and eigenfunctions of the Laplace operator on a rectangle $\prod [0, \pi l_j] \subset \mathbb{R}^d$ with Dirichlet boundary conditions. *Hint*. For all parts first do the case $l_j = 1$.
   (ii) Show that the number of eigenvalues greater than $-\Lambda^2$ is equal to the number of points in the lattice $\prod ((l_j)^{-1}\mathbb{Z})$ which lie inside the ball of radius Λ in $\mathbb{R}^d$.
   (iii) Conclude that as $n \to \infty$, $\lambda_n n^{-2/d}$ converges to a limit and compute the limit.
   (iv) Observe that the limit depends only on d and the d-dimensional measure of the rectangle.

   DISCUSSION. Part (iv) is a special case of Weyl's Theorem on the asymptotic distribution of the eigenvalues which was motivated by Plank's law for black body radiation.

The next problem shows that the algebraic growth $n^{2/d}$ is valid for all of our self-adjoint elliptic eigenvalue problems. The key is the minimax principle for the nth eigenvalue (see Reed and Simon [RS1] and Courant and Hilbert [CH1])

$$\lambda_n = \max_{\dim(V)=n} \min\{(Lv, v) \colon v \in V\}$$

the maximum being over n-dimensional linear subspaces of $\mathring{H}^1(\Omega)$.

3. (i) Suppose that $\Omega_1 \subset \Omega \subset \Omega_3$ are bounded open subsets of $\mathbb{R}^d$ and L satisfies (5.6.1)–(5.6.4) and $L = L^t$ on the large set Ω_3. Prove that

   $$\lambda_n(L, \Omega_3) \le \lambda_n(L, \Omega) \le \lambda_n(L, \Omega_1),$$

   where $\lambda_n(L, \mathcal{O})$ denotes the nth eigenvalue of L on $\mathcal{O}$ with Dirichlet boundary conditions on $\partial\mathcal{O}$.
   (ii) Prove that if L and K are symmetric, they satisfy (5.6.1)–(5.6.4) on Ω, and $(Lu, u) \ge (Ku, u)$ for all $u \in \mathring{H}^1(\Omega)$, then $\lambda_n(L, \Omega) \ge \lambda_n(K, \Omega)$.
   (iii) By comparing L to $\beta_j\Delta + \gamma_j$ on rectangles Ω_j contained in and containing Ω, show that there are constants $c_1 > c_2 > 0$ such that for n large

   $$-c_1 n^{2/d} < \lambda_n(L, \Omega) < -c_2 n^{2/d}.$$

4. Suppose that L is a self-adjoint operator on an open set Ω defined by an operator satisfying (5.6.1)–(5.6.4). In addition, suppose that $\lambda_{j+1} < \lambda_j$ are successive eigenvalues of L. Prove the following theorem:

Theorem 8 (C. Dolph). *If $f \in C^1(\mathbb{R})$ and*

$$\lambda_{j+1} < \inf f'(t) \le \sup f'(t) < \lambda_j, \tag{5}$$

then for any $g \in H^{-1}(\Omega)$ there is a unique solution to the semilinear Dirichlet problem

$$Lu - f(u) = g, \qquad u \in \mathring{H}^1(\Omega).$$

Hint. Apply Theorem 5.6.5.

DISCUSSION. The hypothesis (5) is called a *nonresonance condition.* In a rough sense, it says that the nonlinear term does not interact with the spectrum. The term resonance comes from the study of *nonlinear oscillations.* This is the special case of $\Omega = [0, l]$ with periodic boundary conditions, where one is seeking l-periodic solutions of second-order ordinary differential operators.

See [N] for a nice exposition of problems at resonance.

The next problem discusses waves propagating in $\mathbb{R}_t \times \Omega$ according to the damped wave equation

$$u_{tt} - c^2 \Delta u + a u_t = 0 \quad \text{in } \mathbb{R} \times \Omega, \quad a \in \mathbb{R}_+. \tag{6}$$

Interaction with the boundary is described by the Dirichlet boundary condition (see also Problems 3.7.2, 3.7.3, and 5.2.1)

$$u = 0 \quad \text{on } \mathbb{R} \times \partial\Omega. \tag{7}$$

5. Prove

Theorem

(i) *If $f \in \mathring{H}^1(\Omega)$ and $g \in L^2(\Omega)$, then there is a unique generalized solution $u \in C(\mathbb{R} : \mathring{H}^1(\Omega)) \cap C^1(\mathbb{R} : L^2(\Omega))$ satisfying (6) and (7).*

(ii) *Show that the energy is exponentially decreasing as $t \to \infty$ in the sense that there are positive constants c, α such that for all such u and all $t \ge 0$*

$$\int u_t^2(t, x) + |\nabla_x u(t, x)|^2 \, dx \le c e^{-2\alpha t} \int u_t^2(0, x) + |\nabla_x u(0, x)|^2 \, dx.$$

DISCUSSION. Since the au_t term represents frictional resistance it is reasonable to expect that α would be an increasing function of a. In fact, as friction tends to infinity, one finds that α tends to zero. This counterintuitive result should be revealed by your analysis. The cause, *overdamping*, is also present for the simple damped spring, $y'' + ay'' + ky = 0$, $y'(0) = 0$, for which the energy, $(y')^2 + ky^2$, decays slowly if a is large.

§5.8. Tangential Regularity for the Dirichlet Problem

This and the next section are concerned with the differentiability of solutions of the Dirichlet problem when the data are regular. The goal is to show that the solutions u have two more derivatives than the right-hand side f, provided that the operator L and domain Ω are smooth. In particular, if $f \in C^\infty(\bar{\Omega})$ then $u \in C^\infty(\bar{\Omega})$, so u is a classical solution of the Dirichlet problem. Starting in §1.1, we have emphasized that a gain of m derivatives for an operator of order m is peculiar to elliptic equations.

This regularity property is related to the origins of elliptic equations as equations describing steady states of dissipative physical processes. These processes have the effect of smoothing out irregularities. By the time the steady state is reached the solution is as smooth as the data permits. The easiest example of this sort is steady states for the heat equation, where the solutions are regular thanks to the smoothing property of the heat equation. The smoothing property is one aspect of its dissipative character.

Suppose that Ω lies on one side of its C^∞ boundary, precisely

$$\bar{\Omega} \hookrightarrow \mathbb{R}^d \text{ is a compact submanifold with boundary.} \tag{1}$$

This is equivalent to compactness together with the existence of a defining function $\rho \in C^\infty(\mathbb{R}^d : \mathbb{R})$ with $\rho = 0$ and $\nabla\rho \neq 0$ on $\partial\Omega$. A *coordinate patch* is an open set $\mathcal{O}$ in $\mathbb{R}^d$ and a diffeomorphism $\eta: \mathcal{O} \to \mathbb{R}^d_y$ such that $\eta(\mathcal{O} \cap \Omega) = \eta(\mathcal{O}) \cap \{y_1 > 0\}$, $\eta(\mathcal{O} \cap \partial\Omega) = \eta(\mathcal{O}) \cap \{y_1 = 0\}$. If $\eta(x) = (y_1(x), \ldots, y_d(x))$, the y_i are called local coordinates in $\mathcal{O} \cap \bar{\Omega}$. $\bar{\Omega}$ is covered by a finite set of coordinate patches.

Suppose, in addition, that L satisfies (5.6.1)–(5.6.4) so

$$a_{ij}, a_i, a_0 \text{ belong to } C^\infty(\bar{\Omega}). \tag{2}$$

The proof of regularity has two steps. First, we show that u is differentiable in directions tangent to the boundary. A vector field $V = \sum v_j(x)\partial_j$ is a *tangential vector field* if $v_j \in C^\infty(\bar{\Omega})$, and for all $x \in \partial\Omega$, $V(x) \in T_x(\partial\Omega)$. That is, V is tangent to the boundary. Two equivalent descriptions are, $\partial\Omega$ is characteristic for the partial differential operator V, and $V\rho = 0$ for defining functions ρ. These vector fields play a crucial role. The next result gives some important properties.

Proposition 1

(i) *If V_1 and V_2 are tangential fields, then so is the commutator $[V_1, V_2]$.*

(ii) *If $\eta: \mathcal{O} \to \mathbb{R}^d_y$ is a coordinate patch and V is a tangential vector field supported in $\mathcal{O} \cap \bar{\Omega}$, then in local coordinates, V is a linear combination of the fields $y_1\partial_1, \partial_2, \ldots, \partial_d$ with coefficients smooth in $\{y_1 \geq 0\} \cap \eta(\mathcal{O})$.*

(iii) *There is a finite set of tangential fields, $V_1, \ldots, V_M$, such that V is tangential if and only if V is a linear combination of the V_j with coefficients in $C^\infty(\bar{\Omega})$.*

PROOF. (i) A vector field is tangential if and only if for any $\psi \in C^\infty(\bar{\Omega})$ with $\psi|_{\partial\Omega} = 0$ one has $V\psi|_{\partial\Omega} = 0$. Using this characterization one sees that $V_i V_j(\psi)|_{\partial\Omega} = 0$, and therefore that $[V_1, V_2]\psi|_{\partial\Omega} = 0$.

(ii) In local coordinates y flattening the boundary to $y_1 = 0$, write $V = \sum v_j(y)\partial/\partial y_j$. It suffices to show that $v_1(y) = 0$ whenever $y_1 = 0$. Since the function y_1 vanishes on the boundary and V is tangent to the boundary it follows that $Vy_1 = 0$ when $y_1 = 0$, which is the desired relation.

(iii) Any vector field can be written as a finite sum, $\sum \varphi_j V$, where $\{\varphi_j\}$ is a finite partition of unity subordinate to a covering $\{\mathcal{O}_j\}$ by coordinate patches.

If $\bar{\mathcal{O}}_j \subset \Omega$, $\varphi_j V$ is a combination of the fields $\varphi_j(\eta^{-1})_* \partial/\partial y_k$, $k = 1, 2, \ldots, d$, where $\eta: \mathcal{O} \to \mathbb{R}^d$ is the coordinate map.[1] These fields are supported in the interior of Ω, so are tangential.

If $\mathcal{O}_j$ is a boundary patch with the boundary flattened to $\{y_1 = 0\}$, then for tangential V, $\varphi_j V$ is described in local coordinates as in (ii). Write $v_1 = y_1 w$ with w supported in the same subset of $\bar{\mathcal{O}}$ as w. Then $\varphi_j V$ is a combination of the tangential fields

$$\varphi_j(\eta^{-1})_*\left(y_1 \frac{\partial}{\partial y_1}\right) \quad \text{and} \quad \varphi_j(\eta_j^{-1})_*\left(\frac{\partial}{\partial y_k}\right), \qquad 2 \le k \le d. \qquad \square$$

Definition. Suppose that $s \in \mathbb{N}$ and B is either $\mathring{H}^1(\Omega)$, $L^2(\Omega)$, or $H^{-1}(\Omega)$. Then $B^s_{\tan}$ is the set of $u \in B$ such that whenever $N \le s$ and $V_1, \ldots, V_N$ are tangential vector fields then $V_1 V_2, \ldots, V_N u \in B$.

Aside. Spaces of distributions whose regularity is unchanged by the application of tangential vector fields are called *conormal*. They play an important role in a variety of problems in linear and nonlinear partial differential equations. They are a special case of what are called *Lagrangian distributions*. The associated Lagrangian submanifold of the cotangent bundle is the conormal variety of $\partial\Omega$. The wavefront set (in the sense of Hormander) of conormal distributions belong to this submanifold (see [H2]).

Examples. 1. Let $\Omega =]0, 1[\subset \mathbb{R}$. Then $u \equiv \ln(x)$ belongs to $L^2(\Omega)^s_{\tan}$ for all s. Note that u is smooth in the interior corresponding to the fact that tangential derivatives are, in fact, all derivatives at interior points. However, u is not smooth up to the boundary. The key is that applying xd/dx leaves u unchanged.

2. With the same Ω, u, the function $x(x-1)u$ belongs to $\mathring{H}^1(\Omega)^s_{\tan}$ for all s. The function $x^\alpha(1-x)^\alpha u$ belongs to $\mathring{H}^1(\Omega)^1_{\tan}$ if and only if $\alpha > \frac{1}{2}$ (Problem 1).

3. In the interior of Ω all derivatives are tangential so that if $u \in L^2(\Omega)^s_{\tan}$ and $x \in \Omega$, then $u \in H^s(x)$ (see §5.3 for definition).

4. Analogous definitions work on M, a smooth compact manifold with boundary. If $\partial M = \phi$, then all vector fields are tangential and $L^2(M)^s_{\tan} = H^s(M)$. Thus tangential regularity is full regularity. This is relevant for the Laplace–Beltrami operator and Hodge Laplacian described in §5.4. The proofs of this section work with only minor modifications in these contexts.

[1] If $\Psi: \mathcal{O}_1 \to \mathcal{O}_2$ is a smooth map from an open set in $\mathbb{R}^{d(1)}_x$ to an open set in $\mathbb{R}^{d(2)}_y$ and $W = \sum w_j(x)\partial/\partial x_j$ is a vector field on $\mathcal{O}_1$, then the *push forward* $\Psi_* W$ is a vector field on $\mathcal{O}_2$ defined by choosing a curve $\gamma: \mathbb{R} \to \mathcal{O}_1$ with $\gamma(0) = x$ and $\gamma'(0) = W(x)$. Then $\Psi_* W(\Psi(x)) \equiv (\Psi \circ \gamma)'(0)$. Equivalently, $(\Psi_* W)u \equiv W(u \circ \Psi)$. The expression in coordinates is

$$(\Psi_* W)|_{\Psi(x)} = \sum_{j,k} w_j(x) \frac{\partial \Psi_k}{\partial x_j}(x) \frac{\partial}{\partial y_k}.$$

Let $\mathscr{V} \equiv (V_1, \ldots, V_M)$ be a generating set as in Proposition 1(iii). For $\beta \in \mathbb{N}^M$, let $\mathscr{V}^\beta \equiv V_1^{\beta_1} V_2^{\beta_2} \ldots V_M^{\beta_M}$. Proposition 1(i) implies that taking the products in a different order results in a differential operator which differs from $\mathscr{V}^\beta$ by a sum of terms, each a product of fewer than $|\beta|$ tangential fields. Thus u belongs to $B^s_{\tan}$ if and only if $\mathscr{V}^\beta u$ belongs to B for all β with $|\beta| \leq s$. It follows that $B^s_{\tan}$ is a Hilbert space with norm

$$\|u\|^2_{B^s_{\tan}} \equiv \sum_{|\beta| \leq s} \|(\mathscr{V})^\beta u\|^2_B. \tag{3}$$

The proof of completeness is left as an exercise.

Lemma 2.

(i) *If $Q(x, D)$ is a differential operator of degree 1 with smooth coefficients on $\bar{\Omega}$, then Q maps $\mathring{H}^1(\Omega)^s_{\tan}$ continuously to $L^2(\Omega)^s_{\tan}$ and $L^2(\Omega)^s_{\tan}$ continuously to $H^{-1}(\Omega)^s_{\tan}$.*

(ii) *If $P(x, D)$ is a differential operator of degree 2 with smooth coefficients on $\bar{\Omega}$, then P maps $\mathring{H}^1(\Omega)^s_{\tan}$ continuously to $H^{-1}(\Omega)^s_{\tan}$.*

PROOF. Assertion (ii) is an immediate consequence of (i). The latter is proved by induction on s.

Suppose that $s = 0$. That Q maps L^2 to H^{-1} is Proposition 5.4.5. The other half is elementary.

Suppose next that $s \geq 1$ and that the result is known for $s - 1$. If $|\alpha| = s$, write $\mathscr{V}^\alpha = \mathscr{V}^\beta V_j$ with $|\beta| = s - 1$. It suffices to show that $\mathscr{V}^\alpha Q$ maps $\mathring{H}^1(\Omega)^s_{\tan}$ to $L^2(\Omega)$ and $L^2(\Omega)^s_{\tan}$ to $H^{-1}(\Omega)$. Write

$$\mathscr{V}^\alpha Q = \mathscr{V}^\beta V_j Q = \mathscr{V}^\beta Q V_j + \mathscr{V}^\beta [V_j, Q].$$

For the first term notice that V_j maps $\mathring{H}^1(\Omega)^s_{\tan}$ to $\mathring{H}^1(\Omega)^{s-1}_{\tan}$ and $L^2(\Omega)^s_{\tan}$ to $L^2(\Omega)^{s-1}_{\tan}$. By the inductive hypothesis, $\mathscr{V}^\beta Q$ maps the target spaces to $L^2(\Omega)$ and $H^{-1}(\Omega)$, respectively.

Consider next the second term. Since the commutator $[V_j, Q]$ is a first-order operator, the inductive hypothesis shows that the second term maps $\mathring{H}^1(\Omega)^{s-1}_{\tan}$ to $L^2(\Omega)$ and $L^2(\Omega)^{s-1}_{\tan}$ to $H^{-1}(\Omega)$, which is more than we need since $B^s_{\tan} \subset B^{s-1}_{\tan}$. □

The main result of this section is the following.

Theorem 3 (Tangential Regularity Theorem). *If $u \in \mathring{H}^1(\Omega)$, $s \in \mathbb{N}$, and $Lu \in H^{-1}(\Omega)^s_{\tan}$, then $u \in \mathring{H}^1(\Omega)^s_{\tan}$. In addition, there is a constant $c = c(s)$ such that for all such u*

$$\|u\|_{\mathring{H}^1(\Omega)^s_{\tan}} \leq c(\|u\|_{L^2(\Omega)} + \|Lu\|_{H^{-1}(\Omega)^s_{\tan}}). \tag{4}$$

This theorem shows that if u is the solution of a Dirichlet problem and $f = Lu$ has s tangential derivatives in $H^{-1}(\Omega)$, then u has s tangential derivatives in $\mathring{H}^1(\Omega)$. Note the gain of two derivatives. In the next section we show that both the interior and boundary Elliptic Regularity Theorems are consequences of this basic result.

PROOF. The first step of the proof is to derive (4) as an *a priori* estimate.

Lemma 4. *For any $s \in \mathbb{N}$ there is a constant c so that for all $u \in \mathring{H}^1(\Omega)^s_{\tan}$,* (4) *holds.*

The difference between this and the desired result is that we assume $u \in (\mathring{H}^1)^s_{\tan}$.

PROOF OF LEMMA 4. It suffices to show that the slightly weaker inequality

$$\|u\|_{\mathring{H}^1(\Omega)^s_{\tan}} \le c_s(\|u\|_{L^2(\Omega)^s_{\tan}} + \|Lu\|_{H^{-1}(\Omega)^s_{\tan}}) \tag{5}$$

holds. Given (5) for all s, one estimates

$$\|u\|_{L^2(\Omega)^s_{\tan}} \le c\|u\|_{\mathring{H}^1(\Omega)^{s-1}_{\tan}} \le c_{s-1}(\|u\|_{L^2(\Omega)^{s-1}_{\tan}} + \|Lu\|_{H^{-1}(\Omega)^{s-1}_{\tan}}).$$

Repeating this process s times shows that (5) implies (4).

The proof of (5) is by induction on s beginning with $s = 0$. Let $L_\sigma \equiv -1 + \sum \partial_i((a_{ij} + a_{ji})/2)\partial_j$, so $L - L_\sigma \equiv Q$ is a differential operator of order 1 with coefficients smooth on $\bar{\Omega}$. In §5.4 we showed that L_σ is an isomorphism of $\mathring{H}^1(\Omega)$ to $H^{-1}(\Omega)$. Thus

$$\|u\|_{\mathring{H}^1} \le c\|L_\sigma u\|_{H^{-1}} \le c(\|Lu\|_{H^{-1}} + \|Qu\|_{H^{-1}}) \le c(\|Lu\|_{H^{-1}} + \|u\|_{L^2}),$$

the last inequality follows from Lemma 2 since Q is of order 1. This proves (5) when $s = 0$.

Next suppose that the result is known for $s - 1$ with $s \ge 1$. For $1 \le k \le M$, apply the inductive hypothesis to $V_k u \in (\mathring{H}^1)^{s-1}_{\tan}$ to find

$$\|V_k u\|_{(\mathring{H}^1)^{s-1}_{\tan}} \le c_{s-1}(\|LV_k u\|_{(H^{-1})^{s-1}_{\tan}} + \|V_k u\|_{(L^2)^{s-1}_{\tan}}). \tag{6}$$

Now

$$LV_k u = V_k Lu + [L, V_k]u, \tag{7}$$

and the commutator is a differential operator of order 2 with smooth coefficients. Therefore Lemma 2 implies that

$$\|[L, V_k]u\|_{(H^{-1})^{s-1}_{\tan}} \le c\|u\|_{(\mathring{H}^1)^{s-1}_{\tan}} \le c(\|u\|_{L^2(\Omega)^{s-1}_{\tan}} + \|Lu\|_{H^{-1}(\Omega)^{s-1}_{\tan}}), \tag{8}$$

the last estimate using the inductive hypothesis.

The first term on the right of (7) has $(H^{-1})^{s-1}_{\tan}$ norm dominated by a constant times the $(H^{-1})^s_{\tan}$ norm of Lu. This combined with (6) and (8) completes the inductive proof. □

Theorem 3 is also proved by induction on s. Lemma 4 proves the case $s = 0$. For $s \ge 1$, suppose the result known for $s - 1$. The elegant idea of Nirenberg is to apply Lemma 4 to difference quotients $\delta_j^h u$ approaching $V_j u$. In this way, one derives $(\mathring{H}^1)^{s-1}_{\tan}$ estimates for $\delta^h_j u$ which are uniform in h as h tends to zero. That $V_j u \in \mathring{H}^1(\Omega)^{s-1}_{\tan}$ follows. The details of this argument are presented in the next paragraphs. The proof is long because there are many technical results which must be developed. The ideas are useful in a variety of other contexts.

If V is a tangential vector field, let Φ_t be the one-parameter group of diffeomorphisms of $\bar{\Omega}$ generated by V. The fact that V is tangential implies that $\bar{\Omega}$ and $\partial\Omega$ are invariant under the flow of V.

Example 4 following Proposition 2 of the Appendix shows that for any diffeomorphism $\Phi: \Omega \to \Omega$, the map $C_0^\infty(\Omega) \ni \psi \mapsto \psi \circ \Phi$ extends uniquely to a sequentially continuous map of $\mathscr{D}'(\Omega)$ to itself given by

$$\langle T \circ \Phi, \varphi \rangle = \langle T, |\det D\Phi^{-1}|\varphi \circ (\Phi^{-1})\rangle. \tag{9}$$

Lemma 5. *Suppose that $\Phi: \bar{\Omega} \to \bar{\Omega}$ is a diffeomorphism.*

(i) *For any $s \in \mathbb{N}$, the map $T \mapsto T \circ \Phi$ is a bounded linear map of $B^s_{\tan}$ to itself where B is either $\mathring{H}^1(\Omega)$, $L^2(\Omega)$, or $H^{-1}(\Omega)$.*

(ii) *For the same s and B, if Φ_t is the flow of a tangential field V, then for $u \in B^s_{\tan}$, $u \circ \Phi_t$ is a continuous function of t with values in $B^s_{\tan}$.*

Proof of (i). The proof is by induction on s. Suppose first that $s = 0$.

For $B = L^2(\Omega)$, the result follows from the boundedness of $D\Phi^{-1}$.

For $B = \mathring{H}^1(\Omega)$, choose $g_n \in C_0^\infty(\Omega)$ converging to T in $\mathring{H}^1(\Omega)$. Then $g_n \circ \Phi \in C_0^\infty(\Omega)$ and converges to $T \circ \Phi$ in $\mathscr{D}'(\Omega)$. The desired result then follows from the estimate

$$\|g \circ \Phi\|_{\mathring{H}^1(\Omega)} \le c\|g\|_{\mathring{H}^1(\Omega)}, \qquad \text{for all} \quad g \in C_0^\infty(\Omega).$$

This estimate is an immediate consequence of the formula for $\partial(g \circ \Phi)$ given by the chain rule.

For $B = H^{-1}(\Omega)$, reason by duality. If $T \in H^{-1}(\Omega)$ and $\varphi \in C_0^\infty(\Omega)$

$$\begin{aligned} |\langle T \circ \Phi, \varphi\rangle| &= |\langle T, |\det D\Phi^{-1}|\varphi \circ (\Phi^{-1})\rangle| \\ &\le \|T\|_{H^{-1}(\Omega)} \||\det D\Phi^{-1}|\varphi \circ (\Phi^{-1})\|_{\mathring{H}^1(\Omega)} \end{aligned}$$

Explicitly computing the derivatives of $|\det D\Phi^{-1}|\varphi \circ (\Phi^{-1})$ yields the bound

$$\le c\|T\|_{H^{-1}}\|\varphi\|_{\mathring{H}^1(\Omega)}.$$

This proves the desired estimate for $T \circ \Phi$ in $H^{-1}(\Omega)$.

Next consider $s \ge 1$ assuming the result for $s - 1$. For any $u \in B$ and tangential field $W = \sum w_j(x)\partial_j$

$$W(u \circ \Phi)(x) = \sum w_j(x)\partial_j(u(\Phi(x)) = \sum w_j(x)\frac{\partial u}{\partial x_k}(\Phi(x))\frac{\partial \Phi_k}{\partial x_j}(x) = (Zu)(\Phi(x)), \tag{10}$$

where Z is the vector field on Ω defined by

$$Z(\Phi(x)) = \sum w_j(x) \frac{\partial \Phi_k(x)}{\partial x_j} \frac{\partial}{\partial x_k} \equiv \Phi_* W|_{\Phi(x)}, \tag{11}$$

Z is the *push forward* of W by Φ.

Since W is tangential and Φ maps $\partial\Omega$ to itself, $\Phi_* W \equiv Z$ is also tangential, and so $Zu \in B^{s-1}_{\tan}$ with norm bounded by a multiple of the $B^s_{\tan}$ norm of u. By induction, the $B^{s-1}_{\tan}$ norm of $(Zu) \circ \Phi$ is bounded by a multiple of the $B^{s-1}_{\tan}$ norm of Zu. Combining these two assertions yields the desired boundedness.

For use in the proof of (ii), note that an induction on $|\alpha|$ starting with (10) for $|\alpha| = 1$ proves

$$(\mathscr{V})^\alpha(u \circ \Phi) = (\Phi_* \mathscr{V})^\alpha u|_{\Phi(x)}, \qquad \Phi_* \mathscr{V} \equiv (\Phi_* V_1, \ldots, \Phi_* V_M). \tag{12}$$

Proof. (ii) By induction again.

For $s = 0$ and $B = L^2$ or $\mathring{H}^1$, the proof is almost identical to the proof of Lemma 5.5.2. The proof for $s = 0$ and $B = H^{-1}(\Omega)$ is by a duality argument to show that the $H^{-1}(\Omega)$ norm of $T \circ \Phi_t - T \circ \Phi_{t'}$ is $O(|t - t'|)$ (Problem 2).

For $s > 0$ note that $(\Phi_t)_* V_j$ is a smoothly varying family of tangential vector fields. The proof of Proposition 1(iii) shows that there are smooth functions $a_{jk}(t, x)$ on $\mathbb{R} \times \bar{\Omega}$ such that

$$(\Phi_t)_* V_j = \sum_{k=1}^{M} a_{jk}(t, x) V_k.$$

Using this in identity (12) with $\Phi = \Phi_t$ shows that $(\mathscr{V})^\alpha(u \circ \Phi_t)$ is a combination of tangential derivatives of $u \circ \Phi_t$ with smooth coefficients. The continuity for $s > 0$ is then a consequence of the $s = 0$ continuity. □

Definition. If Φ_t is the flow of a tangential field V let $\delta^h_V u \equiv (u \circ \Phi_{-h} - u)/h$ be the associated difference operator.

The difference operator converges to V in the sense that for all $u \in \mathscr{D}(\Omega)$ (resp. $\mathscr{E}(\Omega)$, $\mathscr{D}'(\Omega)$), $\delta^h_V u$ converges to Vu in $\mathscr{D}(\Omega)$ (resp. $\mathscr{E}(\Omega)$, $\mathscr{D}'(\Omega)$) as h tends to zero.

Lemma 6. *Suppose that* $\mathbb{N} \ni s \geq 1$ *and* $B = \mathring{H}^1(\Omega)$ *or* $L^2(\Omega)$ *or* $H^{-1}(\Omega)$, *and* $u \in B^{s-1}_{\tan}$. *Then* $u \in B^s_{\tan}$ *if and only if the set of distributions* $\{\delta^h_{V_k} u\colon 0 < h < 1$ *and* $1 \leq k \leq M\}$ *is a bounded subset of* $B^{s-1}_{\tan}$.

Proof. For the if part (which is the half needed in the sequel) note that for $u \in B^s_{\tan}$, V tangential, and any α, $\mathscr{V}^\alpha \delta^h u \rightharpoonup \mathscr{V}^\alpha Vu$ in $\mathscr{D}'(\Omega)$. If $|\alpha| \leq s - 1$, the distributions on the left are bounded in the Hilbert space B. Thus there is a subsequence which converges weakly in B to a limit b. This implies that the subsequence converges to b in $\mathscr{D}'(\Omega)$ and therefore that $\mathscr{V}^\alpha Vu = b \in B$. Thus $u \in B^s_{\tan}$.

To prove the only if part, compute using the chain rule

$$\frac{d}{dt} u \circ \Phi_t = \sum (\partial_j u) \circ \Phi_t)(v_j \circ \Phi_t) = (Vu) \circ \Phi_t. \tag{13}$$

The right-hand side is a continuous function of t with values in $B^{s-1}_{\tan}$. Integrating yields

$$\delta^h u = \frac{u \circ \Phi_{-h} - u}{h} = \frac{-1}{h}\int_{-h}^{0} \frac{d}{dt}(u \circ \Phi_t)\, dt = \frac{-1}{h}\int_{-h}^{0} (Vu) \circ \Phi_t\, dt. \quad (14)$$

Use (12) to show

$$\mathscr{V}^\beta \delta^h u = \frac{-1}{h}\int_{-h}^{0} (((\Phi_t)_* \mathscr{V})^\beta Vu) \circ \Phi_t\, dt.$$

Since $u \in B^s_{\tan}$, the integrand is a continuous function of t with values in B provided $|\beta| \le s - 1$. Furthermore, the B norm of the integrand is bounded by a constant times the $B^s_{\tan}$ norm of u, the constant independent of $0 < h < 1$. This shows that the left-hand side is bounded in B. □

To prove Theorem 3 we estimate $\delta^h_V u$ in $\mathring{H}^1(\Omega)^{s-1}_{\tan}$ for tangential fields V. The strategy is to use the case $s - 1$ of Lemma 4. It suffices to show that for $0 < h < 1$, $L\delta^h_V u$ is bounded in $(H^{-1})^{s-1}_{\tan}$ and $\delta^h_V u$ is bounded in $L^2(\Omega)$.

The second assertion follows from the fact that $u \in \mathring{H}^1(\Omega) \subset L^2(\Omega)^1_{\tan}$.

Since $Lu \in H^{-1}(\Omega)^s_{\tan}$, Lemma 6 shows that $\delta^h_V Lu$ is bounded in $H^{-1}(\Omega)^{s-1}_{\tan}$.

The key step is to show that the commutators $[L, \delta^h_V]u$ are bounded in $H^{-1}(\Omega)^{s-1}_{\tan}$. For each h, the difference operator δ^h_V is bounded from $\mathring{H}^1(\Omega)^{s-1}_{\tan}$ (resp. $H^{-1}(\Omega)^{s-1}_{\tan}$) to itself, and L maps $\mathring{H}^1(\Omega)^{s-1}_{\tan}$ to $H^{-1}(\Omega)^{s-1}_{\tan}$ continuously so each term of the commutator maps $\mathring{H}^1(\Omega)^{s-1}_{\tan} \to H^{-1}(\Omega)^{s-1}_{\tan}$ continuously. Individually, they converge to LV and VL which are not bounded as maps $(\mathring{H}^1)^{s-1}_{\tan} \to (H^{-1})^{s-1}_{\tan}$. Thus, the individual terms in the commutators are not bounded independent of $0 < h < 1$.

The arguments above and to follow are presented globally in $\bar{\Omega}$ in a coordinate free way. One could equally well have used partitions of unity and local arguments where the fields $y_1\partial_1, \partial_2, \ldots, \partial_d$ would play a central role. Those more comfortable with computations in local coordinates should have little difficulty translating to that form.

The next lemma is the crucial commutator estimate.

Lemma 7. *If $P(x, D)$ is a differential operator of degree 2 with smooth coefficients on $\bar{\Omega}$ and V is a tangential vector field, then for any $s \in \mathbb{N}$ the commutators $\{[P, \delta^h_V]: |h| < 1\}$ are uniformly bounded as maps of $\mathring{H}^1(\Omega)^s_{\tan}$ to $H^{-1}(\Omega)^s_{\tan}$.*

So as not to lose the thread of the argument we complete the proof of Theorem 3 assuming Lemma 7.

End of Proof of Theorem 3. Recall that the strategy is to prove that $u \in \mathring{H}^1(\Omega)^s_{\tan}$ by showing that $\delta^h_V u$ is bounded in $\mathring{H}^1(\Omega)^{s-1}_{\tan}$ for $0 < h < 1$ and $V \in \{V_1, \ldots, V_M\}$. Using the case $s - 1$ of Lemma 4, this was reduced to estimating the $H^{-1}(\Omega)^{s-1}_{\tan}$ norm of $L(\delta^h_V u)$. Lemma 6 estimates $\delta^h_V Lu$ and

Lemma 7 estimates the commutator $[\delta_V^h, L]u$. One obtains

$$\|\delta_V^h u\|_{\mathring{H}^1(\Omega)_{\tan}^{s-1}} \leq \text{const. times the right-hand side of (4)} \tag{15}$$

for $0 < h < 1$. Lemma 6 implies that $u \in \mathring{H}^1(\Omega)_{\tan}^s$.

The proof of Lemma 6 shows that there is a subsequence $h_n \to 0$ such that $w_n \equiv \delta_V^{h_n} u \rightharpoonup Vu$ in $\mathring{H}^1(\Omega)_{\tan}^{s-1}$. Using (15) and the lower semicontinuity of norm with respect to weak convergence yields

$$\| Vu\|_{\mathring{H}^1(\Omega)_{\tan}^{s-1}} \leq \liminf \|\delta_V^h u\|_{\mathring{H}^1(\Omega)_{\tan}^{s-1}} \leq c(\text{right-hand side of (4)}).$$

Summing over all $V \in \{V_1, \ldots, V_M\}$ and adding the estimate from the $s - 1$ case of the theorem yields estimate (4). □

Lemma 7 is proved using the next two lemmas.

Lemma 8. *Suppose that $s \in \mathbb{N}$, $a \in C_0^\infty(\mathbb{R}^d)$, V is a smooth compactly supported vector field on $\mathbb{R}^d$, and Φ_t is the one-parameter group of diffeomorphisms generated by V. For $h \neq 0$, let $\delta^h u \equiv (u \circ \Phi_{-h} - u)/h$ be the associated difference operators.*

(i) *The maps $u \mapsto u \circ \Phi_h, |h| < 1$ are uniformly bounded from $H^s(\mathbb{R}^d)$ to itself.*
(ii) *The commutators $\{[a, \delta^h]: |h| < 1\}$ are uniformly bounded from $H^s(\mathbb{R}^d)$ to itself.*
(iii) *For any $1 \leq j \leq d$, the commutators $\{[\partial_j, \delta^h]: |h| < 1\}$ are uniformly bounded from $H^s(\mathbb{R}^d)$ to $H^{s-1}(\mathbb{R}^d)$.*

PROOF. (i) For $u \in C_0^\infty(\mathbb{R}^d)$ compute, suppressing h momentarily,

$$\frac{\partial u \circ \Phi}{\partial x_j} = \sum \frac{\partial u}{\partial x_i}(\Phi(x)) \frac{\partial \Phi}{\partial x_j^i}.$$

By induction one shows that for $|\alpha| \leq s$, $\partial^\alpha(u \circ \Phi_h)$ is a finite sum of terms, each of which is a product of $\partial^{\leq s} u(\Phi(x))$ times a finite product of derivatives of the components of Φ_h. As these derivatives of Φ_h are uniformly bounded for $|h| < 1$

$$\|u \circ \Phi_h\|_{H^s(\mathbb{R}^d)} \leq c\|u\|_{H^s(\mathbb{R}^d)} \qquad \text{for all} \quad u \in C_0^\infty(\mathbb{R}^d) \quad \text{and} \quad |h| < 1.$$

The desired result is then a consequence of the density of $C_0^\infty(\mathbb{R}^d)$ in $H^s(\mathbb{R}^d)$.

(ii) Compute for $u \in \mathcal{S}(\mathbb{R}^d)$

$$\begin{aligned} h[a, \delta^h]u &= a(u \circ \Phi_{-h} - u) - ((au) \circ \Phi_{-h} - au) \\ &= a(u \circ \Phi_{-h}) - (au) \circ \Phi_{-h} = (a - a \circ \Phi_{-h})(u \circ \Phi_{-h}). \end{aligned}$$

Since the maps $u \mapsto u \circ \Phi_h$ are uniformly bounded from $H^s(\mathbb{R}^d)$ to itself, it suffices to show that multiplication by $(a - a \circ \Phi_{-h})/h$ is uniformly bounded on $H^s(\mathbb{R}^d)$. These difference quotients converge to $-Va$ uniformly together with all derivatives. In addition, for $|h| < 1$, the supports are contained in a compact set independent of h so that the difference quotients converge to $-Vu$ in $\mathcal{S}(\mathbb{R}^d)$. The desired result is then a consequence of Proposition 2.6.4.

(iii) Compute for $u \in \mathscr{S}(\mathbb{R}^d)$ and $\partial \equiv \partial_j$

$$h\partial\delta^h u = \partial(u \circ \Phi_{-h} - u) \qquad \text{and} \qquad h\delta^h \partial u = (\partial u) \circ \Phi_{-h} - \partial u.$$

Thus

$$h[\partial, \delta^h]u = \partial(u \circ \Phi_{-h}) - (\partial u) \circ \Phi_{-h}.$$

Now

$$\partial_j(u \circ \Phi_{-h}) = \sum \left(\frac{\partial u}{\partial x_k} \circ \Phi_{-h}\right)\left(\frac{\partial(\Phi_{-h})_k}{\partial x_j}\right),$$

so

$$h[\partial, \delta^h]u = \sum \left(\frac{\partial u}{\partial x_k} \circ \Phi_{-h}\right)\left(\frac{\partial(\Phi_{-h})_k}{\partial x_j} - \delta_{k,j}\right), \tag{16}$$

where $\delta_{k,j}$ is the Kronecker delta.

Note that since $\Phi_0(x) = x$, $\partial(\Phi_h)_k/\partial x_j = \delta_{k,j}$ when $h = 0$. Thus the Fundamental Theorem of Calculus yields

$$\frac{\partial(\Phi_h)_k/\partial x_j - \delta_{k,j}}{h} = \frac{1}{h}\int_0^1 \frac{d}{d\theta}\frac{\partial(\Phi_{\theta h})_k}{\partial x_j}\, d\theta \equiv \varphi_{hkj}. \tag{17}$$

The θ derivatives inside the integral are equal to h times t derivatives of Φ_t, so the above expression shows that for $|h| < 1$, the φ's belong to $C_0^\infty(\mathbb{R}^d)$, have support in a fixed compact set, and have partial derivatives bounded independent of h. Proposition 2.6.4 shows that multiplication by the φ's is uniformly bounded from $H^s(\mathbb{R}^d)$ to itself. Then the expression (16), together with the fact that the family $\circ\, \Phi_h$ is uniformly bounded from H^s to itself, completes the proof of Lemma 8. □

Lemma 9. *Suppose that V and W are tangential vector fields on $\bar{\Omega}$, Φ_t is the flow of V, and δ_V^h is the associated difference operator, and B is either $\mathring{H}^1(\Omega)$, $L^2(\Omega)$, or $H^{-1}(\Omega)$. Then, for any $1 \geq s \in \mathbb{N}$, the operators $\{[W, \delta_V^h] : |h| < 1\}$ are uniformly bounded from $B_{\tan}^s$ to $B_{\tan}^{s-1}$.*

Proof. The definition of δ_V^h yields

$$\delta_V^h(Wu) = \frac{(Wu) \circ \Phi_{-h} - Wu}{h},$$

$$W(\delta_V^h u) = \frac{W(u \circ \Phi_{-h}) - Wu}{h} = \frac{((\Phi_{-h})_* Wu) \circ \Phi_{-h} - Wu}{h}.$$

Subtracting the first from the second yields

$$[W, \delta_V^h]u = h^{-1}(((\Phi_{-h})_* Wu) - Wu) \circ \Phi_{-h}.$$

Now $(\Phi_{-h})_* W$ is a smoothly varying family of tangential vector fields. Thus $(\Phi_{-h})_* W = \sum a_j(h, x) V_j$ with $a_j \in C^\infty(\mathbb{R} \times \bar{\Omega})$. Then

$$[W, \delta_V^h] = \left(\sum h^{-1}(a_j(h, x) - a_j(0, x)) V_j \right) \circ \Phi_{-h}.$$

For $|h| \le 1$, the coefficients $h^{-1}(a_j(h, \cdot) - a_j(0, \cdot))$ belong to a bounded set in $C^\infty(\bar{\Omega})$. Thus the family in paraentheses is a bounded family of maps from $B^s_{\tan}$ to $B^{s-1}_{\tan}$. Since $\circ \Phi_{-h}$ is a bounded family from $B^{s-1}_{\tan}$ to itself the desired result follows. □

Finally, we can combine the above ingredients to prove Lemma 7.

PROOF OF LEMMA 7. It suffices to show that for $|\alpha| \le s$ the family $\{\mathscr{V}^\alpha[P, \delta_V^h]: |h| < 1\}$ is uniformly bounded from $\mathring{H}^1(\Omega)^s_{\tan}$ to $H^{-1}(\Omega)$.

The proof is by induction on s. To prove the case $s = 0$ extend the coefficients of P and V to elements in $C_0^\infty(\mathbb{R}^d)$. Then Lemma 8 shows that the family $\{[P, \delta_V^h]: |h| < 1\}$ is bounded from $H^1(\mathbb{R}^d)$ to $H^{-1}(\mathbb{R}^d) = H^1(\mathbb{R}^d)'$. Restricting to the closed subspace $\mathring{H}^1(\Omega)$ yields the desired result.

Next suppose that $s \ge 1$ and the result is known for $s - 1$. For $|\alpha| = s$, write $\mathscr{V}^\alpha = \mathscr{V}^\beta V_j$ with $|\beta| = s - 1$ and $1 \le j \le M$. Then

$$\mathscr{V}^\alpha[P, \delta_V^h] = \mathscr{V}^\beta V_j[P, \delta_V^h] = \mathscr{V}^\beta[P, \delta_V^h]V_j - \mathscr{V}^\beta[[P, \delta_V^h], V_j]. \tag{18}$$

For the first term on the right, note that V_j is bounded from $\mathring{H}^1(\Omega)^s_{\tan}$ to $\mathring{H}^1(\Omega)^{s-1}_{\tan}$, and by induction $\mathscr{V}^\beta[P, \delta_V^h]$ is uniformly bounded from $\mathring{H}^1(\Omega)^{s-1}_{\tan}$ to $H^{-1}(\Omega)$. Thus, to complete the proof, it suffices to show that $[[P, \delta_V^h], V_j]$ is uniformly bounded from $\mathring{H}^1(\Omega)^s_{\tan}$ to $H^{-1}(\Omega)^{s-1}$.

Use Jacobi's identity to write

$$[[P, \delta_V^h], V_j] = -[[\delta_V^h, V_j], P] - [[V_j, P], \delta_V^h]. \tag{19}$$

Lemma 9 shows that $[\delta_V^h, V_j]$ is uniformly bounded from $B^s_{\tan}$ to $B^{s-1}_{\tan}$ with $B = \mathring{H}^1(\Omega)$ and $B = H^{-1}(\Omega)$. This, together with the fact that P is bounded from $\mathring{H}^1(\Omega)^r_{\tan}$ to $H^{-1}(\Omega)^r_{\tan}$ for $r = s$ and $r = s - 1$, suffices to show that each term in the commutator of $[\delta_V^h, V_j]$ with P is uniformly bounded from $\mathring{H}^1(\Omega)^s_{\tan}$ to $H^{-1}(\Omega)^{s-1}$.

Since $[V_j, P]$ is a differential operator of order 2 with smooth coefficients on $\bar{\Omega}$, the inductive hypothesis shows that the second term on the right in (19) is uniformly bounded from $\mathring{H}^1(\Omega)^{s-1}_{\tan}$ to $H^{-1}(\Omega)^{s-1}$, which is a stronger conclusion than needed. □

This completes the proof of Theorem 3.

The same sort of Tangential Regularity Theorem is true for the Neumann problem, that is, $Lu \in H^{-1}(\Omega)^s_{\tan}$ implies $u \in H^1(\Omega)^s_{\tan}$. Note that one has H^1 and not $\mathring{H}^1$ and that the boundary condition is expressed as in (5.4.17). The proof must be modified since $u \circ \Phi_h$ need not satisfy the Neumann condition $\partial(u \circ \Phi_h)/\partial n = 0$. What one does is, take as a test function $\varphi \circ \Phi_h$ in (5.4.17) and subtract the resulting expression from (5.4.17). After some manipulation and estimation of commutators as in this section, one ends up with uniform estimates for a $\delta^h u$ in $H^1(\Omega)$. In this way, one gets the case $s = 1$. The general case is an inductive argument (see Agmon [A]).

Other methods for deriving elliptic regularity do not rely on the use of coercive quadratic forms but rest more on Fourier analysis. The reader is referred to the treatises of Taylor [Ta] and Hormander [H2] for these methods.

PROBLEMS

1. For $1 \geq s \in \mathbb{N}$, prove that $x^\alpha(1-x)^\alpha \ln(x)$ belongs to $\mathring{H}^1(]0, 1[)^s$ if and only if $\alpha > \frac{1}{2}$.
2. Give a detailed proof of the $s = 0$ case of Lemma 5(ii).
3. Suppose that $\Omega =]0, 1[\subset \mathbb{R}$. Prove that $C_0^\infty(]0, 1[)$ is dense in $B_{\tan}^s$ for $B = L^2(\Omega)$, $\mathring{H}^1(\Omega)$, $H^{-1}(\Omega)$.
   DISCUSSION. The same result is valid for any nice subset Ω of $\mathbb{R}^d$. The proof of that more general case should be clear upon combining your solution of this problem with the methods of §5.5.
4. Suppose that $\rho \in C^\infty(\bar{\Omega})$, $\rho|_{\partial\Omega} = 0$, and $\nabla\rho(x) \neq 0$ for $x \in \partial\Omega$.
   (i) Prove that $u \in L^2(\Omega)_{\tan}^1$ if and only if $\rho u \in H^1(\Omega)$.
   (ii) If $1 \geq s \in \mathbb{N}$, prove that $u \in L^2(\Omega)_{\tan}^s$ if and only if $\rho^s u \in H^s(\Omega)$.
   (iii) Formulate and prove $\mathring{H}^1(\Omega)$ and $H^{-1}(\Omega)$ versions of parts (i) and (ii).

§5.9. Standard Elliptic Regularity Theorems

In this section we continue the study of the differentiability of solutions of the Dirichlet problem. In the last section differentiability tangent to the boundary was proved. The results of this section require some simple definitions, motivated by Proposition 2.6.2.

Definition. If $\omega \subset \mathbb{R}^d$ is open and $s \in \mathbb{N}$, then

$$H^s(\omega) \equiv \{u \in L^2(\omega): (\forall |\alpha| \leq s), \partial^\alpha u \in L^2(\omega)\}, \tag{1}$$

$H^s(\omega)$ is a Hilbert space with norm

$$\|u\|_{H^s(\omega)}^2 \equiv \sum_{|\alpha| \leq s} \|\partial^\alpha u\|_{L^2(\omega)}^2. \tag{2}$$

In these definitions, $\partial^\alpha u$ is the distribution derivative. The completeness of H^s is proved as follows. If u_n is a Cauchy sequence then for $|\alpha| \leq s$, $\partial^\alpha u_n$ is a Cauchy sequence in $L^2(\omega)$. Since $L^2(\omega)$ is complete these derivatives converge to limits f_α in $L^2(\omega)$. Then $\partial^\alpha f_0 = f_\alpha \in L^2(\omega)$ so $f_0 \in H^s(\omega)$ and $\partial^\alpha u \to f_\alpha$ in $L^2(\omega)$ so $u_n \to f_0$ in $H^s(\omega)$.

Definition. If $s \in \mathbb{N}$, $\Omega \subset \mathbb{R}^d$, $x \in \bar{\Omega}$, and $u \in \mathscr{D}'(\Omega)$, then u belongs to H^s at x, denoted $u \in H^s(x)$, if there is an $r > 0$ such that $u \in H^s(\Omega \cap B_r(x))$.

Thus $u \in H^s(x)$ means that there is an $r > 0$ such that the distribution derivatives of u of order less than or equal to s are square integrable on $\Omega \cap B_r(x)$.

Consider L satisfying (5.6.1)–(5.6.4) and $\bar{\Omega} \subset \mathbb{R}^d$ a smooth submanifold with boundary. The main results of this section show that if u satisfies the Dirichlet boundary condition, that is, $u \in \mathring{H}^1(\Omega)$, then u is in $H^{s+2}(x)$ at any x where $Lu \in H^s(x)$. The results are straightforward consequences of the Tangential Regularity Theorem proved in the last section.

We begin with a simple result which explains why we expect a gain of two derivatives. An even more special case is Proposition 2.4.6.

Proposition 1. *Suppose that $P(D)$ is a constant coefficient elliptic operator of order m and $s \in \mathbb{R}$,*

$$\text{if } u \in H^s(\mathbb{R}^d) \text{ and } Pu \in H^s(\mathbb{R}^d), \text{ then } u \in H^{s+m}(\mathbb{R}^d). \tag{3}$$

Conversely, if $P(D)$ is an mth order operator such that (3) holds for some $s \in \mathbb{R}$, then P is elliptic.

A similar result is true for variable coefficient operators and local regularity. An mth order $P(x, D)$ is elliptic at x if and only if u is in $H^{s+m}(x)$ whenever u and Pu belong to $H^s(x)$.

Proof of Proposition 1. If P is elliptic, Problem 1.6.7 shows that there are positive constants c_j such that

$$|P(\xi)| \geq c_1\langle\xi\rangle^m - c_2\langle\xi\rangle^{m-1}.$$

Estimate $\langle\xi\rangle^{m-1} \leq \varepsilon\langle\xi\rangle^m + c_\varepsilon$ to see that

$$|P(\xi)| \geq \frac{c_1\langle\xi\rangle^m}{2} - c_3.$$

Then

$$\int_{\mathbb{R}^d} \langle\xi\rangle^{2(s+m)}|\hat{u}(\xi)|^2\, d\xi \leq c \int_{\mathbb{R}^d} \langle\xi\rangle^{2s}(|P(\xi)| + c_3)^2|\hat{u}(\xi)|^2\, d\xi.$$

By hypothesis the right-hand side is finite and it follows that the left-hand side is finite.

Conversely, suppose that P is not elliptic. For $s \in \mathbb{R}$ let $\sigma \equiv s + m - 1$. It suffices to show that there is a u in H^σ such that $Pu \in H^s$, but u is not in $H^{\sigma+1}$.

Since P is not elliptic there is an $\omega \in \mathbb{R}^d$ with $|\omega| = 1$ and $P_m(\omega) = 0$. Then P_m vanishes on the ray $\{r\omega: r > 1\}$. Since ∇P_m is a polynomial of degree $m - 1$, it follows that $|P_m| \leq c\langle\xi\rangle^{m-1}$ on the set of points within unit distance of that ray. Thus $|P(\xi)| \leq c'\langle\xi\rangle^{m-1}$ on the same set.

Suppose φ a nonzero element of $C_0^\infty(|\xi| < 1)$. Define $\hat{u}$ to be a sum of translates of this function in the direction of ω weighted so that u belongs to

H^σ but not to $H^{\sigma+\varepsilon}$ for any $\varepsilon > 0$. For example, take

$$\hat{u}(\xi) \equiv \sum \varphi(\xi - n\omega)(n^{\sigma+1/2} \ln(n))^{-1}.$$

Since $P(\xi)$ is $O(\langle\xi\rangle^{m-1})$ on the support of u we have $Pu \in H^{\sigma-m+1} = H^s$. The proof is complete. □

The next result concerns the regularity of a function at an interior point $\underline{x}$ of an open domain Ω. The differential operator is

$$L \equiv \sum a_{ij}(x)\partial_i\partial_j + \sum a_i(x)\partial_i + a_0(x), \tag{4}$$

$$a_{ij},\, a_i,\, a_0 \text{ belong to } C^\infty(\underline{x} : \mathbb{R}), \tag{5}$$

$$\exists \mu > 0,\ \forall \xi \in \mathbb{R}^d, \qquad \sum a_{ij}(\underline{x})\xi_i\xi_j \geq \mu|\xi|^2. \tag{6}$$

Here $C^\infty(\underline{x})$ means C^∞ on a ball centered at $\underline{x}$.

Theorem 2 (Interior Elliptic Regularity Theorem). *Assume that $\underline{x} \in \Omega$ and that* (4), (5), *and* (6) *hold. If $u \in \mathscr{D}'(\Omega)$ belongs to $H^1(\underline{x})$ and Lu belongs to $H^s(\underline{x})$ for some $0 \leq s \in \mathbb{N}$, then $u \in H^{s+2}(\underline{x})$.*

Proof. We show that if $\sigma \in \mathbb{N}$, $1 \leq \sigma \leq s + 1$, and $u \in H^\sigma(\underline{x})$, then $u \in H^{\sigma+1}(\underline{x})$. Applying this result s times beginning with $\sigma = 1$ yields the desired conclusion.

For $u \in H^\sigma(\underline{x})$, choose $r > 0$ such that the coefficients of L are smooth on $\bar{B}_r(\underline{x})$, the derivatives of u (resp. Lu) up to order σ (resp. s) are square integrable on the ball, and L is elliptic on the ball in the sense that

$$(\forall x \in \bar{B}_r(\underline{x}),\ \xi \in \mathbb{R}^d), \qquad \sum a_{ij}(x)\xi_i\xi_j \geq \frac{\mu|\xi|^2}{2}. \tag{7}$$

Choose $\varphi \in C_0^\infty(B_r(\underline{x}))$ with φ identically equal to 1 on a neighborhood of $\underline{x}$. The strategy is to apply Theorem 5.8.3 to the function φu in the set $B_r(\underline{x})$.

By construction $\varphi u \in H^\sigma(B_r(\underline{x}))$ and φu is compactly supported in the interior so $\varphi u \in \mathring{H}^1(B_r(\underline{x}))$. The crux is to show that $L(\varphi u)$ belongs to $H^{-1}(B_r(\underline{x}))^\sigma_{\tan}$. Compute

$$L(\varphi u) = \varphi Lu + [L, \varphi]u. \tag{8}$$

By the choice of r, the first term belongs to $H^s(\mathbb{R}^d) \cap \mathscr{E}'(B_r(\underline{x}))$. The operator $[L, \varphi]$ is of order 1 and has coefficients supported in supp φ hence strictly in the interior of the ball. Thus the second term belongs to $H^{\sigma-1}(\mathbb{R}^d) \cap \mathscr{E}'(B_r(\underline{x}))$. Since $\sigma \leq s + 1$, both terms belong to $H^{\sigma-1}(\mathbb{R}^d) \cap \mathscr{E}'(B_r(\underline{x}))$, which is included in $H^{-1}(B_r(\underline{x}))^\sigma_{\tan}$.

Theorem 5.8.3 implies that $\varphi u \in \mathring{H}^1(B_r(\underline{x}))^\sigma_{\tan}$. Choose $\psi \in C_0^\infty(B_r(\underline{x}))$ with ψ identically equal to 1 on a neighborhood of supp φ. Then for $1 \leq j \leq d$, $\psi\partial/\partial x_j$ is a tangential derivative in $B_r(\underline{x})$. Thus, for $|\beta| \leq \sigma + 1$, $(\psi\partial)^\beta(\varphi u) \in L^2(B_r(\underline{x}))$. These derivatives are equal to $\partial^\beta(\varphi u)$ and the proof is complete. □

Examples. 1. If $Lu \in C^\infty(\underline{x})$, then $u \in C^\infty(\underline{x})$. To see this note that one can choose φ independent of s to find that $\varphi u \in \bigcap H^s(\mathbb{R}^d)$, and the result follows from the Sobolev Embedding Theorem. Note that the regularity of u near $\underline{x}$ is not influenced by singularities of Lu outside a neighborhood of $\underline{x}$. Similarly, irregularities in the coefficients of L outside a neighborhood of $\underline{x}$ do not influence the regularity of u at $\underline{x}$. This is in sharp contrast to hyperbolic equations were singularities propagate.

2. If $\Omega \subset \mathbb{R}^d$ is open and L satisfies (4), (5), and (6) at each $\underline{x}$ in Ω and if Lu belongs to $C^\infty(\Omega)$, then so does u. Note that one requires neither regularity of the boundary nor uniformity of (5), (6) as one approaches the boundary.

3. The Hodge Decomposition Theorem of differential geometry is a straightforward consequence of the Interior Regularity Theorem applied to the Hodge Laplacian described in §5.4.

4. While $Lu \in H^s(\underline{x}) \Rightarrow u \in H^{s+2}(\underline{x})$ and $Lu \in C^\infty(\underline{x}) \Rightarrow u \in C^\infty(\underline{x})$, it is not true that $Lu \in C^k(\underline{x}) \Rightarrow u \in C^{k+2}(\underline{x})$ (Problem 3).

Remark. A simple partition of unity argument shows that if (4), (5), and (6) hold at all $\underline{x} \in \Omega$, $\omega_1 \subset\subset \omega_2 \subset\subset \Omega$, $u \in H^1(\omega_2)$, and $Lu \in H^s(\omega_2)$, then $u \in H^{s+2}(\omega_1)$. Moreover, there is a constant $c = c(s, \omega_1, \omega_2, L)$ so that, for all such u,

$$\|u\|_{H^{s+2}(\omega_1)} \le c(\|Lu\|_{H^s(\omega_2)} + \|u\|_{H^1(\omega_2)}).$$

The next result allows one to prove regularity up to the boundary, for example, $C^\infty(\bar{\Omega})$. For that one needs to assume that the boundary is smooth and that the coefficients are regular up to the boundary. In addition, one needs to know that boundary conditions are satisfied.

Example. Let $\Omega = B_1(0) \subset \mathbb{R}^2$. Then $u = \mathrm{Re}(1/(z-1)^{1/2})$ is harmonic on Ω, square integrable on Ω, but certainly not regular at $(1, 0)$. One needs appropriate boundary conditions to force regularity. For this u, the Dirichlet data $u|_{\partial\Omega} \in C^\infty(\partial\Omega \backslash (1, 0))$ and regularity at all points other than $(1, 0)$ is forced.

As in Theorem 2, regularity of solutions near x is influenced only by local behavior of L, Lu, and $\partial\Omega$.

Definition. A point $\underline{x} \in \partial\Omega$ is called *regular* if there is an $r > 0$ and a diffeomorphism $\chi: B_r(\underline{x}) \to \mathbb{R}^d_y$ such that $\chi(B_r(x) \cap \Omega) \subset \{y_1 > 0\}$ and $\chi(B_r(\underline{x}) \backslash \bar{\Omega}) \subset \{y_1 < 0\}$.

Examples. 1. $\bar{\Omega} \hookrightarrow \mathbb{R}^d$ is a smooth submanifold with boundary if and only if every boundary point is regular.

2. If Ω is a square in $\mathbb{R}^2$, then the vertices are not regular and the other boundary points are regular.

Definition. If $u \in \mathscr{D}'(\Omega)$ and $x \in \partial\Omega$, then $u \in \mathring{H}^1(x)$ if and only if there is a $\varphi \in C_0^\infty(\mathbb{R}^d)$ such that φ is identically equal to 1 on a neighborhood of x and $\varphi u \in \mathring{H}^1(\Omega)$.

The hypothesis of the next result is that (5) holds with $\underline{x}$ a regular point of $\partial\Omega$. The precise meaning is that there is an $r > 0$ such that the coefficients belong to $C^\infty(\bar{\Omega} \cap \bar{B}_r(\underline{x}))$.

Theorem 3 (Elliptic Regularity at the Boundary). *Suppose that $\underline{x}$ is a regular point of $\partial\Omega$ and that* (4), (5), *and* (6) *hold. If $u \in \mathring{H}^1(\underline{x})$ and $Lu \in H^s(\underline{x})$, $0 \le s \in \mathbb{N}$, then $u \in H^{s+2}(\underline{x})$.*

Proof. Choose r so small that on $B_r(\underline{x}) \cap \bar{\Omega}$ the coefficients of L are C^∞ and satisfy an ellipticity condition analogous to (7), the derivatives of Lu up to order s are square integrable, and there is a local coordinate change on $B_r(\underline{x})$ as in the definition of regular boundary point. Choose $\varphi \in C_0^\infty(B_r(\underline{x}))$ such that φ is identically equal to 1 on a neighborhood of $\underline{x}$ and $\varphi u \in \mathring{H}^1(\Omega)$.

The strategy is to apply the Tangential Regularity Theorem 5.8.3 to φu. To do that, we need that φu is defined on a smooth submanifold with boundary on which $L$is a smooth elliptic operator. Choose $r' < r$ so that $\text{supp}(\varphi) \subset\subset B_{r'}(\underline{x})$. All boundary points of the open set $\Omega \cap B_{r'}(\underline{x})$ are regular except those on $\partial\Omega \cap \partial B_{r'}(\underline{x})$ which lie outside the support of φ (Figure 5.9.1). Smooth the boundary near those points to obtain a smooth embedded submanifold ω such that $\omega \subset B_r(\underline{x}) \cap \Omega$, and for some $\varepsilon > 0$

$$\partial\omega \cap \text{supp}(\varphi) \subset \partial\Omega \cap B_{r'}(\underline{x}).$$

In particular, $\varphi u \in \mathring{H}^1(\omega)$.

The Tangential Regularity Theorem is next used to prove that $\varphi u \in \mathring{H}^1(\omega)^{s+1}_{\tan}$. To do that, it suffices to show that if $1 \le \sigma \le s$ and $\varphi u \in \mathring{H}^1(\omega)^{\sigma}_{\tan}$, then $\varphi u \in \mathring{H}^1(\omega)^{\sigma+1}_{\tan}$.

To show that $\varphi u \in \mathring{H}^1(\omega)^{\sigma+1}_{\tan}$ using the Tangential Regularity Theorem, it

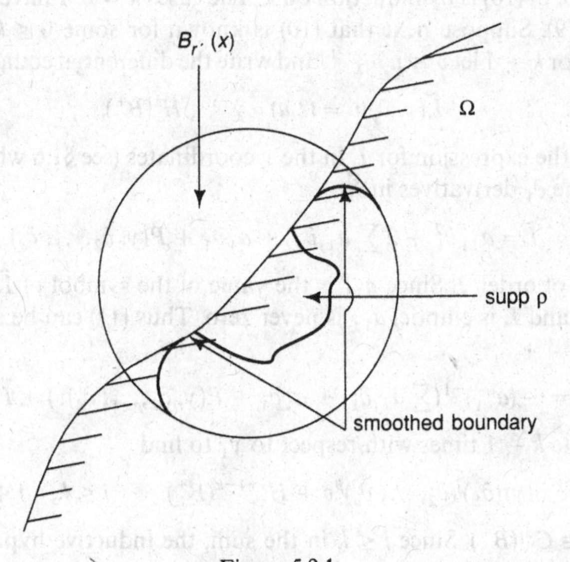

Figure 5.9.1

suffices to verify that $L(\varphi u) \in H^{-1}(\omega)^{\sigma+1}_{\tan}$. Use equation (8) for $L(\varphi u)$. The first term belongs to $H^s(\omega)$ by construction. The commutator $[L, \varphi]$ is a smooth differential operator of order 1 with coefficients supported in $\text{supp}(\varphi)$. Choose $\psi \in C_0^\infty(B_r(\underline{x}))$ with ψ equal to 1 on a neighborhood of $\text{supp}(\varphi)$. Since $\psi u \in \mathring{H}^1(\omega)^\sigma_{\tan}$ it follows that $[L, \varphi]u = [L, \varphi]\psi u \in L^2(\omega)^\sigma_{\tan}$. Thus both terms on the right of (8) belong to $L^2(\omega)^\sigma_{\tan} \subset H^{-1}(\omega)^{\sigma+1}_{\tan}$, since if one applies $\sigma + 1$ tangential derivatives, the first σ maps $L^2(\omega)^\sigma_{\tan}$ to $L^2(\omega)$ and the last maps $L^2(\omega)$ to $H^{-1}(\omega)$. This completes the proof that $\varphi u \in \mathring{H}^1(\omega)^{s+1}_{\tan}$.

This regularity has the correct number of derivatives, namely $s + 2$, but $s + 1$ of them are restricted to be tangential derivatives. The proof is completed by using the differential equation to express arbitrary derivatives in terms of tangential derivatives. This idea bears the intimidating name *partial hypoellipticity at the boundary.* It is little more than the calculation of the Cauchy–Kowaleskaya Theorem. We perform the calculation in the local coordinates y provided by the mapping χ in the definition of regular point. Without loss of generality suppose that $\chi(\underline{x}) = 0$. Choose $\rho > 0$ so small that $B_\rho(0) \subset\subset \chi(\text{int}(\{\varphi = 1\})$. Let $B^+ \equiv B_\rho(0) \cap \{y_1 > 0\}$. The tangential regularity proved above shows that

$$(y_1\partial_1, \partial_2, \ldots, \partial_d)^\alpha(u \circ \chi^{-1}) \in H^1(B^+) \qquad \text{provided} \quad |\alpha| \leq s + 1. \tag{9}$$

The strategy from here is to prove, for $k = 0, \ldots, s + 2$,

$$(\partial_2, \ldots, \partial_d)^\alpha \partial_1^k(u \circ \chi^{-1}) \in L^2(B^+) \qquad \text{if} \quad |\alpha| + k \leq s + 2. \tag{10}$$

This suffices to show that $u \circ \chi^{-1} \in H^{s+2}(B^+)$. However, since χ is a diffeomorphism on a neighborhood of the closure of B^+, the $L^2(\chi^{-1}(B^+))$ norms of the derivatives of u up to order s are bounded by a multiple of the $H^{s+2}(B^+)$ norm of $u \circ \chi^{-1}$. Thus $u \in H^{s+2}(\chi^{-1}(B^+))$. It remains to prove (10).

The proof of (10) is by induction on k. The cases $k = 0, 1$ have already been proved in (9). Suppose next that (10) is known for some $0 \leq k \leq s + 1$. To prove (10) for $k + 1$ let $v \equiv u \circ \chi^{-1}$, and write the differential equation for v as

$$\tilde{L}(y, \partial_y)v = (Lu) \circ \chi^{-1} \in H^s(B^+), \tag{11}$$

where $\tilde{L}$ is the expression for L in the y coordinates (see §1.6 where $\eta = \chi^{-1}$). Separate the ∂_1 derivatives in $\tilde{L}$

$$\tilde{L} = a_{11}\partial_1^2 + 2\sum a_{1j}\partial_{1j} + a_1\partial_1 + P(y, \partial_2, \ldots, \partial_d), \tag{12}$$

where P is of order 2. Since a_{11} is the value of the symbol of $\tilde{L}$ evaluated at $(1, 0, \ldots, 0)$ and $\tilde{L}$ is elliptic, a_{11} is never zero. Thus (11) can be solved for $\partial_1^2 v$ to give

$$\partial_1^2 v = -(a_{11})^{-1}(\sum a_{1j}\partial_{1j} + a_1\partial_1 + P(y, \partial_2, \ldots, \partial_d)v) + H^s(B^+). \tag{13}$$

Differentiate $k - 1$ times with respect to y_1 to find

$$\partial_1^{k+1}v = \sum c_{j,\beta}(y)(\partial_1)^j(\partial_2, \ldots, \partial_d)^\beta v + H^{s+1-k}(B^+), \qquad j \leq k, \quad j + |\beta| \leq k + 1,$$

where $c_{j,\beta} \in C^\infty(\bar{B}^+)$. Since $j \leq k$ in the sum, the inductive hypothesis shows that the right-hand side belongs to $H^{s+1-k}(B^+)$. Thus, if $|\alpha| \le s + 1 - k$, then $(\partial_2, \ldots, \partial_d)^\alpha \partial_1^{k+1} v \in L^2(B^+)$. This proves the case $k + 1$ of (10) which completes the inductive argument. □

Corollary 4. *Suppose that $0 \le s \in \mathbb{N}$, $\bar{\Omega} \subset \mathbb{R}^d$ is a smooth embedded submanifold with boundary, and L satisfies (5.6.1)–(5.6.4). Then $u \in H^{s+2}(\Omega)$ whenever $u \in \mathring{H}^1(\Omega)$ and $Lu \in H^s(\Omega)$. Moreover, there is a $c \in \mathbb{R}$ so that for all such u*

$$\|u\|_{H^{s+2}(\Omega)} \le c(\|Lu\|_{H^s(\Omega)} + \|u\|_{\mathring{H}^1(\Omega)}).$$

Proof. Theorem 3 shows that for each $\underline{x} \in \bar{\Omega}$ there is an open ball $B_{r(\underline{x})}(\underline{x})$ such that $u \in H^{s+2}(B_r)$. Cover the compact set $\bar{\Omega}$ by a finite number of such balls to prove that $u \in H^{s+2}(\Omega)$.

The estimate (13) can be proved by retracting the proof of Theorem 3, keeping track of the estimates at all steps. However, it is worth noting that the estimate (13) is a consequence of the regularity result proved in the first paragraph. Define a normed linear space X by

$$X \equiv \{u \in \mathring{H}^1(\Omega): Lu \in H^s(\Omega)\}, \qquad \|u\|_X \equiv \|Lu\|_{H^s(\Omega)} + \|u\|_{\mathring{H}^1(\Omega)}.$$

X is complete since if u_n is a Cauchy sequence in X, then Lu_n is a Cauchy sequence in H^s and u_n is a Cauchy sequence in $\mathring{H}^1$. By completeness of these two spaces it follows that u_n converges to a limit u in $\mathring{H}^1$ and Lu_n converges to a limit f in H^s. Then $Lu = f$, so $u \in X$ and $u_n \to u$ in X.

The first part of the corollary shows that $X \hookrightarrow H^{s+2}$. Call the inclusion map $\imath$. The inequality (13) is equivalent to the continuity of $\imath$.

Since X and H^{s+2} are complete and $\imath$ is everywhere defined, continuity follows if we show that the graph of $\imath$ is closed. Thus, it suffices to show that if $u_n \to u$ in X and $\imath u_n = u_n \to v$ in H^{s+2}, then $\imath u = v$. Note that in L^2, u_n converges to both u and v. Thus $u = v$. Since $\imath u = u$, this ends the proof. □

If $u \in \mathring{H}^1$ and $Lu \in C^\infty(\bar{\Omega})$, this corollary proves that $u \in H^s(\Omega)$ for all $s \in \mathbb{R}$. One would like to know that u is smooth up to the boundary in the sense that $u \in C^\infty(\bar{\Omega})$. This follows from a smooth bounded set version of the Sobolev Embedding Theorem.

Theorem 5 (Sobolev Embedding Theorem). *Suppose that $\bar{\Omega} \hookrightarrow \mathbb{R}^d$ is a compact smooth submanifold with boundary. If $\mathbb{N} \ni s > k + d/2$, then every element of $H^s(\Omega)$ is equal to a function in $C^k(\bar{\Omega})$. In addition, there is a constant $c = c(s, k, \Omega)$ such that*

$$\|u\|_{C^k(\bar{\Omega})} \le c\|u\|_{H^s(\Omega)} \qquad \textit{for all} \quad u \in H^s(\Omega). \tag{14}$$

The equality of this theorem is in the sense of distributions, that is, there is an element of $C^k(\bar{\Omega})$ which defines the same distribution.

Examples. In these examples suppose that Ω and the differential operators satisfy the hypotheses of Theorem 5.

1. If $f \in C^\infty(\bar{\Omega})$, then the $\mathring{H}^1(\Omega)$ solution of the Dirichlet problem constructed in §5.3 belongs to $C^\infty(\bar{\Omega})$. In particular, it satisfies the differential equation and boundary condition in the classical sense.

2. The eigenfunctions φ_j of self-adjoint operators L, as in §5.7, belong to $C^\infty(\bar{\Omega})$.

3. The unbounded operator defined on $L^2(\Omega)$ by the differential operator in §5.7 has

$$\mathscr{D}(L) = \mathring{H}^1(\Omega) \cap H^2(\Omega),$$

$$\mathscr{D}(L^k) = \{u \in H^{2k}(\Omega): L^j u \in \mathring{H}^1(\Omega) \text{ for } 0 \leq j \leq k-1\}. \tag{15}$$

Thus convergence in the graph norm of $\mathscr{D}(L^k)$ is equivalent to $H^{2k}(\Omega)$ convergence and stronger than $C^m(\bar{\Omega})$ convergence whenever $m < 2k - d/2$. These results complement the conclusions of Theorems 5.7.5–5.7.7.

There are two distinct strategies for proving inequality (14). The first is to use the Fundamental Theorem of Calculus to express u as an integral of suitable derivatives of u and then use Hölder's inequality or a variant (see, e.g. Courant and Hilbert [CH, Vol. 2]). The second is to extend u to an element of $H^s(\mathbb{R}^d)$ and then apply Theorem 2.6.7 to the extension. Following this strategy, Theorem 5 is an immediate consequence of the next result which is of independent interest.

Theorem 6. *If $\bar{\Omega} \hookrightarrow \mathbb{R}^d$ is a smooth compact submanifold with boundary and $s \in \mathbb{N}$, then there is a bounded linear operator E: $H^s(\Omega) \to H^s(\mathbb{R}^d)$ such that $Eu = u$ on Ω.*

The proof of this result has two steps. $C^\infty(\bar{\Omega})$ is proved to be dense in $H^s(\bar{\Omega})$, then the operator E is constructed on $C^\infty(\bar{\Omega})$.

Theorem 7. *If $\bar{\Omega} \hookrightarrow \mathbb{R}^d$ is a smooth compact submanifold with boundary and $s \in \mathbb{N}$, then $C^\infty(\bar{\Omega})$ is dense in $H^s(\Omega)$.*

Proof of Theorem 7. Choose V a compactly supported smooth vector field on $\mathbb{R}^d$ which is transverse to $\partial\Omega$ and points toward the interior of Ω at $\partial\Omega$. Let Φ_t be the flow generated by V and let $\Omega_t \equiv \Phi_t(\Omega)$. The proof of Lemma 5.8.5(i) shows that $\circ\, \Phi_t$ is an isomorphism from $H^s(\Omega_t)$ to $H^s(\Omega)$ with inverse given by $\circ\, \Phi_{-t}$. The norm bounded uniformly for $|t| \leq 1$.

For $u \in L^2(\Omega)$

$$\|u \circ \Phi_t - u\|_{L^2(\Omega)} \to 0 \qquad \text{as} \quad t \to 0+.$$

This is obvious if $u \in C(\bar{\Omega})$. The density of $C(\bar{\Omega})$ in $L^2(\Omega)$, together with the uniform boundedness, completes the proof.

Identity (5.8.13) shows that

$$\partial^\alpha(u \circ \Phi_t) = (((\Phi_t)_*\partial_1, (\Phi_t)_*\partial_2, \ldots, (\Phi_t)_*\partial_d)^\alpha u) \circ \Phi_t.$$

It then follows from the $L^2(\Omega)$ case that for $u \in H^s(\Omega)$

$$\|u \circ \Phi_t - u\|_{H^s(\Omega)} \to 0 \quad \text{as} \quad t \to 0+.$$

Since $\Omega \subset\subset \Omega_{-t}$, $\operatorname{dist}(\Omega, \partial\Omega_{-t}) \equiv \delta > 0$. Let $j_\eta \equiv \eta^{-d} j(x/\eta)$ be a standard smooth approximate delta supported in $|x| \le \eta$. Then for any $w \in L^2(\Omega_{-t})$ and $\eta < \delta$, the values of $j_\eta * w$ on Ω are determined by the values of w in Ω_{-t} and $j_\eta * w \to w$ in $L^2(\Omega)$ as $\eta \to 0$. Differentiating, it follows that for $w \in H^s(\Omega_{-t})$, $j_\eta * w \to w$ in $H^s(\Omega)$.

Given $u \in H^s(\Omega)$ and $\varepsilon > 0$, choose $t > 0$ such that the $u \circ \Phi_t - u$ has $H^s(\Omega)$ norm less than $\varepsilon/2$. Then choose $\eta > 0$ so that $j_\eta * (u \circ \Phi_t) - u \circ \Phi_t$ has $H^s(\Omega)$ norm less than $\varepsilon/2$. Then $(j_\eta * (u \circ \Phi_t))|_\Omega$ is the desired $C^\infty(\bar{\Omega})$ approximation of u. □

Proof of Theorem 6. Decomposing u with a finite partition of unity for $\bar{\Omega}$, it suffices to extend elements of $H^s(\Omega)$ which have support in a fixed compact set K in a coordinate patch $\chi: B_r(\underline{x}) \to \mathbb{R}^d$ with $\underline{x} \in \partial\Omega$.

It suffices to extend $u \circ \chi^{-1}$ from $H^s(y_1 > 0)$ to $H^s(\mathbb{R}^d)$, for if v is such an extension we may choose $\psi \in C_0^\infty(\chi(B_r))$ with ψ identically one on a neighborhood of $\chi(K)$. Then $(\psi v) \circ \chi$ extends u to an element of $H^s(\mathbb{R}^d)$ supported in $B_r\{\underline{x}\}$.

The extension operator on $\{y_1 > 0\}$ is given by

$$(Ew)(y_1, y') \equiv \sum_{0 \le j \le s} a_j w(-jy_1, y') \qquad \text{for} \quad y_1 \le 0.$$

The real coefficients a_j are chosen so that E maps $C^s(y_1 \ge 0)$ to $C^s(\mathbb{R}^d)$. This is achieved by forcing equality of the derivatives $\partial_1^k Ew(0\pm, y')$ for $k \le s$. This holds if and only if

$$\sum_j (-j)^k a_j = 1 \qquad \text{for} \quad 0 \le k \le s.$$

This set of $s + 1$ equations for the $s + 1$ unknowns a_j has as coefficient the Vandermonde matrix which is invertible, so the a_j are uniquely determined.

The extension operator E so defined maps $C^s(y_1 \ge 0) \cap H^s(y_1 > 0)$ to $C^s(\mathbb{R}^d) \cap H^s(\mathbb{R}^d)$ and

$$\|Ew\|_{H^s(\mathbb{R}^d)} \le c\|w\|_{H^s(y_1 > 0)} \qquad \text{for all} \quad w \in C^s(y_1 \ge 0) \cap H^s(y_1 > 0).$$

The proof of Theorem 7 with $V = \partial/\partial y_1$ shows that $C_{(0)}^\infty(y_1 \ge 0)$ is dense in $H^s(y_1 \ge 0)$, so the set of w as above is dense and E extends uniquely to the desired operator. □

The next examples show that some regularity is required of Ω in order for the above results to be true.

Examples. 1. Let $\Omega =]-1, 0[\cup]0, 1[$ in $\mathbb{R}$. Every element of $H^1(\mathbb{R})$ is continuous at $x = 0$ but the function $u \equiv \operatorname{sgn} x$ belongs to $H^1(\Omega)$. Clearly, there is no extension of u to an element of $H^1(\mathbb{R})$.

2. In §5.6 it was shown that $H^1(\Omega)$ is compactly embedded in $L^2(\Omega)$ whenever Ω is bounded and an extension operator as in Theorem 6 exists. Thus the examples in and before Problem 5.6.2 are sets Ω without extension operators.

Example 1 suggests that very narrow inward pointing spikes pose an obstruction to the existence of extension operators. A celebrated theorem of Calderon shows that if every point of the boundary can be touched from the exterior with an open cone of fixed size in the exterior then an extension operator exists (see Agmon [A]). A theorem of Meyers and Serrin shows that $C^\infty(\Omega) \cap H^s(\Omega)$ is dense in $H^s(\Omega)$ for any open Ω. These approximants need not be well-behaved near $\partial\Omega$, so are markedly less useful than the $C^\infty(\bar{\Omega})$ approximants which need not exist for wild domains.

Problems

1. Prove (15) and (16) and the assertions about $\mathscr{D}(L^k)$ convergence from that example. *Hint.* Use Corollary 4 and Theorem 5.

When Ω is not regular, solutions of the Dirichlet problem need not be smooth up to the boundary. The next problem gives an important example.

2. Suppose that $0 < \theta < 2\pi$ and ω is the wedge $\{z \in \mathbb{C}\backslash 0 : 0 < \arg z < \theta\}$. Then $u = \operatorname{Im}(z^{\pi/\theta}) \in C(\bar{\omega})$ is harmonic in ω and satisfies Dirichlet boundary conditions at $\partial\omega$.
   (i) For what values of s is $u \in H^s(0)$?
   (ii) For which θ is $u \in \mathring{H}^1(0)$?
   (iii) Construct an example of a bounded open Ω and a $u \in \mathring{H}^1(\Omega)$ with $\Delta u \in C^\infty(\bar{\Omega})$ such that $u \notin C^\infty(\bar{\Omega})$. *Hint.* Truncate u away from 0 and close the open end of ω.

   Discussion. For such wedge-like regions an analogue of the Tangential Regularity Theorem is valid where the fields V must be tangent to the boundary even at the singular point. This implies that the fields vanish at the singular point (e.g. $x\partial_x + y\partial_y$).

3. Verify that $u = xy(\ln r)^\beta$ with $r \equiv (x^2 + y^2)^{1/2}$ and $0 < \beta < 1$ satisfies $\Delta u \in C(\mathbb{R}^2)$ and $u \notin C^2(\mathbb{R}^2)$, showing that the "gain of two" regularity theorem are false in the C^k category.

   Discussion. Elliptic gain of two is correct in the Hölder spaces $C^{k+\alpha}$ for $\alpha \in]0, 1[$ (see Bers, John and Schecter [BJS] and Gilbarg and Trudinger [GT]).

§5.10. Maximum Principles from Potential Theory

The study of elliptic equations in the previous sections was based almost exclusively on L^2 methods. The basic estimates were proved by integration by parts. This section is devoted to pointwise estimates which rest on so-called *maximum principles*. These methods are very powerful and flexible but are nevertheless useful almost exclusively for scalar equations of second order. Their failure for systems and higher-order equations renders the analysis of such problems more difficult. We begin by describing two classical results which are the precursors of the general maximum principle of E. Hopf.

When $x \in \mathbb{R}$, functions with nonnegative second derivative are convex (thumb down). This shows that

Proposition 1

(i) *If $u \in C^2(]a, b[)$ and $d^2u/dx^2 > 0$, then u can have no local maximum.*
(ii) *If $d^2u/dx^2 \geq 0$ and there is an $\hat{x} \in]a, b[$ such that $u(\hat{x}) \geq u(x)$ for all $x \in]a, b[$, then u is constant.*

The convexity of u in Proposition 1 asserts that for any interval $I \equiv \{|x - x_0| < r\} \subset\subset]a, b[$, u at the center x_0 is smaller than the average of u on ∂I, that is,

$$u(x_0) \leq \frac{u(x_0 + r) - u(x_0 - r)}{2}.$$

This subaverage property has a generalization to functions with nonnegative Laplacian. Such functions are called *subharmonic.*

Theorem 2. *If $u \in C^2(B_r(x_0)) \cap C(\bar{B}_r(x_0))$ is subharmonic, that is, $\Delta u \geq 0$ in $B_r(x_0)$, then the value of u at the center of the ball is less than or equal to the average value of u over the boundary of the ball. That is,*

$$u(x_0) \leq \frac{1}{|\partial B|} \int_{\partial B} u \, d\sigma, \qquad |\partial B| \equiv \int_{\partial B} 1 \, d\sigma.$$

PROOF. For $0 < \rho < r$, let $B_\rho \equiv \{|x - x_0| < \rho\}$. Greens' identity reads

$$\int_{B_\rho} v\Delta u - u\Delta v \, dx = \int_{\partial B_\rho} \frac{\partial u}{\partial n} v - u \frac{\partial v}{\partial n} \, d\sigma.$$

Take $v \equiv 1$ so $\Delta v = 0 = \partial v/\partial n$ to conclude that $\int_{\partial B_\rho} \partial u/\partial n \, d\sigma \geq 0$.

For $\rho > 0$, let $I(\rho) \equiv \int_{\partial B_\rho} u \, d\sigma$. Then

$$\begin{aligned} \partial_\rho I(\rho) &= \partial_\rho \int_{\partial B_\rho} u \, d\sigma = \partial_\rho \int_{S^{d-1}} u(x_0 + \rho\omega)\rho^{d-1} \, d\omega \\ &= \int_{\partial B_\rho} \frac{\partial u}{\partial n} \, d\sigma + (d-1) \int_{S^{d-1}} u\rho^{d-2} \, d\omega \\ &\geq \frac{d-1}{\rho} \int_{S^{d-1}} u\rho^{d-1} \, d\omega = \frac{d-1}{\rho} I(\rho). \end{aligned}$$

Thus the derivative of I with respect to ρ satisfies $I' \geq (d-1)I/\rho$.

Multiply this differential inequality by the integrating factor ρ^{1-d} to find

$$(\rho^{1-d}I)' = \rho^{1-d}I' + (1-d)\rho^{-d}I = \rho^{1-d}\left(I' - \frac{(d-1)I}{\rho}\right) \geq 0.$$

Thus $\rho^{1-d}I(\rho)$ is an increasing function of ρ for $0 < \rho < r$.

Let ω_d denote the $d - 1$ dimensional area of the unit sphere in $\mathbb{R}^d$ so

$\rho^{d-1}\omega_d = |\partial B_\rho|$. Then the quantity $\rho^{1-d}I(\rho)/\omega_d$ is equal to the average value of u over ∂B_p. By continuity of u, $\rho^{1-d}I(\rho)/\omega_d \to u(x_0)$ as $\rho \to 0$.

We conclude that $u(x_0) \leq \rho^{1-d}I(\rho)/\omega_d$. Let ρ increase to r to prove the theorem. □

Corollary 3 (Mean Value Property). *If $u \in C^2(B_r(x_0)) \cap C(\bar{B}_r(x_0))$ is harmonic in B_r, then the value of u at the center of the ball is equal to the average value of u over the boundary of the ball. These values are also equal to the average value of u over the ball.*

PROOF. To prove the first assertion, apply the previous theorem to u and $-u$. Alternatively, retrace the steps of the proof to see that $\rho^{1-d}I(\rho)/\omega_d$ is independent of ρ.

For the second, express the integral over the ball as an integral of integrals over spheres

$$\int_{B_r} u\,dx = \int_0^r \left(\int_{\partial B_\rho} u\,d\sigma\right) d\rho = \int_0^r \omega_d u(x_0)\rho^{d-1}\,d\rho = u(x_0)|B_r|. \qquad \square$$

Application 1. Newton's Theorem on the attraction of spheres.

This corollary can be used to show that the gravitational field in the exterior of a homogeneous spherical shell is equal to the field of a point charge located at the center with mass equal to the mass of the spherical shell. This is one of the important results of Newton's *Principia*. It allows one to replace spherical planets by point masses without committing any error. The field of the spherical shell of radius R with center $\underline{x}$ is equal to

$$\varphi(x) = \int_{\partial B_R(\underline{x})} \frac{c}{|x - y|}\,d\sigma(y),$$

where the constant c is the product of the mass per unit surface area and the gravitational constant. Let $u(y) \equiv c/|x - y|$. Then if x is in the exterior of the ball the function u is a harmonic function of $y \in \bar{B}_R(\underline{x})$ (Problem 4.6.1). Then $\varphi(x)/|\partial B_R|$ is equal to the average of u over the boundary of the ball. The mean value property asserts that this is equal to u at the center. Thus $\varphi(x) = c|\partial B_R|/|x - \underline{x}|$ which is the desired result.

Application 2. Derivative estimates.

It is typical of elliptic equations that one can bound the size of derivatives in terms of the size of the solution. The classical example is holomorphic functions for which the formula

$$\frac{du(z)}{dz} = \left(\frac{-1}{2\pi i}\right) \oint_{|z-\zeta|=r} \frac{u(\zeta)}{(z - \zeta)^2}\,d\zeta$$

shows that if u is a bounded holomorphic function on Ω, then

$$|u'(z)| \leq \tfrac{1}{2}\pi \frac{\|u\|_{L^\infty(\Omega)}}{\text{dist}(z, \partial\Omega)}. \tag{1}$$

Corollary 4. *If $u \in C^2(\Omega) \cap C(\bar{\Omega})$ is harmonic, that is, $\Delta u = 0$, then for $1 \leq j \leq d$ and all $x \in \Omega$*

$$\left|\frac{\partial u(x)}{\partial x_j}\right| \leq d \frac{\|u\|_{L^\infty(\Omega)}}{\text{dist}(x, \partial\Omega)}. \tag{2}$$

PROOF. Suppose that $r < \text{dist}(\underline{x}, \partial\Omega)$. Then $\partial_j u$ is harmonic so

$$\partial_j u(x) = \frac{1}{|B_r|} \int_{B_r(x)} \partial_j u(y)\, dy = \frac{1}{|B_r|} \int_{\partial B_r(x)} u(y) n_j(y)\, d\sigma(y).$$

Since $|n_j| \leq 1$ and $|\partial B_r| = d|B_r|/r$, this yields

$$\left|\frac{\partial u(x)}{\partial x_j}\right| \leq d \frac{\|u\|_{L^\infty(\Omega)}}{r}.$$

Since this is true for all $r < \text{dist}(x, \partial\Omega)$, (2) follows. □

Application 3. A maximum principle.

Corollary 5. *Suppose that Ω is a connected open subset of $\mathbb{R}^d$ and $u \in C^2(\Omega)$ is subharmonic, that is, $\Delta u \geq 0$.*

(i) *If there is an $\hat{x} \in \Omega$ such that $u(\hat{x}) \geq u(x)$ for all $x \in \Omega$, then u is constant.*
(ii) *If Ω is bounded and u is continuous on $\bar{\Omega}$, then $u \leq \max_{\partial\Omega} u$.*

PROOF. Let $m \equiv u(\hat{x})$. Then $\{x \in \Omega: u(x) = m\}$ is closed in Ω since u is continuous.

If $u(x_0) = m$, choose $r > 0$ such that $B_r(x_0) \subset\subset \Omega$ so $u \leq m$ in $B_r(x_0)$. Theorem 2 and the local maximality yield the two inequalities

$$m = u(x_0) = \int_{B_r} \frac{u\, dx}{|B_r|} \leq \int_{B_r} \frac{m\, dx}{|B_r|} = m.$$

Thus u must equal m almost everywhere in B_r. By continuity $u \equiv m$ in B_r.

Thus $\{x \in \Omega: u(x) = m\}$ is both open and closed in Ω and we conclude that $u \equiv m$, since Ω is connected.

Assertion (ii) is an immediate consequence of (i). □

A physical example reveals how reasonable this result is. Consider the equilibrium position of a membrane stretched over Ω and maintained at height $g(x)$ at $\partial\Omega$. If the only forces acting push downward, the height $v(x)$ at equilibrium satisfies

$$\Delta v \geq 0 \quad \text{in } \Omega \qquad \text{and} \qquad u = g \quad \text{on } \partial\Omega.$$

The corollary implies that $u \leq \max g$ throughout $\bar{\Omega}$. Thus with such forces the highest point is a point of support at the boundary.

EXAMPLE. If $u \in C^2(\Omega) \cap C(\bar{\Omega})$ is harmonic, then

$$\min_{\partial\Omega} u \leq u \leq \max_{\partial\Omega} u. \tag{3}$$

This follows upon applying part (ii) to u and to $-u$, both of which are subharmonic.

Using the corollary we can solve the classical Dirichlet problem when the boundary data are continuous.

Theorem 6. *If $\bar{\Omega} \hookrightarrow \mathbb{R}^d$ is a smooth compact embedded manifold with boundary and $g \in C(\partial\Omega)$, then there is one and only one harmonic function $u \in C^\infty(\Omega) \cap C(\bar{\Omega})$ such that $u = g$ on $\partial\Omega$.*

PROOF. Choose $g_n \in C^\infty(\partial\Omega)$ with $g_n \to g$ uniformly on $\partial\Omega$. Let $u_n \in C^\infty(\bar{\Omega})$ be the harmonic functions with $u_n = g_n$ on $\partial\Omega$. Estimate (3) applied to $u_n - u_m$ shows that

$$\|u_n - u_m\|_{L^\infty(\bar{\Omega})} \leq \|g_n - g_m\|_{L^\infty(\partial\Omega)},$$

so $\{u_n\}$ is a Cauchy sequence in $C(\bar{\Omega})$. Let $u \in C(\bar{\Omega})$ be the limit.

On $\partial\Omega$, u is equal to the uniform limit of the g_n, so $u = g$ on $\partial\Omega$.

Estimate (2) applied to $u_n - u_m$ shows that

$$|\nabla u_n(x) - \nabla u_m(x)| \leq \frac{c\|g_n - g_m\|_{L^\infty(\partial\Omega)}}{\text{dist}(x, \partial\Omega)}$$

so u_n is a Cauchy sequence in $C^1(\Omega)$. It follows that the limit u belongs to $C^1(\Omega)$. In particular, $u \in H^1(x)$ for all $x \in \Omega$.

Since u_n converges to u in $\mathcal{D}'(\Omega)$ we have $\Delta u = \mathcal{D}'\text{-}\lim \Delta u_n = 0$. The Elliptic Regularity Theorem implies that $u \in C^\infty(\Omega)$. This proves existence.

Uniqueness follows from (3) applied to the difference $u_1 - u_2$ of two solutions. □

PROBLEMS

The next two problems give alternate proofs of the mean value property of harmonic functions.

1. If $\Delta u = 0$ and $B_r(y) \subset\subset \Omega \subset \mathbb{R}^2$, we may translate coordinates reducing to the case $y = 0$. Define v_φ by $v_\varphi(r, \theta) = u(r, \theta + \varphi)$ (polar coordinates). By rotation invariance of Δ, v_φ is harmonic. Let v be the harmonic function $\int_0^{2\pi} v_\varphi \, d\varphi$.
   (a) Prove that v is rotation invariant. Conclude that for $0 < |x| \leq r$, $v = a \ln r + b$ for $a, b \in \mathbb{R}$ (see Problem 4.6.1).

(b) Prove that $a = 0$ and $b = 2\pi u(0)$, thereby proving that $u(0) = \int u(r, \varphi)\, d\varphi/2\pi$ which is the mean value property.

DISCUSSION. To make this work in dimensions higher than 2 one must sum the rotates of u over all rotations. The measure is Haar measure on the orthogonal group.

2. The "standard" proof of the mean value property considers $\int u\Delta v - v\Delta u\, dx$ over the domain $B_r(y)\backslash B_\varepsilon(y)$, with $v = r^{2-d}$ when $d \neq 2$ and $v = \ln r$ when $d = 2$. Prove the mean value property by performing this computation and passing to the limit $\varepsilon \to 0$.

The function r^{2-d} in the above proof is a centerpiece of potential theory ($\ln r$ when $d = 2$). The reason is that it is a fundamental solution of the Laplace equation.

3. Prove that for $d \geq 3$, $\Delta r^{2-d} = \delta/\omega_d$, where δ is the Dirac measure and ω_d is the $d-1$ area of the sphere S^{d-1}. *Hint.* r^{2-d} is locally integrable, so for $\psi \in C_0^\infty(\mathbb{R}^d)$,

$$\langle r^{2-d}, \Delta\psi\rangle = \lim_{\varepsilon \to 0} \int_{\mathbb{R}^d \backslash B_\varepsilon(0)} |x|^{2-d} \Delta\psi(x)\, dx.$$

Then compute as in the previous problem.

DISCUSSION. The functions r^{2-d} and $\ln(r)$ appear in the solution of Problem 4.6.1. A different proof of $\Delta r^{2-d} = \delta/\omega_d$ is given in Problem 4.6.3. Yet another proof, valid only for $d = 3$, is given in Problem 4.6.2.

4. Prove that there is a constant $c = c(\alpha, d)$ such that if $u \in C^\infty(\Omega)$ is harmonic, then

$$|\partial^\alpha u(x)| \leq \frac{c\, \|u\|_{L^\infty(\Omega)}}{(\operatorname{dist}(x, \partial\Omega))^{|\alpha|}}. \tag{4}$$

The estimate from this problem gives an alternate proof of the Interior Elliptic Regularity Theorem for harmonic functions.

5. Prove that if $u \in L^\infty(\Omega)$ satisfies $\Delta u = 0$ in the sense of distributions then $u \in C^\infty(\Omega)$. *Hints.* Let j_ε be a standard approximate delta function. Show that $u_\varepsilon \equiv j_\varepsilon * u$ is a well-defined harmonic function on the set of points in Ω whose distance to the boundary is at least ε. Apply estimate (4) to the restriction of u_ε to compact subsets of Ω and then use Arzela–Ascoli.

§5.11. E. Hopf's Strong Maximum Principles

This section presents a far-reaching generalization of the maximum principles of the last section. We find results for second-order elliptic operators of the form

$$L = \sum a_{ij}(x) \frac{\partial^2}{\partial x_i\, \partial x_j} + \sum b_i(x) \frac{\partial}{\partial x_i} + c(x) \equiv M + c(x). \tag{1}$$

Thus, M denotes the terms of order 1 and 2. Thanks to the equality of mixed partials, we may suppose without loss of generality that $a_{ij} = a_{ji}$. For the first result, we suppose that

$$a_{ij}, b_i \in L^\infty(\bar{\Omega} : \mathbb{R}) \qquad \text{and} \qquad a_{ij} = a_{ji}, \tag{2}$$

$$\sum a_{ij}(x)\xi_i\xi_j \geq 0, \qquad \text{for all} \quad \xi \in \mathbb{R}^d, \quad x \in \bar{\Omega}. \tag{3}$$

Proposition 1. *Suppose that* (1), (2), *and* (3) *hold and that* $\Omega \subset \mathbb{R}^d$ *is open. Then, if* $u \in C^2(\Omega)$ *satisfies* $Mu > 0$, *then* u *cannot have a local maximum in* Ω.

PROOF. If u had a local maximum at $\hat{x}$ the first derivatives of u would vanish at $\hat{x}$ and the matrix $[\partial^2 u(\hat{x})/\partial x_i \, \partial x_j]$ of second derivatives would be negative semidefinite. We then use the following algebraic lemma.

Lemma 2. *If* A_{ij} *and* B_{ij} *are both positive semidefinite symmetric matrices, then* $\sum_{i,j} A_{ij} B_{ij} \geq 0$.

PROOF. The key observation is that

$$\sum_j A_{ij} B_{ij} = (AB^t)_{i,i} = (AB)_{i,i},$$

so $\sum_{i,j} A_{ij} B_{ij} = \mathrm{tr}(AB)$.

Choose an orthogonal transformation, $\mathcal{O}$, so that $\mathcal{O} A \mathcal{O}^{-1} = \mathrm{diag}(\lambda_i)$ with $\lambda_i \geq 0$. Then

$$\mathrm{tr}(AB) = \mathrm{tr}(\mathcal{O} A \mathcal{O}^{-1} \mathcal{O} B \mathcal{O}^{-1}) = \sum \lambda_i \beta_i,$$

where β_i is the ith diagonal element of $\mathcal{O} B \mathcal{O}^{-1}$. Then $\beta_i \geq 0$ since $\mathcal{O} B \mathcal{O}^{-1}$ is positive semidefinite. □

Returning to the proof of Proposition 1, we see that the lemma implies that $Mu(\hat{x}) \leq 0$ if u has a local maximum at $\hat{x}$. □

Corollary 3. *Suppose that* (1), (2), *and* (3) *hold and that* $\Omega \subset \mathbb{R}^d$ *is a bounded open set. If* $u \in C^2(\Omega) \cap C(\bar{\Omega})$ *satisfies* $Mu > 0$, *then* $\max_\Omega (u) \leq \max_{\partial\Omega} (u)$.

The next results and proofs are due to E. Hopf. We follow the exposition of Bers, John, and Schechter [BJS]. For the remainder of the section hypothesis (3) is strengthened to

$$a_{ij} \in C(\bar{\Omega}) \text{ and } \exists \mu > 0, \ \sum a_{ij}(x) \xi_i \xi_j \geq \mu |\xi|^2, \quad \text{for all} \quad x \in \bar{\Omega}, \quad \xi \in \mathbb{R}^d. \tag{4}$$

This implies that M is uniformly elliptic. □

Theorem 4. *Suppose that* (1), (2), *and* (4) *hold and that* Ω *is a connected open subset of* $\mathbb{R}^d$. *If* $u \in C^2(\Omega) \cap C(\bar{\Omega})$ *satisfies* $Mu \geq 0$ *in* Ω *and there is an* $\hat{x} \in \Omega$ *such that* $u(\hat{x}) \geq u(x)$ *for all* $x \in \Omega$, *then* u *is constant.*

This theorem asserts that if u is not constant then u achieves its maximum value in $\bar{\Omega}$ on $\partial\Omega$ and not in the interior. The next result shows that if $\hat{x}$ is a boundary point where the maximum is attained, then the outward normal derivative at $\hat{x}$ is strictly positive. The fact that $u(\hat{x})$ is maximal implies that the derivative is nonnegative.

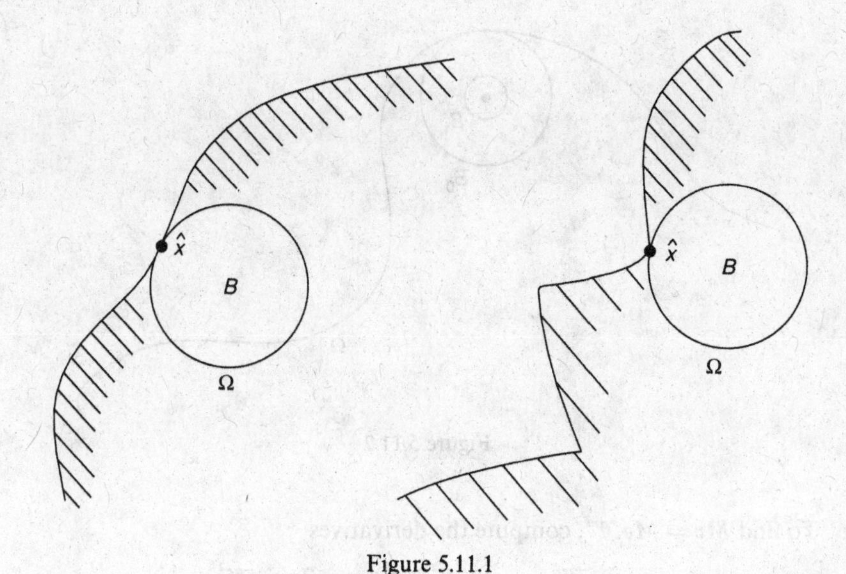

Figure 5.11.1

Theorem 5. *Suppose that $u \in C^2(\Omega) \cap C^1(\bar{\Omega})$ satisfies $Mu \geq 0$ in Ω and attains its maximum value at a point $\hat{x} \in \partial\Omega$ and there is a Euclidean ball, $B \subset \Omega$ with $\bar{B} \cap \partial\Omega = \hat{x}$, then either u is identically constant or $\partial u(\hat{x})/\partial n > 0$.*

Figure 5.11.1(a) shows that if Ω is regular at $\hat{x}$ according to the definition in §5.9, then there is such a Euclidean ball. The derivative $\partial/\partial n$ in the conclusion is the directional derivative in the direction from the center of the ball to $\hat{x}$. This agrees with the standard definitions when Ω is regular at $\hat{x}$. Figure 5.11.1(b) shows that Ω can have inward pointing irregularities and still satisfy the hypothesis. Note that when Ω is irregular, as in Figure 5.11.1(b), choosing different balls yields a cone of outward normal directions and u must be strictly increasing in all of those directions.

Proof of a Weakened Version of Theorem 5. We first prove Theorem 5 under the additional hypothesis

$$u(x) < u(\hat{x}), \qquad \text{for all} \quad x \in \Omega. \tag{5}$$

Choose concentric balls $B_1 \subset\subset B_0 \subset \Omega$ such that $\overline{B_0} \cap \partial\Omega = \hat{x}$ (Figure 5.11.2). Translate coordinates so that the center of the balls is the origin and let $r \equiv |x|$.

Consider the function

$$v(x) \equiv e^{-\alpha r^2} - e^{-\alpha (r_0)^2}, \qquad \alpha > 0, \quad r_0 \equiv \text{radius of } B_0.$$

Then

$$v = 0 \quad \text{on } \partial B_0, \qquad v > 0 \quad \text{in } B_0, \qquad \text{and} \qquad \frac{\partial v}{\partial n} < 0 \quad \text{on } \partial B_0.$$

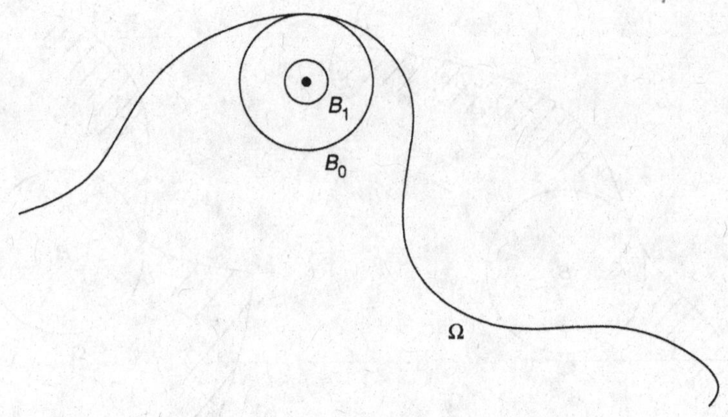

Figure 5.11.2

To find $Mv = Me^{-\alpha r^2}$, compute the derivatives

$$\partial_i e^{-\alpha r^2} = -\alpha e^{-\alpha r^2}\partial_i(r^2) = -2\alpha x_i e^{-\alpha r^2},$$

and

$$\partial_i \partial_j e^{-\alpha r^2} = +4\alpha^2 x_i x_j e^{-\alpha r^2} - 2\delta_{ij}\alpha e^{-\alpha r^2}.$$

Thus

$$Mv \geq (4\alpha^2(\sum a_{ij}x_i x_j) - C\alpha)e^{-\alpha r^2}.$$

Since the sum is at least as large as $\mu|x|^2$, which is bounded below on the complement of B_1, it follows the $Mv > 0$ in $\bar{B}_0\backslash B_1$ if α is sufficiently large. Fix such an α.

Since u is strictly less than $u(\hat{x})$ on ∂B_1 we may choose $\varepsilon > 0$ so that

$$\varepsilon v + u < u(\hat{x}) \quad \text{on } \partial B_1.$$

Then hypothesis (5) shows that $\max(\varepsilon v + u)$ occurs in $B_0\backslash \bar{B}_1$ or at $\hat{x}$.

Since $M(\varepsilon v + u) > Mu \geq 0$ in $\bar{B}_0\backslash B_1$, Proposition 1 implies that the maximum of $\varepsilon v + u$ in $\bar{B}_0\backslash B_1$ must occur on the boundary.

The maximum at the boundary must occur at $\hat{x}$, so $\partial/\partial n(\varepsilon v + u)(\hat{x}) \geq 0$. Thus at $\hat{x}$, $\partial u/\partial n \geq -\varepsilon\, \partial v/\partial n > 0$. □

Proof of Theorem 4 from the Weakened Theorem 5. If m is the maximum of u on $\bar{\Omega}$ and $u(x_0) = m$ for some $x_0 \in \Omega$, we must show that u is constant. Let $S \equiv \{x \in \Omega: u(x) = m\} \ni x_0$. Then S is closed since u is continuous. It suffices to prove that S is open.

If $x_1 \in S$ let $d \equiv \operatorname{dist}(x_1, \partial\Omega)$. It suffices to show that $B(x_1, d/2) \subset S$. If $x_2 \in B(x_1, d/2)$, then

$$\delta \equiv \operatorname{dist}(x_2, S) \leq \operatorname{dist}(x_2, x_1) = \frac{d}{2}.$$

It suffices to show that $\delta = 0$.

If $\delta > 0$, consider the restriction of u to $\bar{B}(x_2, \delta)$. Since $\bar{B}$ is contained in the interior of Ω, u is twice differentiable on the closure of B. By definition of δ there is an $\hat{x} \in \partial B(x_2, \delta) \cap S$, so $u(\hat{x}) = m$ and $u < m$ on $B(x_2, \delta)$. By the weakened form of Theorem 4, $\partial u/\partial v > 0$ at $\hat{x}$. However, $\hat{x}$ is an interior point of Ω at which u is maximal so $\nabla u(\hat{x}) = 0$. This contradiction shows that δ must vanish. □

PROOF OF THEOREM 5. Theorem 4 implies that if u is not constant, then $u|_\Omega$ is strictly less than $\max\{u(x): x \in \bar{\Omega}\}$. The hypotheses of the weakened Theorem 5 are satisfied and it follows that $\partial u(\hat{x})/\partial v > 0$. □

Theorem 5 required u to be differentiable on the closure of Ω. However, the proof works for $u \in C^2(\Omega) \cap C(\bar{\Omega})$, provided that the conclusion is interpreted in the sense

$$0 < \left(\frac{\partial}{\partial n}\right)^{-} u(\hat{x}) \equiv \liminf_{h \to 0-} \frac{u(\hat{x} + hn) - u(\hat{x})}{h}.$$

The maximum principles have an enormous variety of applications. Several are given below. The book of Protter and Weinberger [PW] is devoted to them and is highly recommended as easy and enjoyable reading.

Our first application is a comparison theorem. Note that terms of order 0 are allowed provided that they have the right sign. To keep track of the signs remember that M, with positive definite coefficient matrix like Δ, acts like a negative operator. The differential operator (1), with

$$c \in C(\bar{\Omega}) \quad \text{and} \quad c(x) \le 0 \quad \text{for all} \quad x \in \bar{\Omega}, \tag{6}$$

is then expected to be negative. This is borne out both by the sign of the eigenvalues of L as part of a self-adjoint boundary value problem and the next theorem.

Theorem 6 (Comparison Theorem). *Suppose that hypotheses* (1), (2), (4), *and* (6) *hold. If $u, v \in C^2(\Omega) \cap C(\bar{\Omega})$ satisfy $Lu \le Lv$ and $u \ge v$ on $\partial\Omega$, then $u \ge v$ in $\bar{\Omega}$.*

PROOF. If $\omega \equiv \{x \in \Omega: u < v\}$, it suffices to show that ω is empty. In ω

$$M(u - v) = L(u - v) - c(u - v) \le 0 \qquad \text{and} \qquad u - v \ge 0 \quad \text{on } \partial\omega.$$

If ω is nonempty, Theorem 5 implies that $u - v \ge 0$ in ω contradicting the definition. □

EXAMPLES/APPLICATIONS. 1. If $\Delta u \le 0$ in Ω and $u \ge 0$ on $\partial\Omega$, then $u \ge 0$ in Ω.
Proof. $v = 0$ is a lower comparison function.

In this way we recover Corollary 5.10.4. A physical interpretation of this result, complementing the one after Corollary 5.10.4, is the following. The equilibrium temperature in a domain Ω, subject to a nonnegative time-independent heat source and kept at nonnegative temperature at the boundary, is strictly positive in the interior.

Let $u_+ := \max\{u, 0\}$ and $u_- := \min\{u, 0\}$. Then using $\min u_-$ as a lower comparison and $\max u_+$ as a upper comparison yields the following corollary.

Corollary 7. *If hypotheses* (1), (2), (4), *and* (6) *hold and* $u \in C^2(\Omega) \cap C(\bar{\Omega})$ *satisfies* $Lu = 0$, *then for all* x *in* $\bar{\Omega}$,

$$\min_{\partial\Omega} u_- \leq u(x) \leq \max_{\partial\Omega} u_+. \tag{7}$$

2. The Fredholm alternative proved in §5.6 shows that to prove the solvability of Dirichlet's problem it is sufficient to prove a uniqueness theorem. The maximum principle is one of the best tools for that purpose.

For example, if $c \leq 0$ in Ω, $Lu = 0$ in Ω, and $u = 0$ on $\partial\Omega$, then Corollary 7 applies and shows that $u \equiv 0$, proving the desired uniqueness and therefore unique solvability of the Dirichlet problem, provided $c \leq 0$.

3. In the same way, uniqueness of solutions of the Neumann problem follows from Theorem 5. Suppose again that (6) is satisfied and that $Lu = 0$ in Ω and $\partial u/\partial n = 0$ at the boundary. Then a nonconstant u cannot have a positive maximum at the boundary, since in a neighborhood of such a point one would have $Mu = Lu - cu \geq 0$, so Theorem 5 would imply that $\partial u/\partial n > 0$ at such a maximum point. Applying the same reasoning to $-u$ shows that a nonconstant solution u cannot have a negative minimum at the boundary. The conclusion is that u is constant. There can be nonzero constant solutions only if $c(x)$ is identically zero. This proves uniqueness when $c \leq 0$ and is not identically zero. The Fredholm alternative then proves solvability. *The Neumann problem is uniquely solvable provided* $c(x) \leq 0$ *and* $c(x)$ *is not identically equal to zero.* If $c \leq 0$ and $a \geq 0$ are not both identically zero, the same argument proves uniqueness for the *Robin Problem* with boundary condition $\partial u/\partial n + a(x)u = 0$.

4. If u is a solution of the nonlinear Dirichlet problem

$$\Delta u - u^3 = f \leq 0 \quad \text{in } \Omega \qquad \text{and} \qquad u \geq 0 \quad \text{on } \partial\Omega,$$

then $u > 0$ in Ω.
Proof. Let $c(x) = -u^2(x) \leq 0$ and apply the Comparison Theorem with $v = 0$.

5. Find upper and lower bounds for the solution u of the Dirichlet problem

$$a(x)u - \Delta u = f \quad \text{in } \Omega, \qquad u = g \quad \text{on } \partial\Omega,$$

where $A \geq a(x) \geq a > 0$.
Solution. To estimate u from above use a comparison function w which is constant, $w \equiv \beta \geq 0$. Then

$$a(x)w - Aw \geq a\beta \quad \text{in } \Omega.$$

If $\beta \geq \sup f_+/a$ and $\beta \geq \sup g_+$, the Comparison Theorem shows that $u \leq w$.

Thus

$$u \leq \max\left\{\frac{\sup(f_+)}{a}, \sup g_+\right\}.$$

Similarly,

$$\min\left\{\frac{\inf(f_-)}{a}, \inf g\right\} \leq u.$$

In particular,

$$\|u\|_{L^\infty} \leq \min\{a^{-1}\|f\|_{L^\infty(\Omega)}, \|g\|_{L^\infty}\}.$$

6. Derive analogous estimates when $a(x) \geq 0$ but is not necessarily strictly positive. Here the comparison functions are more tricky. Use constants to dominate the boundary values. For the inhomogeneous term, f, we take advantage of the fact that $\Delta|x|^2 = 2d$.

Choose $x_0 \in \Omega$ and $R > 0$ so that $\Omega \subset B_R(x_0)$. Let

$$w \equiv \alpha(R^2 - |x - x_0|^2) + \beta, \qquad \alpha \geq 0, \quad \beta \geq 0.$$

Take $\beta = \sup g_+$, so that $v \geq u$ on $\partial\Omega$. Then

$$(a(x) - \Delta)w = a(x)w + 2\alpha d \geq 2\alpha d,$$

so if $\alpha = \sup(f_+)/2d$, the Comparison Theorem shows that $w \geq u$. In this way, we find

$$\inf(g_-) + \frac{\inf(f_-)}{2d}(R^2 - |x - x_0|^2) \leq u \leq \sup(g_+) + \frac{\sup(f_+)}{2d}(R^2 - |x - x_0|^2).$$

7. The example

$$\frac{d^2}{dx^2}u + u = 0, \qquad u = \sin x, \qquad \Omega =]0, \pi[,$$

which has a positive strict local maximum at $x = \pi/2$, shows that the hypothesis $c \leq 0$ cannot be dropped from Corollary 7.

8. The example

$$\frac{d^4}{dx^4}u = 0, \qquad u = x(x - 1)(x + 1), \qquad \Omega =]0, 1[,$$

which has a positive strict local maximum at $x = \frac{1}{2}$, shows that the theory of this section does not extend to higher-order elliptic operators.

9. The solution $u = \cosh x$ of $d^2u/dx^2 - u = 0$ with $u(\pm 1) = \cosh 1$ and $\Omega :=]-1, 1[$, shows that the conclusion (7) cannot be strengthened to $\min_{\partial\Omega} u \leq u(x) \leq \max_{\partial\Omega} u$.

Problems

The maximum principle is one of the most incisive tools for studying nonlinear problems. Here we give three standard applications.

1. **Theorem.** *If f is a bounded continuous function on $\mathbb{R}$, there is a constant c depending only on $\|f\|_{L^\infty(\mathbb{R})}$ and Ω, so that any $u \in C^2(\Omega) \cap C(\bar{\Omega})$ satisfying $\Delta u + f(u) = 0$ in Ω and $u = 0$ on $\partial\Omega$ satisfies $\|u\|_{L^\infty(\Omega)} \le c$.*

2. **Theorem.** *If $f: \mathbb{R} \to \mathbb{R}$ is nondecreasing and continuously differentiable and u, $v \in C^2(\Omega) \cap C(\bar{\Omega})$ satisfy*

$$-\Delta u + f(u) \ge -\Delta v + f(v) \quad \text{in } \Omega, \qquad u \ge v \quad \text{on } \partial\Omega,$$

then $u \ge v$ in Ω. If equality holds at an interior point, then $u \equiv v$.

Hint. Subtract the equations for u and v and write $f(u) - f(v) = c(x)(u - v)$ with

$$c(x) = \int_0^1 f'(v(x) + \theta(u(x) - v(x)))\, d\theta.$$

DISCUSSION. This idea of subtracting then using the mean value theorem to get a linear equation for the difference of two solutions to nonlinear problems is surprisingly useful. It also appears in the hint for Problem 3.6.4. Note that the coefficients of the linear problem depend on u. A related method is to differentiate nonlinear equations to obtain linear equations for the derivatives of the unknown. The coefficients of the resulting linear equation depend on the unknown function.

3. Prove that if $u \in C^\infty(\Omega) \cap C(\bar{\Omega})$ satisfies the equation of constant mean curvature H (see Problem 5.1.4) with $H \le 0$, and u attains its maximum or minimum value at an interior point, then u is constant. In particular, minimal surfaces cannot have interior local extrema.

DISCUSSION. Many other applications to quasi-linear problems from mechanics and geometry can be found in the books of Protter and Weinberger [PW] and Gilbarg and Trudinger [GT].

APPENDIX

A Crash Course in Distribution Theory

This appendix presents some of the elementary notions of distribution theory. Beginning in Chapter 2 these ideas will be used extensively. More detailed brief introductions can be found in [H1, pp. 1–17], [R, pp. 135–162], and [Sc2, pp. 71–140]. More complete treatments are [Don, Sc1, GS, H2], though that level of coverage is not needed.

The Theory of Distributions grew from many disparate sources. One is the treatment of impulsive forces. Newton's second law asserts that the rate of change of momentum is equal to the force applied, $dp/dt = F$. Consider an intense force which acts over a very short interval of time $\underline{t} < t < \underline{t} + \Delta t$. An example is the force exerted by the strike of a hammer. The impulse, I, is defined as $I \equiv \int F(t)\, dt$ so

$$p(\underline{t} + \Delta t) = p(\underline{t}) + I.$$

The exact shape of $F(t)$ does not enter. In the limit, as Δt tends to zero, we arrive at an idealized force which acts instantaneously to give rise to a jump I in the momentum p. Formally, the force law satisfies

$$F = 0 \quad \text{for } t \neq 0 \qquad \text{and} \qquad \int F(t)\, dt = I. \tag{1}$$

This idealized impulsive force is denoted $I\delta_{\underline{t}}$, and $\delta_{\underline{t}}$ is called *Dirac's delta function* though no function can satisfy (1). The idealized equation of motion is $dp/dt = I\delta_{\underline{t}}$. The solution satisfies $p(t+) - p(t-) = I$. Such idealizations are quite successful in a variety of problems of mechanics and electricity.

The mathematical framework developed by L. Schwartz in the 1940s has an additional motivation from mechanics. It has long been realized that if $u(x)$ is a physical observable that depends on x, then it is impossible to measure point values of u since any measuring device has finite size. This is true even in classical physics but the problem becomes more serious in Quantum Field Theory where the point values of the field may not exist. The Ehrenfests suggested that smeared or averaged fields, $\int u(x)\varphi(x)\,dx$, with regular weight functions φ are well defined and meaningful. Both observations about the measurement process suggest the importance of averages $\int u(x)\varphi(x)\,dx$. The observable u is "observed" as a linear functional

$$\varphi \mapsto \int u(x)\varphi(x)\,dx. \tag{2}$$

An example is the impulsive force field $F = I\delta_{\underline{t}}$ of (1) which corresponds to the functional

$$\varphi \mapsto I\varphi(\underline{t}) = \text{“}\int F(t)\varphi(t)\,dt\text{”}.$$

This example suggests that the test functions φ must at least be continuous. Examples of dipoles and more general multipoles, $\varphi \mapsto \partial^\alpha\varphi(\underline{x})$, suggest that the test functions be infinitely differentiable to permit commonly considered observables. In order for observables which are large at infinity to give finite answers, it is reasonable to impose that the test functions be required to have compact support, that is, $\operatorname{supp}(\varphi) \equiv \operatorname{cl}\{x: \varphi(x) \neq 0\}$ is compact. In summary, we are led to the idea that an observable u on an open subset $\Omega \subset \mathbb{R}^d$ is a linear functional on $C_0^\infty(\Omega)$. Following Schwartz, $C_0^\infty(\Omega)$ is denoted $\mathscr{D}(\Omega)$ and $C^\infty(\Omega)$ is denoted $\mathscr{E}(\Omega)$.

Only those functionals which give nearby results for nearby tests can represent reproducible measurements. Thus we are led to assume that the linear functionals are continuous.

Definition. A distribution on an open $\Omega \subset \mathbb{R}^d$ is a linear map $l\colon \mathscr{D}(\Omega) \to \mathbb{C}$, which is continuous in the sense that if $\{\varphi_n\} \subset \mathscr{D}(\Omega)$ satisfies

$$\text{there is a compact } K \subset \Omega \text{ such that for all } n,\ \operatorname{supp}(\varphi_n) \subset K, \tag{3}$$

and

$$\text{there is a } \varphi \in \mathscr{D}(\Omega) \text{ such that for all } \alpha \in \mathbb{N}^d,\ \partial^\alpha\varphi_n \text{ converges uniformly to } \partial^\alpha\varphi, \tag{4}$$

then $l(\varphi_n) \to l(\varphi)$. The set of all distributions on Ω is denoted $\mathscr{D}'(\Omega)$. When φ_n, φ satisfy (3), (4) we say that φ_n converges to φ in $\mathscr{D}(\Omega)$.

The action of a distribution $l \in \mathscr{D}'(\Omega)$ on a test function $\varphi \in \mathscr{D}(\Omega)$ will often be denoted $\langle l, \varphi\rangle$. The set $\mathscr{D}'(\Omega)$ is clearly a complex vector space.

EXAMPLES. 1. If $u \in L^1_{\text{loc}}(\Omega)$, then there is a natural distribution l_u defined by $\langle l_u, \varphi\rangle \equiv \int u(x)\varphi(x)\,dx$. In this sense, the distributions are generalizations of functions and are sometimes called *generalized functions.* Two locally integrable functions define the same distribution if and only if the functions are equal almost everywhere. We say that a distribution l is a locally integrable function and write $l \in L^1_{\text{loc}}(\Omega)$ if $l = l_u$ for some $u \in L^1_{\text{loc}}(\Omega)$. Similarly, we say that l is a continuous (resp. C^∞) function if $l = l_u$ for a $u \in C(\Omega)$ (resp. $C^\infty(\Omega)$). For example, the distribution defined by the characteristic function of the irrationals, $\chi_{\mathbb{R}\setminus\mathbb{Q}}$, is a C^∞ function. In fact, this distribution is equal to the function 1.

2. If μ is a Radon measure on Ω, then $\langle l_\mu, \varphi\rangle \equiv \int \varphi(x)\,d\mu(x)$ defines a distribution. If M is an embedded k-dimensional submanifold in Ω and $d\sigma$ is the k-dimensional area in M, then $\varphi \mapsto \int_M \varphi\,d\sigma$ is a distribution.

3, 4. If $\underline{x} \in \Omega$, then $\langle l, \varphi\rangle \equiv \varphi(\underline{x})$ is a distribution denoted $\delta_{\underline{x}}$ and called the Dirac delta at $\underline{x}$. When $\underline{x}$ is not mentioned it is assumed to be the origin. More generally, $\langle l, \varphi\rangle \equiv \partial^\alpha\varphi(\underline{x})$ is a distribution.

The fact that an infinite number of derivatives are needed in (4) might seem excessive. In fact, for any fixed distribution and compact $K \subset \Omega$, a finite number suffice.

Proposition 1. *A linear map $l\colon \mathscr{D}(\Omega) \to \mathbb{C}$ belongs to $\mathscr{D}'(\Omega)$ if and only if for every compact subset $K \subset \Omega$ there is an integer $n(K, l)$ and a $c \in \mathbb{R}$ such that for all $\varphi \in \mathscr{D}(\Omega)$ with support in K*

$$|\langle l, \varphi\rangle| \le c\|\varphi\|_{C^n}, \qquad \|\varphi\|_{C^n} \equiv \sum_{|\alpha|\le n} \max|\partial^\alpha\varphi|. \tag{5}$$

Proof. The "if" part is clear. To prove "only if" suppose that (5) is violated for a compact K. For each integer n, choose $\varphi_n \in \mathscr{D}(\Omega)$ with support in K such that

$$|\langle l, \varphi_n\rangle| > 1 \qquad \text{and} \qquad \|\varphi_n\|_{C^n} < 1/n.$$

Then φ_n satisfy (3), (4) with $\varphi = 0$, but $\langle l, \varphi_n\rangle$ does not converge to zero so l is not a distribution. □

Definitions. If there is a c such that (5) holds, l is said to be of order n on K. If l is of order n on every compact $K \subset \Omega$, then that l is of order n on Ω. The set of distributions of order n is naturally identified with the dual of $C^n_0(\Omega)$. It is sometimes denoted $C^{-n}(\Omega)$. A distribution is of *finite order* on Ω if it is of order n on Ω for some n.

Examples. 1–4. The examples above are of order 0, 0, 0, and $|\alpha|$.

5. The functional on $\mathscr{D}'(\mathbb{R})$, defined by

$$l(\varphi) \equiv \lim_{\varepsilon\to 0} \int_{|x|>\varepsilon} \frac{\varphi(x)}{x}\,dx \equiv \text{P.V.} \int \frac{\varphi(x)}{x}\,dx,$$

is called the *principal value of* $1/x$. It is a distribution of order 1 on $\mathbb{R}$ (Problem 14).

6. The functional

$$l(\varphi) \equiv \sum \left(\frac{d}{dx}\right)^n \varphi\left(\frac{1}{n}\right)$$

defines a distribution on $]0, 1[$ which is not of finite order. It follows that there is no distribution on $\mathbb{R}$ whose restriction to $\mathscr{D}(]0, 1[)$ is equal to l.

Definition. *A sequence of distributions $l_n \in \mathscr{D}'(\Omega)$ converges to $l \in \mathscr{D}'(\Omega)$ if and only if for every test function $\varphi \in \mathscr{D}(\Omega)$, $l_n(\varphi) \to l(\varphi)$. This convergence is denoted $l_n \rightharpoonup l$ or $\mathscr{D}'\text{-}\lim l_n = l$.*

EXAMPLES. 1. If $j \in \mathscr{D}(\mathbb{R}^d)$ with $\int j\, dx = 1$, let $j_\varepsilon(x) = \varepsilon^{-n} j(x/\varepsilon)$. Then $j_\varepsilon \rightharpoonup \delta_0$. A more general result is proved in Proposition 2.2.3.

2. We will show in Proposition 4 that every distribution is the $\mathscr{D}'(\Omega)$ limit of a sequence of elements in $\mathscr{D}(\Omega)$. This suggests again the interpretation as generalized functions.

3. The definition of P.V.$(1/x)$ shows that $\mathscr{D}'\text{-}\lim \chi_{n|x|>1}(1/x) = \text{P.V.}(1/x)$.

The great utility of distributions rests largely on the fact that the standard operations of calculus extend to $\mathscr{D}'(\Omega)$. In particular, one can differentiate distributions. For the study of differential equations that fact is particularly important.

The recipe for defining operations on distributions is nearly always the same: pass the operator onto the test function. For example, for $l \in \mathscr{D}'(\mathbb{R}^d)$ the translate of l by the vector y, denoted $\tau_y l$, is defined as follows. If l were equal to the function u, then

$$\langle \tau_y l, \varphi \rangle = \int u(x-y)\varphi(x)\, dx = \int u(z)\varphi(z+y)\, dz = \langle l, \tau_{-y}\varphi \rangle.$$

This motivates the definition, $\langle \tau_y l, \varphi \rangle \equiv \langle l, \tau_{-y}\varphi \rangle$. It is easy to verify that $\tau_y l$ so defined is a distribution and that the definition agrees with $\tau_y u$ when $l = l_u$.

To differentiate a distribution l on $\mathbb{R}^d$, form the difference quotients which should converge to $\partial l/\partial x_j$. Let $e_j \equiv (0, \ldots, 0, 1, 0, \ldots, 0)$ be the jth standard basis element in $\mathbb{R}^d$. The difference quotients are given by

$$\left\langle \frac{\tau_{-he_j} l - l}{h}, \varphi \right\rangle \equiv \left\langle l, \frac{\tau_{he_j}\varphi - \varphi}{h} \right\rangle. \tag{6}$$

The test functions on the right converge to $-\partial\varphi/\partial x_j$, so the continuity of l implies that the right-hand side of (6) converges to $\langle l, -\partial\varphi/\partial x_j \rangle$. This suggests that $\langle \partial l/\partial x_j, \varphi \rangle$ be defined by $\langle l, -\partial\varphi/\partial x_j \rangle$. This defines a distribution and if $u \in C^1(\Omega)$ and $l = l_u$, then the derivatives of l are equal to the distributions $l_{\partial u/\partial x_j}$. Thus the operator $\partial/\partial x_j$ on $\mathscr{D}'$ is an extension of $\partial/\partial x_j$ on $\mathscr{D}$.

As an example consider $H(x) \equiv \chi_{[0,\infty[}(x)$, the *Heaviside function* on $\mathbb{R}$. The difference quotient $(\tau_{-he_j} H - H)/h$ is equal to the function $h^{-1}\chi_{[0,h[}$ which converges to δ in the sense of distributions. Thus $dH/dx = \delta$. Note that the difference quotients converge to zero almost everywhere. Since H is not a constant, zero is surely not the desired derivative. The pointwise limit gives the wrong answer and the distribution derivative is the right answer.

The operations on distributions discussed above are special cases of a general algorithm. The following version appears in unpublished notes of P.D. Lax.

Proposition 2. *Suppose that L is a linear map from $\mathscr{D}(\Omega_1)$ to $\mathscr{D}(\Omega_2)$, which is sequentially continuous in the sense that $\varphi_n \to \varphi$ implies $L(\varphi_n) \to L(\varphi)$. Suppose, in addition, that there is an operator L', sequentially continuous from $\mathscr{D}(\Omega_2)$ to $\mathscr{D}(\Omega_1)$, which is the transpose of L in the sense that $\langle L(\varphi), \psi\rangle = \langle \varphi, L'(\psi)\rangle$ for all $\varphi \in \mathscr{D}(\Omega_1)$, $\psi \in \mathscr{D}(\Omega_2)$. Then the operator L extends to a sequentially continuous map of $\mathscr{D}'(\Omega_1)$ to $\mathscr{D}'(\Omega_2)$ given by*

$$\langle L(l), \psi\rangle \equiv \langle l, L'(\psi)\rangle \quad \text{for all} \quad l \in \mathscr{D}'(\Omega_1) \quad \text{and} \quad \psi \in \mathscr{D}(\Omega_2). \tag{7}$$

The uniqueness of the extension is proved in Proposition 8.

PROOF. The sequential continuity of L' shows that $L(l)$ defined in (7) is a distribution. If $l = l_\varphi$ for some $\varphi \in \mathscr{D}(\Omega_1)$, then

$$\langle L(l), \psi\rangle \equiv \langle l, L'(\psi)\rangle = \int_{\Omega_1} \varphi(x)L'(\psi)(x)\,dx = \int_{\Omega_2} L(\varphi)(x)\psi(x)\,dx, \tag{8}$$

the last equality from the hypothesis that L' is the transpose of L. Thus $L(l)$ is the distribution associated to $L(\varphi)$ which proves that L defined by (7) extends $L|_{\mathscr{D}}$.

Finally, if $l_n \rightharpoonup l$ in $\mathscr{D}'(\Omega_1)$, it follows immediately from (7) that $L(l_n) \rightharpoonup L(l)$, proving the sequential continuity of L. □

EXAMPLES. 1. If $a(x) \in C^\infty(\Omega)$, $(\equiv \mathscr{E}(\Omega))$, then the map $L(\varphi) \equiv a\varphi$ is equal to its own transpose. This statement is equivalent to the identity

$$\langle L(\varphi), \psi\rangle = \int (a(x)\varphi(x))(\psi)\,dx = \int (\varphi(x))(a(x)\psi)\,dx = \langle \varphi, L(\psi)\rangle.$$

Thus for $l \in \mathscr{D}'(\Omega)$, al is a well-defined distribution given by $\langle al, \varphi\rangle \equiv \langle l, a\varphi\rangle$.

2. If $\Omega_2 = y + \Omega_1$ and $L = \tau_y$ is translation by y, then $L' = \tau_{-y}$ is sequentially continuous, so for $l \in \mathscr{D}'(\Omega_1)$ the translates of l are well defined by $\langle \tau_y l, \varphi\rangle \equiv \langle l, \tau_{-y}\varphi\rangle$. Similarly, the reflection operator $(\mathscr{R}u)(x) \equiv u(-x)$ is its own transpose, so $\mathscr{R}l$ is a well-defined distribution on the reflection of Ω.

3. If $L = \partial^\alpha$ (see §1.3 for this notation), then integration by parts shows that the transpose is $L' = (-1)^{|\alpha|}\partial^\alpha$ which is sequentially continuous on $\mathscr{D}$, so the derivatives of distributions are defined by

$$\langle \partial^\alpha l, \varphi\rangle \equiv \langle l, (-1)^{|\alpha|}\partial^\alpha \varphi\rangle.$$

Notice that since $\partial_j\partial_k = \partial_k\partial_j$ on $\mathscr{D}$, it follows from the definition of distribution derivative that for any l, $\partial_j\partial_k l = \partial_k\partial_j l$.

Having defined multiplication and derivative we can compute a product rule for $\partial_j(al)$. Omitting the subscript we find

$$\langle \partial(al), \psi \rangle \equiv \langle l, -a\partial\psi \rangle = \langle l, -\partial(a\psi) \rangle + \langle l, (\partial a)\psi \rangle = \langle a\partial l + (\partial a)l, \psi \rangle.$$

Thus we have the familiar rule $\partial(al) = a\partial l + (\partial a)l$ and, by induction, the usual Leibniz formula for $\partial^\alpha(al)$ is valid

The derivatives combined with Example 2, show that if $P(x, D) = \sum a_\alpha(x)\partial^\alpha$ is a linear partial differential operator with coefficients in $\mathscr{E}(\Omega)$, then P maps $\mathscr{D}'(\Omega)$ to itself with $\langle Pl, \varphi \rangle \equiv \langle l, P'\varphi \rangle$. The transpose of P is given by

$$P'\psi = \sum (-1)^\alpha \partial^\alpha(a_\alpha \psi).$$

The transpose is also denoted P^t in this text.

4. If $\eta: \Omega_2 \to \Omega_1$ is a diffeomorphism and $L(\varphi) \equiv \varphi \circ \eta$, then L is sequentially continuous and

$$\langle L(\varphi), \psi \rangle = \int_{\Omega_2} \varphi(\eta(y))\psi(y)\,dy = \int_{\Omega_1} \varphi(x)\psi(\eta^{-1}(x)) \left|\frac{Dy}{Dx}\right| dx,$$

where $|Dy/Dx|$ is the Jacobian determinant of the transformation $y = \eta^{-1}(x)$. Thus the transpose of L is the map $\psi \mapsto \psi(\eta^{-1}(x))|Dy/Dx|$. Therefore, for any $l \in \mathscr{D}'(\Omega_2)$, $l \circ \eta$ is well defined by

$$\langle l \circ \eta, \psi \rangle \equiv \langle l, |\det D\eta^{-1}| \psi \circ \eta^{-1} \rangle.$$

This example is important if one wants to define distributions on a manifold.

5. **Convolution 1.** Suppose that $\Omega = \mathbb{R}^d$ and $\varphi \in \mathscr{D}(\mathbb{R}^d)$. Let L be the operator $L(\psi) = \varphi * \psi$. Leibniz' rule for differentiating under the integral implies that L maps $\mathscr{D}(\mathbb{R}^d)$ continuously to itself. Fubini's Theorem shows that the transpose of L is convolution with $\mathscr{R}\varphi$ (exercise). Thus $\varphi * l$ makes sense for any $l \in \mathscr{D}'(\mathbb{R}^d)$ and is given by $\langle \varphi * l, \psi \rangle \equiv \langle l, (\mathscr{R}\varphi) * \psi \rangle$. As an example, we compute $\varphi * \delta$,

$$\langle \varphi * \delta, \psi \rangle \equiv \langle \delta, (\mathscr{R}\varphi) * \psi \rangle = ((\mathscr{R}\varphi) * \psi)(0) = \int \varphi(y)\psi(y)\,dv = \langle \varphi, \psi \rangle.$$

Thus $\varphi * \delta = \varphi$. The definitions yield

$$\langle \partial^\alpha(\varphi * l), \psi \rangle \equiv \langle \varphi * l, (-\partial)^\alpha \psi \rangle \equiv \langle l, (\mathscr{R}\varphi) * (-\partial)^\alpha \psi \rangle = \langle l, (-\partial)^\alpha((\mathscr{R}\varphi) * \psi) \rangle.$$

A similar sequence of computations shows that the last term is equal to $\langle \varphi * \partial^\alpha l, \psi \rangle$. On the other hand, applying the derivative in the last term to the $\mathscr{R}\varphi$ term of the convolution, and then unraveling, yields $\langle (\partial^\alpha \varphi) * l, \psi \rangle$. In this way, we prove that in the sense of distributions

$$\partial^\alpha(\varphi * l) = \varphi * \partial^\alpha l = (\partial^\alpha \varphi) * l.$$

Applied when $l = \delta$, we find that $\varphi * \partial^\alpha \delta = \partial^\alpha \varphi$.

Convolution on the right, $L(\psi) = \psi * \varphi$, is equal to $\varphi * \psi$ and also extends. In particular, $l * \varphi = \varphi * l$.

Proposition 3. *If $l \in \mathscr{D}'(\mathbb{R}^d)$ and $\varphi \in \mathscr{D}(\mathbb{R}^d)$, then $l * \varphi$ is equal to the C^∞ function whose value at x is $\langle l, \tau_x(\mathscr{R}\varphi)\rangle$.*

PROOF. First observe that if $x_n \to x$, then $\tau_{x_n}(\mathscr{R}\varphi) \to \tau_x(\mathscr{R}\varphi)$ in $\mathscr{D}(\mathbb{R}^d)$, which suffices to show that the function $\gamma(x) \equiv \langle l, \tau_x(\mathscr{R}\varphi)\rangle$ is continuous on $\mathbb{R}^d$.

Similarly, if Δ_j^h is the forward difference operator $(\tau_{-he_j} - I)/h$ approximating ∂_j, then

$$\Delta_j^h \gamma = \langle l, \tau_x \Delta_j^h(\mathscr{R}\varphi)\rangle,$$

and

$$\tau_x \Delta_j^h(\mathscr{R}\varphi) \to \tau_x \partial_j(\mathscr{R}\varphi) \quad \text{in } \mathscr{D}(\mathbb{R}^d) \quad \text{as} \quad h \to 0.$$

This suffices to show that $\gamma \in C^1(\mathbb{R}^d)$ and $\partial_j \gamma = \langle l, \tau_x \partial_j(\mathscr{R}\varphi)\rangle$. By induction on n it follows that for all $n \in \mathbb{N}$, $\gamma \in C^n(\mathbb{R}^d)$, and $\partial^\alpha \gamma = \langle l, \tau_x \partial^\alpha(\mathscr{R}\varphi)\rangle$ for all $|\alpha| \le n$.

Next we show that the distribution defined by γ is equal to the convolution $l * \varphi$. Toward that end, write $\langle \gamma, \psi\rangle$ as a limit of Riemann sums

$$\langle \gamma, \psi\rangle = \lim_{n\to\infty} \sum_{\alpha \in \mathbb{Z}^d} \psi\left(\frac{\alpha}{n}\right) \langle l, \tau_{\alpha/n}(\mathscr{R}\varphi)\rangle n^{-d} = \lim_{n\to\infty} \sum_{\alpha \in \mathbb{Z}^d} \left\langle l, \psi\left(\frac{\alpha}{n}\right) \tau_{\alpha/n}(\mathscr{R}\varphi)\right\rangle n^{-d}.$$

Note that $n^{-d}\sum \psi(\alpha/n)\tau_{\alpha/n}(\mathscr{R}\varphi) \to (\mathscr{R}\varphi) * \psi$ in $\mathscr{D}(\mathbb{R}^d)$ (exercise), so the limit on the right is equal to $\langle l, (\mathscr{R}\varphi) * \psi\rangle = \langle \varphi * l, \psi\rangle$. Thus, $\gamma = \varphi * l$ as distributions. □

Proposition 4. *Suppose that $\chi, j \in \mathscr{D}(\mathbb{R}^d)$, $\int j(x)\,dx = 1$, $\chi(0) = 1$, $j_\varepsilon(x) \equiv \varepsilon^{-d} j(x/\varepsilon)$, and $\chi_\varepsilon(x) \equiv \chi(\varepsilon x)$. Then for any $l \in \mathscr{D}'(\mathbb{R}^d)$, $\chi_\varepsilon l$, $j_\varepsilon * l$, and $\chi_\varepsilon(j_\varepsilon * l)$ converge to l in $\mathscr{D}'(\mathbb{R}^d)$ as ε tends to zero. In particular, any such l is the limit in $\mathscr{D}'(\mathbb{R}^d)$ of elements of $\mathscr{D}(\mathbb{R}^d)$.*

The function χ_ε is a vast plateau of height very close to 1 over a diameter of order $1/\varepsilon$. Thus multiplication by χ_ε is nearly the identity operator. Convolution by j_ε is close to convolution with the Dirac delta which is the identity operator. These two approximation processes, plateau multiplication and convolution with an approximate delta, are simple but remarkably useful methods in analysis.

PROOF OF PROPOSITION 4. We treat only $\chi_\varepsilon(j_\varepsilon * l)$. The definitions yield $\langle \chi_\varepsilon(j_\varepsilon * l), \varphi\rangle \equiv \langle l, (\mathscr{R}j_\varepsilon) * (\chi_\varepsilon \varphi)\rangle$ for any $\varphi \in \mathscr{D}(\mathbb{R}^d)$. The result is then a consequence of the fact that $(\mathscr{R}j_\varepsilon) * (\chi_\varepsilon\varphi) \to \varphi$ in $\mathscr{D}(\mathbb{R}^d)$. The verification of that is an exercise in advanced calculus which is left to the reader. □

If l is a distribution on Ω and ω is an open subset of Ω, then *l is equal to zero on ω* means that for all $\varphi \in \mathscr{D}(\omega)$, $\langle l, \varphi\rangle = 0$.

Definition. The support of $l \in \mathscr{D}'(\Omega)$ is the complement of $\{x \in \Omega\colon l \text{ is equal to zero on a neighborhood of } x\}$. The support is denoted $\operatorname{supp}(l)$. The set of all $l \in \mathscr{D}'(\Omega)$ such that $\operatorname{supp}(l)$ is compact in Ω is denoted $\mathscr{E}'(\Omega)$.

EXAMPLES. 1. The support of every l is a closed subset of Ω.

2. $\operatorname{supp} \delta_{\underline{x}} = \{\underline{x}\}$.

3. If l is equal to a continuous function f, then $\operatorname{supp}(l) = \operatorname{cl}\{x: f(x) \neq 0\}$.

4. $\operatorname{supp}(\partial^\alpha l) \subset \operatorname{supp}(l)$.

Proposition 5. *If $l \in \mathscr{D}'(\Omega)$ and $\varphi \in \mathscr{D}(\Omega)$ have disjoint supports, then $\langle l, \varphi\rangle = 0$.*

PROOF. For each y in $\operatorname{supp}(\varphi)$ choose an open $\omega_y \subset \Omega$ containing y and on which l is equal to zero. Choose a nonnegative $h_y \in \mathscr{D}(\Omega)$ with $\operatorname{supp}(h_y) \subset \omega_y$ and $h_y(y) > 0$. The sets $\{x: h_y(x) > 0\}$ are an open cover of the compact set $\operatorname{supp}(\varphi)$. Thus there is a finite subcover.

Call the corresponding functions $h_1, \ldots, h_m$. Define $\psi_j \equiv \varphi h_j/(h_1 + \cdots + h_m)$ on $\bigcup \{h_j > 0\}$, and $\psi_j \equiv 0$ otherwise.

Since the sum of the h's is positive on $\operatorname{supp}(\varphi)$, ψ_j is smooth and is supported in the union of the ω_j containing $\operatorname{supp}(h_j)$.

Since l is equal to zero on ω_j, $\langle l, \psi_j\rangle = 0$.

Since $\varphi = \sum \psi_j$ the result follows. □

Recall that $\mathscr{E}(\Omega)$ is a complete metric space whose topology is defined as follows. Let $K_1 \subset K_2 \subset \cdots$ be an exhaustion of Ω by compact sets. For each n, define a seminorm $\|\cdot\|_n$ on $\mathscr{E}(\Omega)$ by

$$\|\varphi\|_n \equiv \sum_{|\alpha| \leq n} \max_{K_n} |\partial^\alpha \varphi|.$$

A metric for $\mathscr{E}(\Omega)$ is given by

$$\operatorname{dist}(\varphi, \psi) \equiv \sum \frac{2^{-n}\|\varphi - \psi\|_n}{1 + \|\varphi - \psi\|_n}.$$

A sequence converges in $\mathscr{E}(\Omega)$ if and only if each partial derivative converges uniformly on compact subsets of Ω. An argument like the proof of Proposition 1 shows that a linear functional $k: \mathscr{E}(\Omega) \to \mathbb{C}$ is continuous if and only if there is an n and a c such that for all $\varphi \in \mathscr{E}(\Omega)$, $|\langle k, \varphi\rangle| \leq c\|\varphi\|_n$. In particular, continuous linear functionals on $\mathscr{E}(\Omega)$ are distributions of compact support. The converse is also true.

Proposition 6

(i) *If a distribution $l \in \mathscr{D}'(\Omega)$ has compact support in Ω, then u has finite order.*
(ii) *$l \in \mathscr{D}'(\Omega)$ has compact support if and only if l extends uniquely to a continuous linear functional on $\mathscr{E}(\Omega)$.*

This result explains the notation $\mathscr{E}'(\Omega)$ for the distributions of compact support.

PROOF. (i) Given $l \in \mathscr{D}'(\Omega)$ with compact support, choose $\psi \in \mathscr{D}(\Omega)$ with ψ equal to 1 on a neighborhood of $\operatorname{supp}(l)$. Then Proposition 5 implies that for any $\varphi \in \mathscr{D}(\Omega)$, $\langle l, \varphi\rangle = \langle l, \psi\varphi\rangle$. Let $K = \operatorname{supp}(\psi)$, and choose $n(K, l)$ and c

such that (5) holds. Then, for any $\varphi \in \mathscr{D}(\Omega)$,

$$|\langle l, \varphi \rangle| = |\langle l, \psi\varphi \rangle| \leq c\|\psi\varphi\|_{C^n} \leq c'\|\varphi\|_{C^n}.$$

(ii) The map $\varphi \in \mathscr{E}(\Omega) \mapsto \langle l, \psi\varphi \rangle$ is then a continuous extension of l to $\mathscr{E}(\Omega)$. Since $\mathscr{D}(\Omega)$ is dense in $\mathscr{E}(\Omega)$ (exercise), the extension is unique. □

Distributions of compact support extend to distributions on $\mathbb{R}^d$. The extension by zero is uniquely determined.

Proposition 7. *If $l \in \mathscr{E}'(\Omega)$ there is one and only one $\tilde{l} \in \mathscr{D}'(\mathbb{R}^d)$ such that $l = \tilde{l}$ on $\mathscr{D}(\Omega)$ and $\operatorname{supp} \tilde{l} \subset \Omega$.*

PROOF. If $\tilde{l}$ is such an extension, choose $\psi \in C_0^\infty(\Omega)$ with $\operatorname{supp}(\psi) \subset \Omega$ and ψ equal to 1 on a neighborhood of $\operatorname{supp}(\tilde{l})$. Note that $\operatorname{supp}(\tilde{l})$ contains $\operatorname{supp}(l)$. For any $\varphi \in \mathscr{D}(\mathbb{R}^d)$

$$\langle \tilde{l}, \varphi \rangle = \langle \tilde{l}, \psi\varphi \rangle = \langle l, \psi\varphi \rangle.$$

Thus $\tilde{l}$ is uniquely determined. Furthermore, defining $\tilde{l}$ by $\langle \tilde{l}, \varphi \rangle \equiv \langle l, \psi\varphi \rangle$ proves existence. □

It is common practice to take this extension for granted and, therefore, to consider $\mathscr{E}'(\Omega)$ as a subset of $\mathscr{E}'(\mathbb{R}^d)$. The next propositions use the notions developed above to prove useful results about distributions.

Proposition 8. $\mathscr{D}(\Omega)$ is sequentially dense in $\mathscr{D}'(\Omega)$. In particular, the extensions of the operators L in Proposition 2 are uniquely determined.

PROOF. Choose an exhaustion $\Omega_1 \subset\subset \Omega_2 \subset\subset \cdots$ of Ω. Choose $\psi_k \in \mathscr{D}(\Omega_{k+1})$ with $\psi_k = 1$ on Ω_k. For $l \in \mathscr{D}'(\Omega)$, $\psi_k l \rightharpoonup l$ (exercise). Denote by $l_k \in \mathscr{E}'(\mathbb{R}^d)$, the extension by zero of $\psi_k l$.

Choose $\varepsilon_k \to 0$ with $\varepsilon_k < \operatorname{dist}(\Omega_k, \partial\Omega_{k+1})$. Then $j_{\varepsilon_k} * l_k \in \mathscr{D}(\Omega_{k+1})$. As in the proof of Proposition 4, $j_{\varepsilon_k} * l_k \rightharpoonup l$ in $\mathscr{D}'(\mathbb{R}^d)$ and therefore in $\mathscr{D}'(\Omega)$. □

Proposition 9. *The only distributions on Ω, with support equal to the single point $\underline{x}$, are finite linear combinations of the derivatives of the Dirac delta at $\underline{x}$.*

PROOF. Without loss of generality we may suppose that $l \in \mathscr{E}'(\mathbb{R}^d)$. Translating, if necessary, we may suppose that $\underline{x} = 0$. Then $\langle l, \varphi \rangle = 0$ for all $\varphi \in \mathscr{E}(\mathbb{R}^d)$ which vanish on a neighborhood of 0.

Choose an integer n such that l is of order n. For $\varphi \in \mathscr{E}(\mathbb{R}^d)$, express φ as a sum of its Taylor polynomial of order n at 0 plus a remainder

$$\varphi = \sum_{|\alpha| \leq n} \frac{(\partial^\alpha \varphi(0))x^\alpha}{\alpha!} + r(x).$$

Then

$$\langle l, \varphi \rangle = \langle l, r \rangle + \sum_{|\alpha| \leq n} c_\alpha \partial^\alpha \varphi(0) \qquad \text{where} \quad c_\alpha \equiv \left\langle l, \frac{x^\alpha}{\alpha!} \right\rangle.$$

The proposition follows with $l = \sum c_\alpha(-\partial)^\alpha \delta$ if we can show that $\langle l, r \rangle = 0$.

Choose a function $\zeta \in \mathscr{E}(\mathbb{R}^d)$ such that ζ vanishes on a neighborhood of 0 and ζ is identically equal to 1 for $|x| \geq 1$. Then, for $m = 1, 2, \ldots$, $\langle l, \zeta(mx)r(x)\rangle = 0$.

The difference $\zeta(mx)r(x) - r(x)$ is supported in $|x| \leq 1/m$. To show that $\langle l, r\rangle = 0$, it is sufficient to show that $\langle l, \zeta(mx)r(x) - r(x)\rangle$ tends to zero as m tends to infinity. To do that, it suffices to show that for all α with $|\alpha| \leq n$, $\max|\partial^\alpha(\zeta(mx)r(x) - r(x))|$ tends to zero as $m \to \infty$.

The derivative of the difference is equal to a sum

$$(\zeta(mx) - 1)\partial^\alpha r(x) + \sum_{\beta+\gamma=\alpha, |\beta|>0} c_\beta m^{|\beta|}\partial^\beta\zeta(mx)\partial^\gamma r(x).$$

Since the derivatives of r up to order n vanish at 0 the first term is $O(m^{-n-1+|\alpha|})$. The summands of the second term are $O(m^{|\beta|})O(m^{-n-1+|\gamma|})$. Since $|\beta| + |\gamma| = |\alpha| \leq n$ the result is $O(m^{-1})$ and the proof is complete. □

Proposition 10. *If Ω is connected then the only distributions on Ω, all of whose partial derivatives vanish, are the constants.*

PROOF. It suffices to show, for any connected open $\omega \subset\subset \Omega$, that l is constant on ω (exercise).

Given such an ω, choose $\chi \in C_0^\infty(\Omega)$ such that χ is equal to 1 on a neighborhood of $\bar{\omega}$.

Then χl extends naturally to an element of $\mathscr{E}'(\mathbb{R}^n)$ with support in Ω and with derivatives vanishing on a neighborhood of $\bar{\omega}$.

With j_ε as in Proposition 4, $j_\varepsilon * (\chi l)$ converges to χl in $\mathscr{D}'(\mathbb{R}^d)$. Therefore, as elements of $\mathscr{D}'(\omega)$, $j_\varepsilon * (\chi l) \rightharpoonup l$.

On the other hand, for small ε, $j_\varepsilon * (\chi l)$ is a smooth function whose derivatives vanish in ω. Thus $j_\varepsilon * (\chi l)$ is equal to a constant c_ε in $\mathscr{D}'(\omega)$. Choose $\varphi \in \mathscr{D}(\omega)$ with $\int \varphi\, dx = 1$. Then, as ε tends to zero,

$$c_\varepsilon = \langle j_\varepsilon * (\chi l), \varphi\rangle \to \langle \chi l, \varphi\rangle \equiv c.$$

Since $c_\varepsilon = j_\varepsilon * (\chi l)|_{\mathscr{D}(\omega)}$ this suffices to show that $l = c$ in ω. □

Finally, we take a second look at convolutions to show that $\mathscr{E}'(\mathbb{R}^d) * \mathscr{D}'(\mathbb{R}^d)$ makes sense. From the definitions it is not difficult to show that $\operatorname{supp}(\varphi * l) \subset \operatorname{supp}(\varphi) \cup \operatorname{supp}(l)$. Thus $\varphi * l \in \mathscr{D}(\mathbb{R}^d)$ when $l \in \mathscr{E}'(\mathbb{R}^d)$.

For such an l consider the map $L(\varphi) \equiv \varphi * l$. This is a sequentially continuous map of $\mathscr{D}(\mathbb{R}^d)$ to itself and it has transpose $L'(\psi) = \psi * (\mathscr{R}l)$ (exercise). The transpose being sequentially continuous, it follows that L extends to an operator from $\mathscr{D}'(\mathbb{R}^d)$ to itself. For $u \in \mathscr{D}'(\mathbb{R}^d)$ and $l \in \mathscr{E}'(\mathbb{R}^d)$, $u * l$ is given by $\langle u * l, \varphi\rangle \equiv \langle u, \varphi * (\mathscr{R}l)\rangle$.

Symmetrically, left convolution $l * u$ is defined as an extension of $l *$. One has $l * u = u * l$ for all $l \in \mathscr{E}'(\mathbb{R}^d)$ and $u \in \mathscr{D}'(\mathbb{R}^d)$.

PROBLEMS

Many details in this Appendix were left as exercises. Working them out is good practice. There are two things one needs to do to learn distributions. One must manipulate the definitions in simple proofs and one must gain familiarity with computations with simple distributions. The previous pages provide many exercises of the first sort. The next problems have the second skill as goal.

1. Compute $(d/dx)^k|x|^j$ for $j, k = 1, 2, 3 \ldots$.

2. Compute $(d/dx)^k|\sin x|$ for $k = 1, 2, 3 \ldots$.

3. (i) Let $u(x)$ be the function which is equal to $\ln(x)$ for $x > 0$ and zero for $x \leq 0$. Then u is locally integrable. Compute the distribution derivative du/dx. *Answer.* $\langle du/dx, \psi \rangle = \int_0^1 (\psi(x) - \psi(0))/x\, dx + \int_1^\infty \psi(x)/x\, dx$.
   (ii) Compute the distribution derivative of the function $\ln(|x|) \in L^1_{\text{loc}}(\mathbb{R})$.

4. Find the most general solution $T \in \mathscr{D}'(\mathbb{R})$ of the following equations:
   (i) $xT = 0$; (ii) $x\, dT/dx = 0$; (iii) $x^2 T = \delta$; (iv) $x\, dT/dx = \delta$;
   (v) $dT/dx = \delta$; (vi) $dT/dx + T = \delta$; (vii) $T - (d/dx)^2 T = \delta$.
   Hint. For the problems with δ on the right hand side, find a form for T in $x > 0$ and in $x < 0$. This computes candidates for T which are correct up to a distribution supported at (0).

5. Let μ be the distribution which integrates a test function over the unit ball in $\mathbb{R}^d$. Compute
   (i) $\partial\mu/\partial x_i$; (ii) $\Delta\mu$ where $\Delta \equiv \sum(\partial/\partial x_i)^2$.

6. For $j \in \mathscr{D}(\mathbb{R}^d)$, $j \geq 0$, $j_\varepsilon \equiv \varepsilon^{-d} j(x/\varepsilon)$, sketch $j_\varepsilon * \mu$ with μ as in Problem 5, and ε converging to zero.

7. With $n = 1$, define f by $f(x) = \sin 1/x$ if $x > 0$ and zero otherwise. Sketch $j_\varepsilon * f$ discussing the behavior as ε tends to zero.

8. Let f be the characteristic function of the positive quadrant, $\{x \in \mathbb{R}^2: x_i > 0, i = 1, 2\}$. Compute (i) $\partial_{x_i} f$; (ii) $\partial_{x_1}\partial_{x_2} f$.

9. Let f be the characteristic function of the set $x_1 x_2 > 0$. Perform the same calculations as in Problem 8.

10. Define $f \in L^\infty(\mathbb{R}^2)$ by $f = x_1 x_2/(x_1^2 + x_2^2)$ for $x \neq 0$. Perform the calculations as in Problem 8. *Hint.* Be careful about $x = 0$. The answer must identify the distribution derivative on all of $\mathbb{R}^2$. Away from $x = 0$ the derivatives are given by elementary calculus. The formal second derivatives are not even locally integrable near the origin. Problems 12 and 13 have similar difficulties.

11. Let f be the characteristic function of a nonempty infinite wedge in $\mathbb{R} \times \mathbb{R}$ with vertex at the origin. Find a constant coefficient second-order partial differential operator $P(D)$ such that $P(D)f = \delta$.

12. Prove that $(\frac{1}{2})(\partial/\partial x + i\partial/\partial y)(1/z) = \pi\delta$. *Hint.* Write $\langle 1/z, P^t(D)\varphi \rangle = \lim \int_{|z| > \varepsilon} z^{-1} P^t(D)\varphi\, dx\, dy$. Use Green's Theorem for the integral on $|z| > \varepsilon$.
    DISCUSSION. The differential operator appearing here is denoted $\partial/\partial\bar{z}$. The fact that $\partial z^{-1}/\partial\bar{z} = 0$ for $z \neq 0$ expresses the fact that z^{-1} is holomorphic away from the origin. The formula of this problem is equivalent to the Cauchy integral formula.

13. Take an example from a calculus text illustrating the inequality of mixed partials and compute the distribution derivatives showing how the mixed partials end up being equal.

14. Prove that the principal value of $1/x$ is a distribution of order 1.

15. Show that the functional

$$l(\varphi) = \sum_{1}^{\infty} \frac{1}{n}\left(\varphi\left(\frac{1}{n}\right) - \varphi(0)\right)$$

defines a distribution of order 1. Find supp(l). Show that there does not exist a constant c such that

$$|\langle l, \varphi\rangle| \le c \max_{\text{supp}(l)} (|\varphi(x)| + |\varphi'(x)|).$$

DISCUSSION. This example of L. Schwartz shows that a reasonable conjecture connecting support and order of distributions of compact support is not true.

References

[A] S. Agmon, *Lectures on Elliptic Boundary Value Problems*, with B.F. Jones and G.W. Batten, Van Nostrand, Princeton, NJ, 1965.

[BJS] L. Bers, F. John, and M. Schechter, *Partial Differential Equations*, American Mathematical Society, Providence, RI, 1964.

[CH] R. Courant and D. Hilbert, *Methods of Mathematical Physics*, Interscience, New York, 1953 (vol. 1), 1962 (vol. 2).

[Dolp] C. Dolph, Nonlinear integral equations of Hamerstein type, *Trans. Amer. Math. Soc.* **66** (1949), 289–307.

[Doll] J. Dollard, Scattering into cones I: Potential scattering, *Comm. Math. Phys.* **12** (1969), 193–203.

[Don] W. Donoghue, *Distributions and Fourier Transforms*, Academic Press, New York, 1969.

[Fi] R. Finn, Remarks relevant to minimal surfaces and to surfaces of prescribed mean curvature, *J. d'Analyse Math.* **14** (1965), 139–160.

[Fo] G.B. Folland, *Introduction to Partial Differential Equations*, Princeton University Press, Princeton, NJ, 1976.

[Fr] F.G. Friedlander, On the radiation field of pulse solutions of the wave equation II, *Proc. Roy. Soc.* **279A** (1964), 386–394.

[Gara] P. Garabedian, *Partial Differential Equations*, Wiley, New York, 1964.

[Gard] L. Gärding, Linear hyperbolic partial differential equations with constant coefficients, *Acta. Math.* **85** (1950), 1–62.

[GS] I.M. Gelfand and G.E. Shilov, *Generalized Functions*, vol. 1, trans. by E. Saletan, Academic Press, New York, 1964.

[GT] D. Gilbarg and N. Trudinger, *Elliptic Partial Differential Equations of Second Order*, 2nd ed., Springer-Verlag, New York, 1983.

[He] D. Henry, *Geometric Theory of Semilinear Parabolic Equations*, Springer-Verlag, New York, 1981.

[H1] L. Hörmander, *Linear Partial Differential Operators*, 3rd printing, Springer-Verlag, New York, 1969.

[H2] L. Hörmander, *The Analysis of Linear Partial Differential Operators*, Springer-Verlag, New York, 1982 (vol. 1), 1983 (vol. 2), 1985 (vol. 3), 1985 (vol. 4).

[J] F. John, *Partial Differential Equations*, 4th ed., Springer-Verlag, New York, 1982.

[LL] E. Landesman and A. Lazer, Nonlinear perturbations of linear elliptic boundary value problems at resonance, *J. Math. Mech.* **19** (1970), 609–623.

[La] P.D. Lax, The formation and decay of shock waves, *Amer. Math. Soc. Monthly* **79** (1972), 227–241.

[LP] P.D. Lax and R. Phillips, *Scattering Theory*, Academic Press, New York, 1967.

[Lio] J.L. Lions, *Équations Différentielles Opérationelles et Problèmes aux Limites*, Springer-Verlag, Berlin, 1963.

[Lit] W. Littman, The wave equation and L_p norms, *J. Math. Mech.* **12** (1963), 55–68.

[LS] L. Loomis and S. Sternberg, *Advanced Calculus*, Addison-Wesley, Reading, MA, 1968.

[Ma] A. Majda, *Compressible Fluid Flow and Systems of Conservation Laws in Several Space Variables*, Springer-Verlag, New York, 1984.

[Me] A. Messiah, *Quantum Mechanics*, vol. 1, trans. G. Temmer, Interscience, New York, 1961.

[N] L. Nirenberg, An application of generalized degree to a class of nonlinear problems, Colloque sur l'Analyse Functionelle Tenu à Liège du 14 a 16 Septembre 1970, Vander, Louvain, 1971, 57–74.

[P] E. Picard, *Traité d'Analyse*, 3ed., t. 2, Gauthier-Villars, Paris, 1925.

[PW] M. Protter and H. Weinberger, *Maximum Principles in Differential Equations*, Prentice-Hall, Englewood Cliffs, NJ, 1967.

[RS] M. Reed and B. Simon, *Methods of Mathematical Physics*, vols. 1 and 2, Academic Press, New York, 1972 (vol. 1), 1975 (vol. 2).

[RSzN] F. Riesz and B. Sz-Nagy, *Functional Analysis*, trans. L. Boron, Ungar, New York, 1955.

[R] W. Rudin, *Functional Analysis*, McGraw-Hill, New York, 1973.

[Sc1] L. Schwartz, *Théorie des Distributions*, nouvelle ed., Hermann, Paris, 1966.

[Sc2] L. Schwartz, *Mathematics for the Physical Sciences*, Addison-Wesley, Reading, MA, 1966.

[Sm] J. Smoller, *Shock Waves and Reaction–Diffusion Equations*, Springer-Verlag, New York, 1983.

[Sp] M. Spivak, *Calculus on Manifolds, a Modern Approach to the Classical Theorem of Advanced Calculus*, W.A. Benjamin, New York, 1965.

[Ta] M. Taylor, *Pseudodifferential Operators*, Princeton University Press, Princeton, NJ, 1981.

[Tr] F. Treves, *Basic Linear Partial Differential Equations*, Academic Press, New York, 1975.

[Wh] G.B. Whitham, *Linear and Nonlinear Waves*, Wiley New York, 1974.

[W] D.V. Widder, Positive temperatures on an infinite rod, *Trans. Amer. Math. Soc.* **55** (1944), 85–95.

Index

Graduate Texts in Mathematics

continued from page ii

119 ROTMAN. An Introduction to Algebraic Topology.
120 ZIEMER. Weakly Differentiable Functions: Sobolev Spaces and Functions of Bounded Variation.
121 LANG. Cyclotomic Fields I and II. Combined 2nd ed.
122 REMMERT. Theory of Complex Functions. *Readings in Mathematics*
123 EBBINGHAUS/HERMES et al. Numbers. *Readings in Mathematics*
124 DUBROVIN/FOMENKO/NOVIKOV. Modern Geometry—Methods and Applications. Part III.
125 BERENSTEIN/GAY. Complex Variables: An Introduction.
126 BOREL. Linear Algebraic Groups.
127 MASSEY. A Basic Course in Algebraic Topology.
128 RAUCH. Partial Differential Equations.
129 FULTON/HARRIS. Representation Theory: A First Course. *Readings in Mathematics*
130 DODSON/POSTON. Tensor Geometry.
131 LAM. A First Course in Noncommutative Rings.
132 BEARDON. Iteration of Rational Functions.
133 HARRIS. Algebraic Geometry: A First Course.
134 ROMAN. Coding and Information Theory.
135 ROMAN. Advanced Linear Algebra.
136 ADKINS/WEINTRAUB. Algebra: An Approach via Module Theory.
137 AXLER/BOURDON/RAMEY. Harmonic Function Theory.
138 COHEN. A Course in Computational Algebraic Number Theory.
139 BREDON. Topology and Geometry.
140 AUBIN. Optima and Equilibria. An Introduction to Nonlinear Analysis.
141 BECKER/WEISPFENNING/KREDEL. Gröbner Bases. A Computational Approach to Commutative Algebra.
142 LANG. Real and Functional Analysis. 3rd ed.
143 DOOB. Measure Theory.
144 DENNIS/FARB. Noncommutative Algebra.
145 VICK. Homology Theory. An Introduction to Algebraic Topology. 2nd ed.
146 BRIDGES. Computability: A Mathematical Sketchbook.
147 ROSENBERG. Algebraic K-Theory and Its Applications.
148 ROTMAN. An Introduction to the Theory of Groups. 4th ed.
149 RATCLIFFE. Foundations of Hyperbolic Manifolds.
150 EISENBUD. Commutative Algebra with a View Toward Algebraic Geometry.
151 SILVERMAN. Advanced Topics in the Arithmetic of Elliptic Curves.
152 ZIEGLER. Lectures on Polytopes.
153 FULTON. Algebraic Topology: A First Course.
154 BROWN/PEARCY. An Introduction to Analysis.
155 KASSEL. Quantum Groups.
156 KECHRIS. Classical Descriptive Set Theory.
157 MALLIAVIN. Integration and Probability.
158 ROMAN. Field Theory.
159 CONWAY. Functions of One Complex Variable II.
160 LANG. Differential and Riemannian Manifolds.
161 BORWEIN/ERDÉLYI. Polynomials and Polynomial Inequalities.
162 ALPERIN/BELL. Groups and Representations.
163 DIXON/MORTIMER. Permutation Groups.
164 NATHANSON. Additive Number Theory: The Classical Bases.
165 NATHANSON. Additive Number Theory: Inverse Problems and the Geometry of Sumsets.
166 SHARPE. Differential Geometry: Cartan's Generalization of Klein's Erlangen Program.
167 MORANDI. Field and Galois Theory.
168 EWALD. Combinatorial Convexity and Algebraic Geometry.
169 BHATIA. Matrix Analysis.
170 BREDON. Sheaf Theory. 2nd ed.

9 780387 974729